The Paleoecology of Lower Magdalenian Cantabrian Spain

James T. Pokines

BAR International Series 713
1998

Published in 2019 by
BAR Publishing, Oxford

BAR International Series 713

The Paleoecology of Lower Magdalenian Cantabrian Spain

ISBN 9780860549529 paperback
ISBN 9781407350295 e-book

DOI https://doi.org/10.30861/9780860549529

A catalogue record for this book is available from the British Library

This book is available at www.barpublishing.com

BAR Publishing is the trading name of British Archaeological Reports (Oxford) Ltd. British Archaeological Reports was first incorporated in 1974 to publish the BAR Series, International and British. In 1992 Hadrian Books Ltd became part of the BAR group. This volume was originally published by John and Erica Hedges in conjunction with British Archaeological Reports (Oxford) Ltd / Hadrian Books Ltd, the Series principal publisher, in 1998. This present volume is published by BAR Publishing, 2019.

BAR
PUBLISHING

BAR titles are available from:

BAR Publishing
122 Banbury Rd, Oxford, OX2 7BP, UK
EMAIL info@barpublishing.com
PHONE +44 (0)1865 310431
FAX +44 (0)1865 316916
www.barpublishing.com

TABLE OF CONTENTS

LIST OF TABLES

LIST OF FIGURES

ACKNOWLEDGEMENTS

This work is drawn largely from my Ph.D. dissertation at the Department of Anthropology of the University of Chicago. Thanks go to my dissertation committee for their guidance and forbearance: Dr. Leslie G. Freeman and Dr. Michael Dietler (University of Chicago), Dr. Julian Kerbis Peterhans (Field Museum of Natural History, Chicago) and Dr. Jane Buikstra (University of Arizona). Dr. Joaquín González Echegaray (Instituto para Investigaciones Prehistóricas, Santander, Spain) also provided invaluable support, years of instruction, and access to data. Computing support was supplied by L. Reed Associates.

Several seasons of investigators and field school students spent weeks in the careful collection of this data. These have been led by Carlos Fernández Rodríguez, Monica Angulo Ibañez, Steven Kluskens, Heather Stettler, Marcy Krupa and Lance Reed. I must also thank the Department of Mammals of the Field Museum of Natural History and especially Dr. William Stanley, Dr. Bruce Patterson, Melissa Morales, Michi Solomon and Dr. Steven Goodman for their advice, help and space. Also invaluable was the help, access to data and general saving grace of the staff of the Centro de Investigación y Museo de Altamira (Santillana del Mar, Spain): Carmen de las Heras Martín and Dr. José A. Lasheras Corruchaga.

Many colleagues have contributed indirectly to this work, through their years of advice and encouragement. These include April Nowell, Dr. Gail Larsen Peterkin, Dr. Heather Price, Dr. Anne Pike-Tay, and Dr. Heidi Knecht.

Funding for this research was supplied by National Science Foundation Dissertation Improvement Grant SBR-9408920, the Instituto para Investigaciones Prehistóricas (Santander, Spain), and by the Department of Mammals, Field Museum of Natural History (Chicago, U.S.A.).

Leah

ABSTRACT

The terrestrial mammalian microfauna from the Lower Magdalenian site of El Juyo (Santander, Spain) and related sites are used to reconstruct the paleoenvironment Cantabrian Spain. This lowland cave site is located near the modern coastline and provides 3 m of stratified, undisturbed Lower Magdalenian deposits. Occupation spanned approximately 1,000 years ca. 14,000 B.P. Fine screening has yielded a large sample of small mammal remains. Taphonomic analysis indicates that these remains were accumulated as a result of barn owl (*Tyto alba*) pellet regurgitation within the cave vestibule. The small mammalian fauna is dominated by microtines, primarily of the taxa *Arvicola* cf. *terrestris* and *Microtus oeconomus*, and includes *Microtus nivalis*, *M. agrestis*, *M. arvalis*, *Pitymys lusitanicus*, and *P. pyrenaicus*. The late survival of the now extinct microtine species *Pliomys lenki* is also indicated. Also present are the murids *Apodemus flavicollis* and *A. sylvaticus*, the insectivores *Talpa europaea*, *T. caeca*, *Sorex coronatus*, *S. minutus*, *Neomys fodiens*, *N. anomalus*, and *Erinaceus europaeus*. A single lagomorph species *Oryctolagus cuniculus* is present. Small carnivores are represented by *Mustela nivalis*, *M. putorius* and *Vulpes vulpes*. El Juyo's large faunal species include (in descending order of abundance) *Cervus elaphus*, *Capreolus capreolus*, *Bison bonasus*, *Equus caballus*, *Capra pyrenaica*, *Sus scrofa*, *Panthera leo*, *Canis lupus*, and *Ursus arctos*. Human food remains also include the abundant remains of marine mollusc species *Patella vulgata* and *Littorina littorea* and Atlantic salmon (*Salmo salar*).

A new method of paleoenvironmental reconstruction is presented, based upon the relative proportions of terrestrial mammalian microfauna. Analysis of the modern barn owl diet indicates that the relative biomass contributions of micromammalian prey are highly indicative of the relative habitat proportions within the owl's hunting radius: wetlands, forest, humid meadow, heathland, dry meadow, etc. These data are compared with other sources (pollen, large mammalian fauna, aquatic fauna, sea level changes and macrobotanical remains) to produce a clearer picture of the Lower Magdalenian setting. This analysis is extended to the remaining phases of the Upper Paleolithic.

These sources indicate that the climate during the occupation of El Juyo was very humid and cold, with open coastal heath the dominant terrain type. Small areas of forest and grassland were also present. Rapid climatic deterioration occurred during the deposition of the upper occupation levels. The medium microfaunal species *Erinaceus europaeus* (hedgehog) entered the site as a commensal and was not owl or human prey. The larger microfauna (primarily mustelids) are posited to have entered the cave as the result of human fur trapping activity, since these are represented almost exclusively by foot elements. A larger pattern of fur acquisition is detected for the Cantabrian Upper Paleolithic.

The Upper Paleolithic human ecological adaptations in Cantabrian Spain are examined as they relate to this interpretation of the paleoenvironment. The pattern is an increasing separation of human adaptations from the constraints of environment. These adaptations focused upon the "wild harvesting" of concentrated resources, especially red deer, ibex, salmon, and marine molluscs. These species were harvested with increasing efficiency, despite major fluctuations in local climate and sea level. These changes in wild food procurement in many aspects presage the adaptations leading to food production

CHAPTER 1
INTRODUCTION

This research reconstructs the paleoenvironment of the Lower Magdalenian inhabitants of Cantabrian Spain, the strip of land composed of the four northern coastal regions and provinces of Asturias, Cantabria, Vizcaya and Guipúzcoa (Figure 1.1). The data presented are the terrestrial mammalian microfaunal remains (defined as species with an adult body weight less than 5 kg) from El Juyo, an extremely rich Lower Magdalenian cave site. The microfaunal data from El Juyo and other Upper Paleolithic sites are compared with the other sources of palaeoenvironmental data. These include environmental data derived from large terrestrial fauna, marine fauna, sedimentology, palynology, and chronological synchronization with changes in world ocean level. Each source possesses inherent biases that must be accounted for when extrapolating past environmental conditions from them (Kuniholm 1990). The author maintains that in both the resolution of information contained in them and in the potential to accommodate the biases incurred in their introduction to a site, terrestrial mammalian microfauna are the most accurate information source on past environmental conditions. These biases are best evaluated and controlled for using a taphonomic approach: determining the path these organisms took from the biosphere to the lithosphere (Efremov 1940). An even higher degree of resolution may be achieved using these multiple sources together (Tessier *et al.* 1993).

Paleoenvironmental research is relevant to more than the natural history and evolution of the small mammals themselves. Its relevance to archaeological study is in the establishment of an arena for human ecological interaction. One cannot assess how human responses to their environment changed throughout hominid evolution without establishing what those environments were. Changes in environment also may have prompted changes in human foraging activity, food exploitation technology and landscape use and must be understood before changes in these aspects may be assigned to other causes. Human separation from the constraints of environment during the end of the Upper Paleolithic is also relevant to later major changes, including sedentism, urbanism, social specialization and agriculture.

Cantabrian Spain is a particularly rich region in which to study past human adaptations to changing environment, where human occupation is documented throughout the Lower, Middle, and Upper Paleolithic and Mesolithic periods (González Echegaray and Freeman, 1998; Straus 1985, 1992). The region is physically circumscribed, bordered by the Bay of Biscay (Cantabrian Sea) on the north, the Pyrenees to the east and the Cantabrian Cordillera along its southern edge. Through much of its history these natural obstacles have partially isolated it from adjacent areas. One may thus study regional human ecological adaptations (hunting/foraging behavior, landscape use, and resource acquisition) without placing necessarily artificial borders around an area of interaction. Due to its coastal location with abruptly rising mountains, Cantabrian Spain has multiple ecological zones within easy access of one another, allowing for a variety of adaptive responses and logistic strategies by its human occupants.

Its current biogeographic position is distinct from the remainder of the Iberian Peninsula, where a drier, warmer Mediterranean climate now prevails. The local spectrum of exploitable species in Cantabrian Spain is and was distinctive. During the colder phases of glaciation, climate was also distinct from neighboring France, where true tundra and tundra-dwelling species could be found in abundance (reindeer, arctic fox, snowy owls, wolverine, etc.). Cantabrian Spain thus offered a distinct set of foraging options to its inhabitants. The area's karst terrain has allowed excellent preservation of archaeological materials, albeit in a selective depositional setting: almost all of its Paleolithic occupation levels have been found in limestone caves.

HISTORY OF INVESTIGATION

Paleoenvironmental research traces its origins to the very beginnings of geological investigation in the last half of the 18th century. Uniformitarian thought--that the same geological processes visible today were also in operation long into the past--pushed the conjectured age of the Earth far beyond that held by Biblical scholars. The slow deposition of sediments must indicate that long sequences of sedimentary rock took millions of years to accumulate, compact and erode. Though preceded by other early geologists, James Hutton in his *The Theory of the Earth* (1795) first published the principle of uniformitarianism: the Earth's sedimentary rocks were accumulated by the same processes of sediment deposition visible in contemporary rivers and seas. Charles Lyell later codified these ideas of geological processes in his *Principles of Geology* (1830-33).

Hutton and many other nascent geologists of this early period also noted that glacial transport was a plausible explanation for large boulders displaced far from their parent bedrock (glacial erratics). Other related phenomena that begged explanation were striations on exposed rock surfaces and moraine deposits far from extant alpine glaciers. The main competing theory was that boulders and other debris melted out of drifting icebergs in a universal deluge (increasingly disassociated throughout this period with the notion of a Biblical flood). Holders of this viewpoint included William Buckland, who published many insightful observations in support of this theory in his *Reliquiae Diluvianae* in 1823. Evidence from the study of modern alpine glaciation in Switzerland, however, began to convince multiple investigators that these phenomena resulted from the action of moving glaciers.

The idea that vast ice sheets covered Europe in the past was long in gaining acceptance among the scientific community (summarized by Hallam 1989). Ignace Venetz reported to a

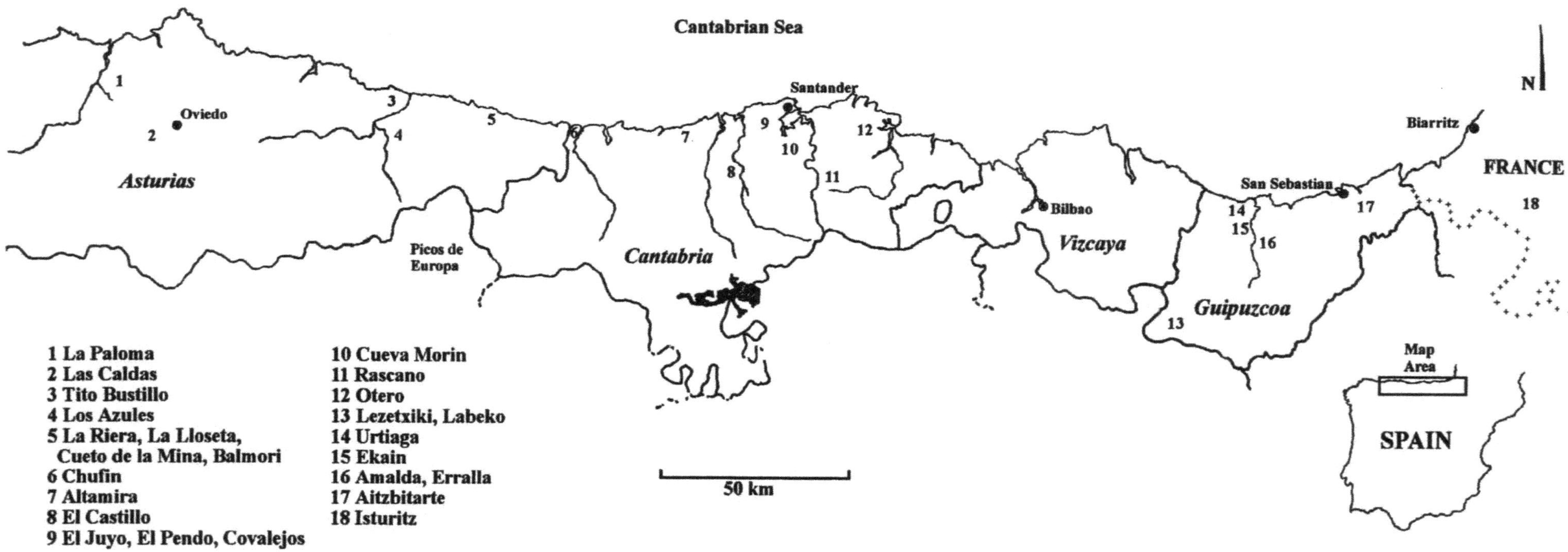

Figure 1.1: Sites Discussed in Chapter 1.

Swiss natural history society in 1821 the theory that glaciers had once covered much more of Europe than previously posited. He was to expand upon this idea throughout the decade, including the scope of how far these alpine glaciers had spread. The notion of an *Eiszeit* (Ice Age) was first published by the botanist Karl Schimper in 1837. This theory was further expanded upon and popularized through the work of two naturalists, Jean Charpentier and (more famously) Louis Agassiz. Agassiz published a substantial work in 1840 concluding that a vast ice sheet once covered Europe, and through much effort he succeeded in personally convincing many colleagues of glacial theory. This theory was to gain acceptance throughout the early half of the 19th century, with perhaps consumating proof in the discovery of the Greenland ice sheet in 1852. The notion had arrived that the Earth's environments were mutable, and in the past vast sheets of ice covered what is now Europe.

Also notable from this period were some early published observations concerning site formation and taphonomic processes. Lyell (1830-33:II:219-227) discussed how animal remains could accumulate in the deposits of caves and fissures by their acting as natural pit traps over long periods. In this manner, these remains often became associated with those of humans. Buckland (1823:93) had previously noted that "the bones of various birds, of moles, water-rats, mice" had been accumulated in limestone caves by the "agency of hawks", while fish bones had been accumulated in the same sites by gulls. This relationship between the type of small faunal remains recovered and the predatory agency of accumulation was to become central in 20th century investigations of taphonomy (including the present work).

Archaeologists of the 19th century began to tie human history into these past glacial environments (summarized by Grayson 1983). The antiquity of stone tools themselves was verified by their associations with extinct fauna. Most notably, de Mortillet (1883) published a stage system of human development. His system is relevant to the origins of paleoenvironmental thought in its correlation of stone tools with supposed unique or dominant faunal associations. Our Stone Age ancestors were theorized to have passed through multiple stages: Great Cave Bear, Elephant/Rhinoceros, Reindeer, Aurochs, and Recent. While later rejected, de Mortillet's scheme did link Stone Age peoples with fauna totally or locally extinct in contemporary temperate Europe. People were posited to have passed through a Reindeer Stage, adapting to a unique and much more climatically rigorous environmental situation. The idea, however hesitant, that human culture was mutable and adaptable to changing environmental conditions had been established.

Serious Paleolithic investigation in Cantabrian Spain commenced as early as 1872 with the excavations by E. de la Predraja in the site of Covalejos, which yielded assemblages (later identified as) Mousterian (González Echegaray and Freeman, 1998). The first explorations of Altamira were carried out in 1876 by Marcelino Sanz de Sautuola, a local amateur prehistorian. This world-renowned cave site has yielded occupation layers from the Solutrean and Magdalenian periods, as well as magnificent parietal art dating from the latter period (Freeman 1987). These early investigations naturally focused upon the stratigraphic succession of archaeological techno-complexes and the parietal art associated with them. It was at this site that the veracity of Paleolithic cave art was established among the scientific community (Cartailhac 1902).

In the early half of the 20th century archaeological focus remained upon establishing the succession of the stone age culture-stratigraphic units of Western Europe. Many important and famous sites were excavated with varying degrees of publication success, such as the immense deposits of El Castillo (Cabrera 1984; Obermaier 1924). This site has yielded 18 m of deposits with archaeological levels spanning virtually the entire prehistory of Spain: Acheulean, Mousterian, Aurignacian, Perigordian, Solutrean, Magdalenian, Azilian, and Neolithic. Typical to the excavation standards of the day, the only paleoenvironmental data collected was that inferred from the large faunal remains and infrequently recovered small fauna. Research focus lay upon the temporal succession of stone and osseous tool industries. At El Castillo the correct temporal placement of the Solutrean industries--between the Aurignacian and Magdalenian industries--was established. Palynological studies were unknown, and the retrieval of small faunal remains was subject to the low probability of being spotted during rather hasty excavation and not a part of any systematic attempt at their collection.

Palynological investigation in Cantabrian Spain began with the analysis of the long paleolithic sequence at El Pendo (Mousterian, Aurignacian, Gravettian, Magdalenian and Azilian Levels). Dug originally by Sautuola in 1878 and many other investigators in subsequent years (including Jesús Carballo in 1926), extensive re-excavations were carried out in 1953-1957, though publication waited until 1980 (González Echegaray). Arlette Leroi-Gourhan (1980) produced a brief analysis of the changing pollen counts. The palynological analysis for the site of Isturitz was published much earlier (Leroi-Gourhan 1959), though this site exists just outside of Cantabrian Spain in French Basque country. Palynological analysis became standard practice for subsequent excavations. These include such sites as Otero, Morín, and La Riera (Arl. Leroi-Gourhan 1966, 1978 and 1986, respectively); Tito Bustillo, Rascaño, Chufín, and Eralla (Boyer-Klein 1976, 1981, 1984, 1985); Los Azules (López 1981); Ekain and Amalda (Dupré 1984, 1990), and Lezetxiki, Labeko and Urtiaga (Sánchez Goñi 1993).

This concern for palynological analysis is one of a sheaf of new interpretive techniques applied to paleolithic investigation. Analysis of sediments and their changes between levels as an indication of paleoenvironment has also been undertaken at least since the 1953-57 re-excavations at El Pendo (Butzer 1980). Integrated research of this kind has grown during the past three decades with the advent of controlled, modern excavations beginning with the re-excavation of Cueva Morín (González Echegaray and Freeman 1971, 1973, 1978) during the late 1960s. New excavations at the site of La Riera (Clark and Richards 1978; Straus and Clark 1978, 1986) were undertaken with the specific project focus of paleoecology. Ongoing investigations at El Castillo (Cabrera 1984; Bischoff *et al.* 1991) and El Pendo are undertaken with the specific goals of more precise excavation of remnant stratigraphy and to gather evidence missed during the earlier, less controlled digs. Such modern techniques are applied to the modern excavations of such sites as Tito Bustillo (Moure Romanillo 1975), Erralla (Altuna 1985a), Amalda (Altuna *et al.* 1990), Ekain (Altuna and Merino 1984), and Rascaño (González Echegaray and Barandiarán 1981). Much contemporary research centers upon utilization of landscape, site location, and site territories with regard to food procurement (Bailey 1983; Bailey and Davidson 1983; Clark 1985). Investigation of decorative motifs on bone implements has indicated the possible seasonal aggregation of social groups at certain loci (Altamira and Cueto de la Mina), thus directly tying together the economic activities of multiple sites (Conkey 1978, 1980). The trend in contemporary Cantabrian paleolithic investigations has thus been integrative research combining multiple specialties, with sites studied in the context of their regional setting and interactions.

Most research prior to the past two decades tended to lump microfauna with large fauna, the latter of which was given the greater priority of analysis and interpretation. The work of Jesús Altuna (1972, 1976, 1978, 1986; Altuna and Mariezkurrena 1980, 1985) spans this separation period of large and small faunal analyses. As a direct result of more refined excavation techniques, research specifically concerning mammalian microfauna (as well as other small fauna such as fish, reptiles, amphibians and birds) becomes practical. While some of these faunal elements were recovered even under earlier, less rigorous excavation regimes (Cabrera 1984), their totals were far smaller than the more archaeologically visible and economically important large fauna (Altuna 1972). Fine-mesh screening of sediments is a minimum requirement for their recovery. Some modern excavations have yielded significantly large microfaunal samples, notably Ekain (n = 1,008 identified remains [Zabala 1984]), Amalda (n = 2,346 [Pemán 1990]), and Eralla (n = 15,710 [Pemán 1985]). Modern excavation techniques are no guarantee of substantial microfaunal samples, as at La Riera where modern re-excavation of remnant stratigraphy yielded a small microfaunal sample (n = 152) (Altuna 1986). Modern recovery techniques have the additional benefit of producing unprecendented large samples of stone and osseous microtools such as microblades, fine points, and fragments of bone needles (Freeman *et al.* 1998).

RESEARCH OBJECTIVES

Three problems are addressed by this study of paleoenvironment. One is the reconstruction of the Lower Magdalenian habitat surrounding El Juyo and by extension the habitat of the numerous other sites existing in the same time range and coastal lowland environmental zone. These sites include the Magdalenian occupations of Altamira (Altuna and Straus 1976; Straus 1976-77), La Riera (Straus and Clark 1986), El Castillo (Cabrera 1984), Tito Bustillo (Moure Romanillo 1975; Moure Romanillo and Cano Herrera 1976, 1978, 1979), Cueva Morín (Altuna 1978), and Ekain (Zabala 1984). By further extrapolation, the ecological setting of upland sites such as Rascaño may be surmised. The basic parameters of this reconstruction are the varying components and proportions of terrestrial mammalian microfauna. These taxa are representative of certain microhabitats and tend to be far more specialized in their choices than are large fauna (Avery 1982a; Bourlière 1975; Delibes de Castro 1985; Fa *et al.* 1982). Changes in latitudinal (or cliserial--Graham 1979), altitudinal (Moreno and Barbosa 1992), degree of precipitation, and vegetation associative makeup of small mammal communities are noted. These changing microfaunal parameters indicate the changing Late Pleistocene environment.

When accumulated by generalist predators as is indicated at El Juyo (some owl species in particular), microfaunal assemblages can give a less nebulous picture of local habitats and their relative proportions (Andrews 1990; Avery 1982a, 1982b, 1987, 1992; Brain 1974; Brain and Brain 1977; Lovari *et al.* 1976; Thackeray 1987; Thackeray and Avery 1990). Biases due to prey selectivity by predators or preference for certain hunting habitats are understood by comparison of fossil assemblages with modern ones accumulated by predators in a variety of ecological zones. Paleoenvironmental reconstruction therefore comes from comparison between microfaunal signals derived from given environments (Avery 1993), and not from the assumption that relative species proportions directly predict relative habitat proportions. A large body of research from habitats throughout the Holarctic exists for comparisons of this type (Brunet-Lecomte and Delibes 1984; Buckley and Goldsmith 1975; Campos 1978; Delibes *et al.* 1984; Glue 1967, 1970, 1974, 1975; Goszczynski 1974; Jaksic *et al.* 1982; Pitelka *et al.* 1955; Southern 1954). The present research, however, presents a new method for calculation directly from the archaeological remains (Chapter 4).

The second research problem is the comparison of this source of environmental data with other sources already available for El Juyo and other sites. Sedimentological analysis suggests increasing instability and an eventual return to colder conditions during the time of deposition of the upper habitation levels (Freeman *et al.* 1988:11). Measurement of the degree of cryoclastic spalling can give a relative indication of an increase in cold conditions. Any actual paleoenvironmental reconstruction, however, is based upon extrapolation from paleotemperature regimes and analogy with present-day cold climates (Butzer 1986). Additionally, human activity may potentially alter sediment deposition in ways that may be mistaken for changes in environment. These include changes in fluvial deposition due to the increased runoff allowed by trampled vegetation outside an occupation site, and changes to the freeze/thaw regime of a cave or rockshelter ceiling by human-caused temperature regulation: erection of barriers to air flow and use of fire in a cave. Sedimentology thus is significantly limited in its potential for inferring changes in climate.

Occupation at El Juyo occurred during the closing stages of the Pleistocene (Oxygen Isotope Stage 2), with sea levels, ocean circulation, and temperatures very different from today's temperate and humid climate (CLIMAP 1976; McIntyre *et al.* 1976; Ruddiman and McIntyre 1976; Shackleton and Opdyke 1976; van Donk 1976). Global sea levels at the last glacial maximum were at least 85 m lower than present-day levels (CLIMAP 1976), with other estimates placing sea level from 90 m to 130 m lower than present (for summaries see Bailey 1983; Butzer 1983). Evidence derived from deep sea core foraminifera suggests that the North Atlantic surface ocean water temperatures at 42° North latitude were around 10° C colder on average ca. 18,000 B.P. than at present, indicating polar water conditions (McIntyre *et al.* 1976). The lowest elevation of terminal moraines left by alpine glaciers during the last glacial maximum in the Cantabrian Cordillera is 550 m above sea level (Clark 1981: 7). Air temperatures at glacial maximum for the coast near El Juyo had a February average of around -1.6° C and an August average of 10.4° C. The nearest modern analog for these weather conditions is the treeless coast of Iceland at 66° N latitude (Butzer 1986:215). During the harshest climatic conditions of El Juyo's later occupation, the terrain around the site had much less forest cover and much more open grassland than modern climate would allow.

One may infer that the climate was significantly colder than today in general throughout Europe during the later (ca. 14,000 B.P.) occupation of El Juyo, with ameliorating climate conditions from the glacial maximum ca. 22,000-18,000 B.P. (Straus 1991b). Local conditions, however, may have varied greatly from region to region. Such is particularly true for Cantabrian Spain, which owes in large part its modern temperate climate to its coastal proximity and the warm waters of the Gulf Stream. Since the southward advance of the polar ice cap into the North Atlantic may have deflected the Gulf Stream abruptly southward (Ruddiman and McIntyre 1976), this source of local climatic warming may have been switched on or off over relatively short periods. Changes in local climate may have been similarly abrupt.

Palynological analysis has been undertaken for El Juyo as well (Boyer-Klein and Arl. Leroi-Gourhan 1987; Arl. Leroi-Gourhan 1994). The problems of pollen taphonomy, differential production and buoyancy have increasingly drawn into question the face value of any pollen diagram (Sánchez Goñi 1993). The propensity of some wind-pollinated taxa (such as *Pinus*) to produce abundant and far-traveling pollen will cause their overrepresentation in a pollen profile (Faegri and Iversen 1989; Prentice 1986). Pollen from ground plants will also be introduced to a site on the bodies of humans and animals, causing overrepresentation of shrubby and herbaceous forms. Foliage transported into the site for food, bedding, medicine, or as an incidental inclusion may also concentrate the pollen of selected species in proportions unlike those in the local surroundings. Research has shown that pollen is reworked and transported by water (Davis 1968; Traverse 1990) and that consequently sorting by size may occur (Fall 1987; Goodwin 1988; Holmes 1990). Pollen and spores of different species are also differentially susceptible to destruction by microbial attack, oxidation, and mechanical forces of corrosion (Havinga 1967). While broadly suggestive of past environmental settings and gross changes in climatic regimes, palynology alone cannot give a complete and convincing picture of the paleoenvironment of Cantabrian Spain, any more than can sedimentology.

The large fauna of El Juyo and other Magdalenian sites have been affected by cultural biases in their selective harvest, making their relative proportions generally unsatisfactory for paleoenvironmental reconstruction. They are useful as indicator species, however, and complement other sources of data. One must note that the principal prey species at El Juyo and most other Cantabrian sites, red deer (*Cervus elaphus*), can thrive across a range of environments throughout the Holarctic and is indicative of no one habitat. Larger species of microfauna (such as lagomorphs or small mustelids) also may have been food or fur resources for humans and must be critically evaluated for evidence of their inclusion in the site having been culturally affected.

A further source of evidence utilized in conjunction with microfauna are the macrobotanical remains recovered from El Juyo. Excellent preservation (due in large part to the sealing of the deposits by flowstone shortly after occupation ceased) combined with comprehensive retrieval practices has allowed unprecedented recovery of carbonized and uncarbonized plant parts and seeds (Crowe 1987; Freeman *et al.* 1988; Freeman *et al.* 1998; Ibañez-Angulo 1991; Stettler 1995). This botanical data does suffer from some of the same biases as pollen from the site: passage through a cultural filter in the presence and relative abundance of taxa, differential potential for transport, and differential susceptibility to destruction. Local plant taxa represented through botanical remains reconfirm the local origin of pollen introduced to the site and as indicator species in themselves. In this role they compliment the microfaunal data from the site.

The final problem under consideration is the nature of the changing adaptations of the human groups inhabiting Cantabrian Spain at the end of the Pleistocene in relation to their environment. Broad changes in food procurement strategy over time are evident in the archaeological levels already excavated. Mousterian and Early Upper Paleolithic levels are characterized by opportunistic hunting and/or scavenging of less dangerous large mammal taxa, such as horse (*Equus caballus*), large bovids and red deer (Freeman 1973, 1975, 1981; Straus 1977, 1992, 1993). Beginning with the Late Upper Paleolithic (Solutrean, Magdalenian, and Azilian culture-stratigraphic units) and accelerating into Mesolithic (Asturian) times, food procurement in Cantabrian Spain intensified in the exploitation of key species, primarily red deer, Atlantic salmon (*Salmo salar*) and marine molluscs (*Patella vulgata* and *Littorina littorea* most commonly). Site location/function began to diversify in ways that suggest the exploitation of other more areally restricted/specialized resources, such as ibex (*Capra pyrenaica*) hunting at the upland site of Rascaño (González Echegaray and Barandiarán 1981). The Lower Magdalenian levels of El Juyo thus span a fascinating period where intensification of some aspects of food procurement (slaughter of whole herds of red deer, increased use of marine resources, broadening of the food base to include previously underutilized species) occurred. A necessary component in understanding these changes in food procurement strategy is the reconstruction of their changing environmental setting.

ARCHAEOLOGICAL CONTEXT

The El Juyo materials fit within the overall context of the Late Upper Paleolithic as represented sequentially by the Solutrean (20,500 to 16,000 B.P.), Lower Magdalenian (16,500 to 13,500 B.P.), Upper Magdalenian (13,500 to 11,000 B.P.), and Azilian (11,000 to 9,000 B.P.) tool industries (following Straus 1992). Some researchers recognize further subdivisions such as a Middle Magdalenian, which corresponds to Magdalenian IV of France (marked by the presence of "proto-harpoons"). This phase has been demarcated at sites including La Paloma (Martínez and Chapa 1980) and Las Caldas (Corchón 1981). Other finer subdivisions akin to those commonly denoted in French Late Magdalenian sites based upon changes in harpoon style (Bordes 1968:161-166) have been applied with limited success to Cantabrian Spain (see Straus 1992:135-138 for discussion). Finer gradations, whether valid or invalid as applied to Cantabrian Spain, are not relevant to the present topic.

Human habitation in this region dates from the Acheulean as represented by the basal deposits at El Castillo, an important cave site also yielding Mousterian, Aurignacian, Gravettian, Solutrean, Lower and Upper Magdalenian, Azilian, and post-Neolithic levels, as well as abundant cave art (Bischoff *et al.* 1991; Cabrera 1984; Moure Romanillo *et al.* 1996). Other Acheulean occurrences consist only of (mostly equivocal) open air sites, thus making this early tool complex poorly represented at the present time. The Mousterian is better represented in Cantabrian Spain, with levels excavated at sites such as El Castillo (Cabrera 1984), Cueva Morín (González Echegaray and Freeman 1978), and El Pendo (González Echegaray 1980) among others.

Early Upper Paleolithic sites in Cantabrian Spain tend to be more numerous and may indicate a continuing trend of slow population increase (Straus 1992). While Chatelperronian levels have been found at only a few sites, as at Cueva Morín, El Pendo and possibly Ekain (Altuna and Merino 1984), Aurignacian levels occur at several sites, notably Cueto de la Mina (Vega del Sella 1916), Rascaño (González Echegaray and Barandiarán 1981) and again at El Castillo, El Pendo, and Cueva Morín. The Early Upper Paleolithic is continued by Upper Perigordian/Gravettian levels at sites such as El Castillo, Amalda (Altuna *et al.* 1990) and Cueva Morín.

The period examined in the present work, the Late Upper Paleolithic (specifically the Lower Magdalenian), shows a significant increase in site abundance. The Solutrean is represented presently by more than 44 sites in Cantabrian Spain (Straus 1992:96), although materials from some early excavations have been lost or dispersed over the years. This list of sites includes El Pendo, El Castillo, Cueva Morín, Rascaño, Cueto de la Mina, Altamira (Altuna and Straus 1976; Cabrera and Gimenez 1991; Freeman 1988; González Echegaray 1988; Straus 1976-77), La Riera (Clark and Richards 1978; Clark and Straus 1983; González Morales 1986; Straus *et al.* 1981; Straus and Clark 1978, 1986), Las Caldas (Corchón 1981), and Balmori (Clark 1974-75). The Solutrean tool industry is characterized by foliate stone points in willow leaf, laurel leaf, or shouldered forms. Additional components of the lithic industry include endscrapers, burins, denticulates and notches, along with low percentages of sidescrapers. Lithic elements generally lacking are backed bladelets, with exceptions occurring at some sites. Its osseous industry is well developed, with antler projectile points generally having circular or oval (ovate) cross sections, and other items including eyed needles of bone or antler.

The Solutrean is succeeded with some temporal overlap by the Lower Magdalenian, again abundantly represented among Cantabrian sites. Total site numbers vary according to how researchers classify equivocal assemblages. Utrilla (1981) compiles 18 definite Lower Magdalenian sites in Cantabria Spain, with many more Magdalenian sites indeterminate as to Lower or Upper phase and many she designates Middle Magdalenian. Straus (1992:129) updates this figure to as high as 52, but notes the probable inflation due to the inclusion of equivocal Magdalenian phase sites and does not recognize a distinct Middle Magdalenian phase in Cantabrian Spain. Lower Magdalenian sites include El Castillo, Altamira, La Riera, Rascaño, La Paloma, Cueto de la Mina, Cueva de la Lloseta (Jordá 1958), and El Juyo (Barandiarán *et al.* 1987; Freeman and González Echegaray 1981, 1982, 1995; Freeman *et al.* 1988; Janssens and González Echegaray 1958; Krupa 1994; Pokines 1993, 1997; Stettler 1995). The Upper Magdalenian is equally well represented at such sites as El Castillo, La Riera, Rascaño, La Paloma, and Tito Bustillo (Moure Romanillo 1975; Moure Romanillo and Cano Herrera 1976, 1978, 1979). The Azilian follows, with levels at Los Azules, La Riera, Rascaño, and La Paloma among other sites.

While each of these archaeological tool complexes derives its name from a locality in France (Solutré, La Madeleine, St. Acheul, etc.), these phases as represented in Cantabrian Spain do not correlate precisely with extant French-based chronologies. The Solutrean is often divided into phases based upon the successive appearance of foliate point types. These phases do not appear applicable to Cantabrian sites, such as at La Riera where shouldered Solutrean points (supposedly a marker of the Upper Solutrean) were found in earlier levels (Straus 1990:437). Similarly, though the Cantabrian Lower Magdalenian generally corresponds to the Magdalenian III phase of the normative French system, available radiocarbon dates indicate a later persistence than in France, with the first two phases (Magdalenian I and II) absent locally (González Echegaray 1960).

The primary division of the Cantabrian Magdalenian occurs among assemblages based upon the presence or absence of laterally barbed harpoon heads made of antler. Lower Magdalenian tool assemblages are primarily characterized by unbarbed projectile points of antler, with quadrangular forms (rectangular, square, trapezoidal, rhomboid) becoming far more prevalent in relation to ovate forms (circular, oval, halfmoon) than was typical of the preceding Solutrean phase. By definition, Lower Magdalenian sites lack harpoons and occur before Upper Magdalenian sites, but there is some overlap in dating between the two phases just as overlap occurs between the Solutrean and the Lower Magdalenian. Elements of Lower Magdalenian lithic tool kits include the usual Upper Paleolithic forms, with an increase in the importance of nucleiform endscrapers and (at some sites) of backed bladelets.

THE LOWER MAGDALENIAN SITE OF EL JUYO

El Juyo is a cave site located approximately 8 km west of the city of Santander and 5 km south of the present-day coastline of the Cantabrian Sea (Barandiarán *et al.* 1987). The present (artificial) cave entrance is located in a sinkhole in the local

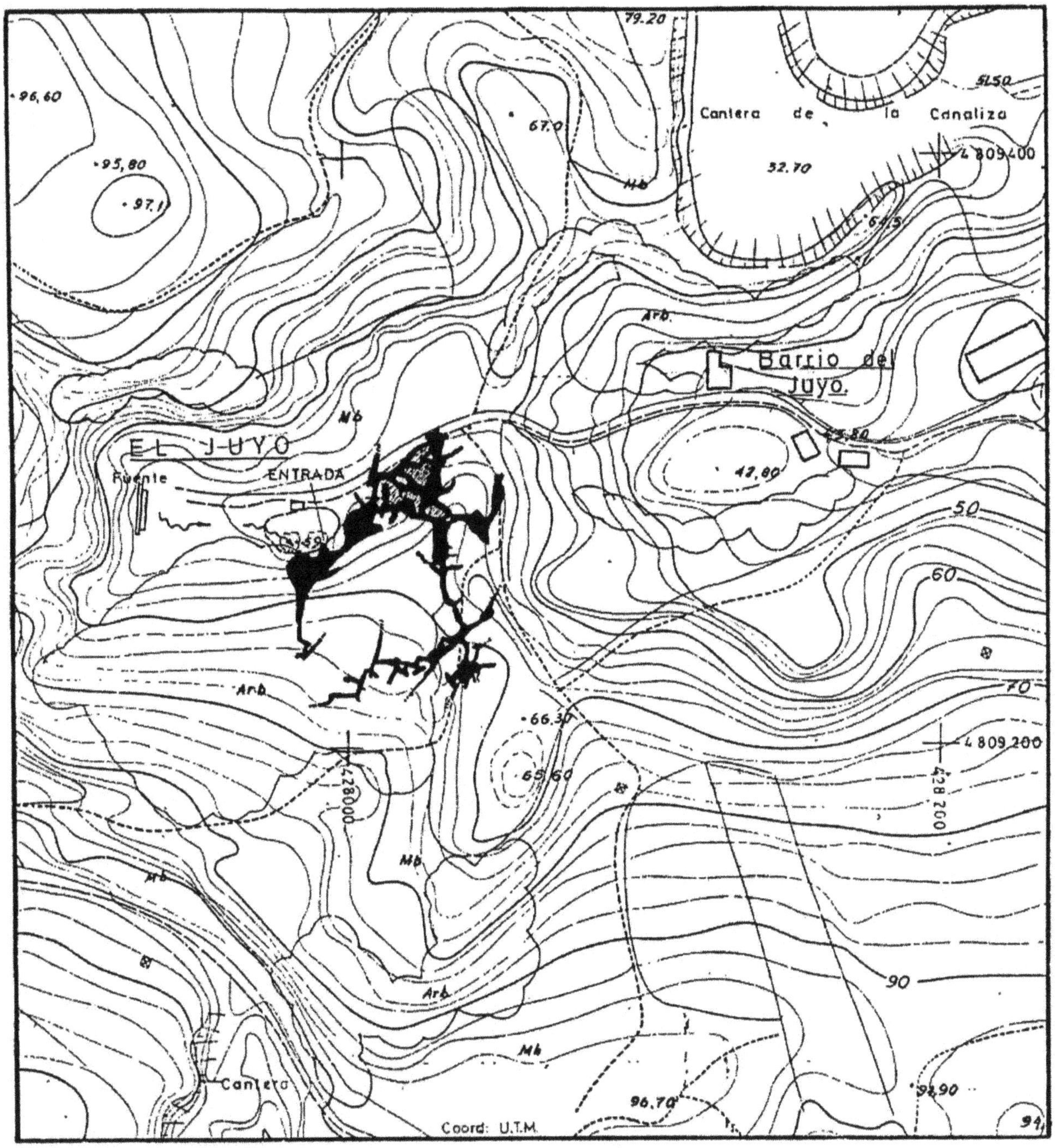

Figure 1.2: Topographic Surroundings of El Juyo (contours at 2 m) (after Barandiarán *et al.* 1987:20).

limestone topography, near the top of a hill offering an excellent view of the surrounding terrain (Figure 1.2). The collapsed prehistoric entrance to El Juyo, like the modern one, is approximately 60 m above sea level, while the adjacent hills reach over 100 m. During the last glacial maximum, the site was slightly further from the coastline (then less than 15 km away). The site was similarly close to the montane region parallel to the north coast of Spain (the Cantabrian Cordillera), which is a westward extension of the Pyrenees. These uplands begin their climb nearly at the present-day coastline itself, serving to circumscribe the approximately 60 km wide coastal strip known informally as Cantabrian Spain (see Straus and Clark 1978; Straus 1992).

Except for traces of sporadic use, including the few artifacts left by what may have been a single Bronze Age explorer and a later (Visigothic) burial in one of the narrow back galleries of the cave, the deposits in El Juyo consist only of Cantabrian Lower Magdalenian materials (Barandiarán *et al.* 1987; Janssens and González Echegaray 1958). These main deposits in the cave's former vestibule were protected by the collapse of the main entrance around 14,000 B.P. and subsequent sealing by a layer of flowstone (Freeman *et al.* 1988:6). The site was rediscovered only in the early 1950s and tested archaeologically for the first time from 1955 to 1957 by Janssens and González Echegaray (1958; see also González Echegaray 1987). New excavations were begun in 1978 and 1979 under the joint direction of Barandiarán, Freeman, González Echegaray and Klein, and in subsequent seasons (1982-83, and 1987-96) by Freeman and González Echegaray (Barandiarán *et al.* 1987; Freeman and González Echegaray 1981; Freeman *et al.* 1988; Klein *et al.* 1981).

The ongoing excavations in the main vestibule cover an area of somewhat more than 40 square meters (see Figure 1.3). About a 3.5 m depth of stratified deposits exists at the site (see Figure 1.4), with an estimated duration of occupation of about 1,000 years (Freeman *et al.* 1988). Charcoal from Levels 4 and 7 has been radiocarbon dated to 13,920+/-240 B.P. and 14,440+/-180 B.P., respectively (Freeman *et al.* 1988:18). The uppermost Lower Magdalenian deposit is Level 4, a thin layer of occupation debris above the nearly sterile flowstone of Level 5. Its lateral variant Level 4S (the sanctuary constructed within El Juyo) intrudes into the lower deposits from Level 4 downward (see Freeman and González Echegaray 1981 for a detailed description of this complexly constructed series of mounds, walls, offering layers, and related features). Level 4 Ramp refers to an area of fill containing materials mixed from other levels, resulting apparently from accumulations of debris removed during the building of the sanctuary. Levels 6, 7, 8, and 9 are the lowermost levels which to date have been exposed over a wide area. Small exposures of Levels 10 and 11 have been made to date. Level 6 is notable for a dug out, stone-lined, multiroomed structure complex yielding much worked and unworked antler, a large lamp in one small lateral room, and an abundance of red deer remains in the main room of the complex (Klein and Cruz Uribe 1987; Klein *et al.* 1981). Level 7 contained two basin-shaped hearths. Level 8 is distinguished by its near midden-like character: it has yielded tens of thousands of shells, mainly of limpets (*Patella vulgata*) and periwinkles (*Littorina littorea*) (Krupa 1994). Fish remains (primarily salmonids) are also abundant. Level 9 is similar in character to Level 8, though mollusc content is generally lower. Levels 10 and 11 lack the shell abundance of the previous two levels and so far contain dense concentrations of large faunal remains and (in the latter level) a large pit.

METHODOLOGY

The microfaunal remains reported here were retrieved by wet-screening after dispersal of clays from the site's deposits as part of a process of full recovery of all archaeological remains (Freeman *et al.* 1988), including flotation for the recovery of botanical remains from all sediments excavated. Screen mesh size employed for the fine fraction is 0.25 mm, one order of magnitude finer than the 0.125 inch (3.2 mm) mesh commonly employed in other modern excavations (Shaffer and Sánchez 1994). The introduction of bias through the loss of remains of smaller species therefore is insignificant. This smaller size mesh used during excavation is also necessary to assure the full recovery of botanical remains. Each sediment subsample was plotted to sector (9 per square meter) and a normal maximum depth of 5 cm. In later processing by the author, dried sediments were reshaken through a 1 mm mesh screen prior to sorting, since it was determined that the finer dust served to obscure the tiniest remains, nor could remains this small (> 0.25 mm and < 1.0 mm) be identified under normal circumstances. Sediments were manipulated using a fine brush. Terrestrial mammalian microfaunal remains and the remains of other taxonomic groups (fish, birds, reptiles and amphibians) and bone and lithic microtools were separated from the fine sediments under constant illumination. The microfauna recovered in the larger screen mesh fraction (2-3 mm) was also incorporated into this analysis, as were the rare instances of mapped microfaunal remains spotted during excavation.

Most species were identified from isolated teeth, since complete or partial tooth rows were rarely recovered. The majority of postcranial remains could only be identified to general microfaunal size class (Small, Medium and Large) and are analized taphonomically in Chapter 2. An exception to this pattern are the postcranial remains of moles (*Talpa*), which are highly robust and identifiable. Isolated postcranial remains of larger species of microfauna (mustelids, hedgehog, etc.) were also sometimes identified. The basic enumeration of microfaunal species therefore is expressed in terms of the Number of Identified Teeth (NIT), with the Minimum Number of Individuals (MNI) derived from (sided) NIT counts. Since teeth are generally among the most durable elements of the vertebrate skeleton, no major loss of information is incurred by this failure to identify postcranial remains to species level.

Chapter 2 also covers the other aspects of taphonomy relevant to this analysis. These include the taphonomic signatures of various predator/scavenger types that prey upon small mammals. These species include diurnal and nocturnal raptors, pack rats, porcupines, felids, mustelids, canids, viverrids and herpestids. The power of natural forces such as moving water, traps formed by sinkholes, and in-den mortality to concentrate bones is also be discussed. These known modern patterns are related to the El Juyo sample in order to define the possible ways in which the bones were introduced

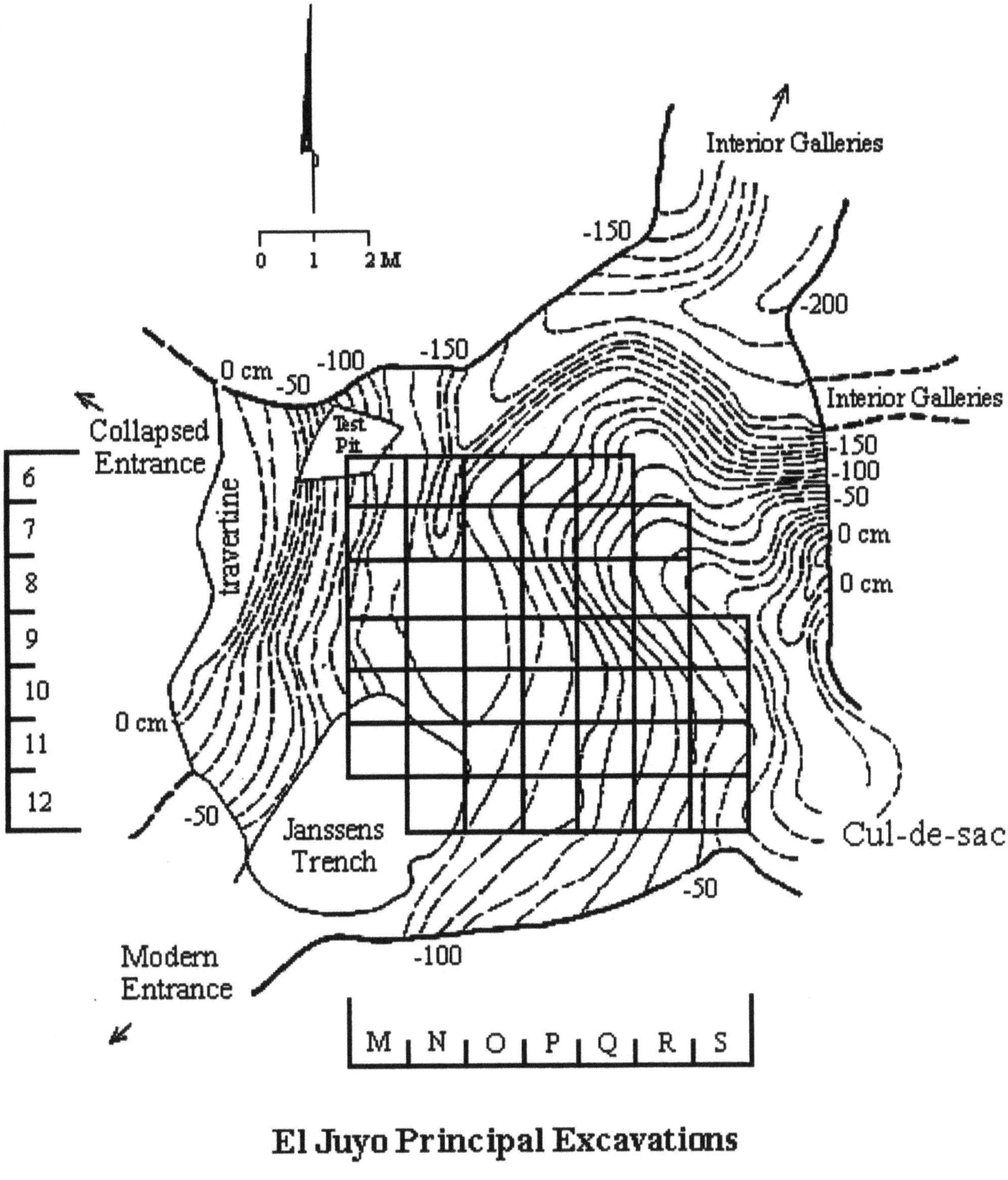

Figure 1.3: Areal Extent of El Juyo Excavations (after Freeman *et al.* 1988:23).

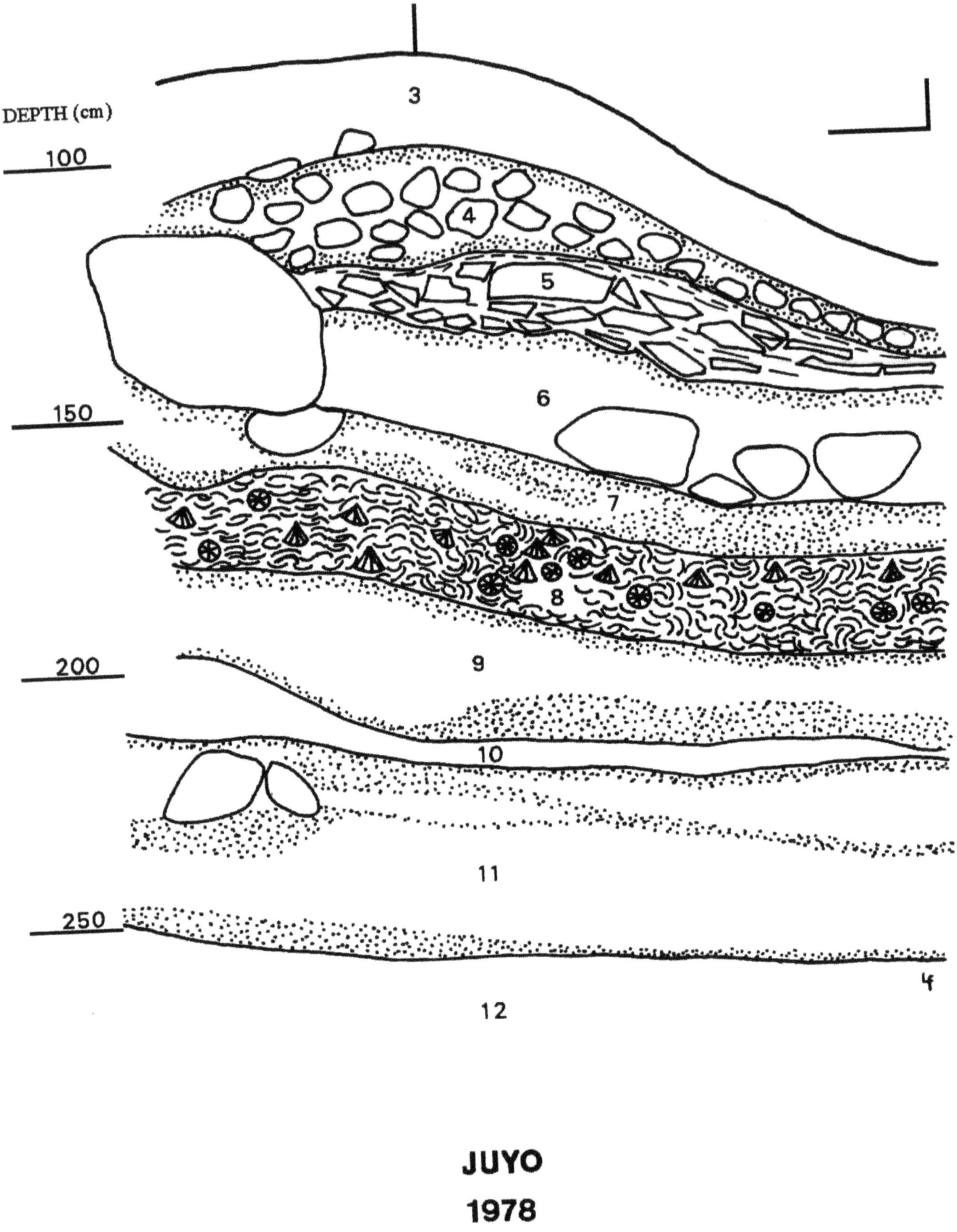

Figure 1.4: Stratigraphic Profile of El Juyo (after Barandiarán *et al.* 1987:50).

to the site. These in turn shape the interpretation of the paleoenvironmental data and allow for greater resolution of reconstruction. These data also illuminate the competing use of the cave by humans, raptors, and other species such as hedgehog or denning carnivores. Spatial data reflect site usage and surface alteration patterns (i.e. structure building) by El Juyo's Magdalenian occupants, since most of these remains were introduced independently of human action. Their natural tendency to collect in low points on the site surface illuminates in some cases the duration of construction, particularly in the Sanctuary area of Level 4.

Species identification also aids in determining the probable accumulators of the microfaunal assemblage. Nocturnal prey species tend to indicate nocturnal predators as an agency of accumulation. The size classes of the accumulated species also aids in primary predator identification, since predators often specialize, hunting within a given size range of prey (Goszczynski 1974, 1977). Determination of agencies of accumulation is of great importance towards the evaluation of possible biases in relative species abundance. Evidence of digestive corrosion and weathering (Andrews 1990; Behrensmeyer 1978) also helps to identify the taphonomic history of the assemblage. Some predators of small fauna (canids and diurnal raptors most relevantly) have a sufficiently low stomach pH to corrode the bones of their prey (Andrews 1990; Clark 1972; Cummings *et al.* 1976; Duke *et al.* 1975; Hoffman 1988; Korth 1979). Such corrosion does not normally occur in the stomachs of nocturnal raptors (owls), which as a group tend to egest very complete and undamaged osseous remains in their pellets (Dodson and Wexlar 1979; Levinson 1982; Pokines and Kerbis Peterhans 1998). In addition, any corrosion on a break surface would indicate that breakage occurred as a result of predation and not as a result of trampling while in the ground or through other postmortem destructive forces. These data aid in the characterization of the relative effects of trampling, water transport and initial deposition upon the dispersal pattern of the microfauna in El Juyo. This information in turn aids in the determination of the origin(s) of the microfauna within the site and their reliability as paleoenvironmental indicators.

Age profiles were derived for each species sufficiently represented to help determine probable agencies of accumulation, since predator species prey upon different age segments of some prey populations. Aging for some species of small mammal is accomplished using tooth wear stages (Adamczewska-Andrzejewska 1967; Delany and Davis 1961; Trout 1978) or estimated from degree of postcranial fusion. Determination of age using length of the maxillary diastema for species with continuously growing teeth (Saint Girons 1973) was not practical, due to the low numbers of these elements that survived and could be assigned to species. Very young small mammals still in a burrow or den are unlikely to be taken by raptors and may be introduced into an archaeological site by terrestrial predators or through natural death in situ.

The location of the microfaunal remains within the site is also discussed in Chapter 2. Concentration near the former cave entrance or under a likely roosting place for a raptor within the twilight zone helps to indicate the agency of introduction into the site. Concentration around an area of food storage or disposal may indicate commensalism by micromammalian species or even consumption by humans, thus indicating a direct human-environment interaction.

Chapter 3 describes the morphological and ecological characteristics of the species encountered in the Paleolithic levels from Cantabrian Spain, as well as those from modern raptor pellet- and trapping-derived collections. This information is presented and summarized to allow the reconstruction of Pleistocene climate based upon the relative abundance of these taxa. The reproductive cycles of these prey species are also informative concerning the patterns of site usage by predators and humans (discussed in Chapter 10).

Chapter 4 presents new data on the modern microfauna of Cantabrian Spain along with a review of studies in the Iberian Peninsula. A sample of owl and hawk pellets from the rural environs of El Juyo has been collected as a baseline for comparison of local environmental change. The area holds many ideal roosting sites, and the author has observed many species of pellet-producing raptor (barn owl *Tyto alba*, peregrine falcon *Falco peregrinus* and buzzard *Buteo buteo*) hunting near El Juyo. Other species of owl common to northern Spain include Scops owl (*Otus scops*), eagle owl (*Bubo bubo*), little owl (*Athene noctua*), tawny owl (*Strix aluco*), long-eared owl (*Otus asio*), and short-eared owl (*Asio flammeus*) (Burton 1973). The osseous remains contained in the regurgitated pellets of these species give a thorough picture of the modern small fauna community in the immediate site area. They also illustrate the radical changes the small mammal community has undergone as a result of foreign species introduction and landscape alteration due to agriculture, pastoralism, urbanization, and forestry. The modern small mammal community in the immediate area around El Juyo and other Cantabrian locations has also been sampled by trapping to indicate the changes in its makeup. The more complete faunal composition and larger sample accumulated from owl pellets allows direct comparison with the microfauna remains from El Juyo, since owls were their primary accumulators during Lower Magdalenian occupation.

As will be discussed, nocturnal raptors are far better accumulators of microfauna within their predatory range than are trap arrays and as such give far more complete information on the relative abundances of the species present. These data serve as a baseline for comparison with the Pleistocene data and as a source of biometric data. They also demonstrate how the modern assemblages derived from owl pellets relate to the modern vegetation and climate, and therefore how assemblages introduced to archaeological sites by owls in the past can be used to reconstruct paleoenvironments.

Chapter 5 details the microfauna recovered from the new excavations at El Juyo. These species include a single molar identified to *Pliomys lenki*, a now-extinct microtine. This remain represents perhaps the latest survival of this species in Cantabrian Spain. Changes in species composition and diversity among the levels is interpreted in a taphonomic framework, including such factors as the ecological habits (selectivity of predation, choice of prey, denning/roosting habits, etc.) of the accumulating species. In this manner the paleoenvironment of El Juyo is examined. This analysis is later extended where possible to other Cantabrian sites yielding sufficient amounts of microfauna, including Amalda (Pemán 1990), Aitzbitarte (Altuna 1972), Ekain (Zabala 1984) and Erralla (Pemán 1985).

Chapter 6 (and Appendix A) present the results of biometrical analysis of the microfauna from El Juyo. These data allow comparison with those derived from other sites of Pleistocene and Holocene Europe for microevolutionary changes. These small mammal microevolutionary changes are critical to Pleistocene chronology as built upon the succession of evolutionary forms (Chaline 1974, 1985; Heinrich 1982, 1987; Hinton 1911, 1926; Janossy 1975; Stuart 1980, 1981, 1982; Tchernov 1968, 1975, 1979, 1982; Von Koenigswald 1973). Size changes may also reflect changes in environmental parameters such as rainfall or temperature (Klein 1991) and will be compared with modern data from small fauna populations existing across varying climatic regimes. Biometrical analysis was also performed, in some cases, to separate the closely related species of a genus where this was not possible on morphological grounds alone. Species identification within a genus may affect paleoenvironmental interpretation.

Chapter 7 discusses the other available sources of paleoenvironmental data from El Juyo. These are correlation with global ocean level chronology, sedimentology, palynology, large terrestrial mammals (red deer, bison, horse, etc.), aquatic resources (fish and molluscs), and macrobotanical remains (carbonized seeds and other plant remains). The limitations and biases of these methods are also discussed. These multiple sources along with the microfaunal data are synthesized in Chapter 8 for El Juyo and the Cantabrian region to give a more complete paleoenvironmental reconstruction of the Lower Magdalenian and the surrounding periods.

Chapter 9 reports the finds of larger microfauna from El Juyo and other Cantabrian Spanish sites. While most of these taxa fall within the 5 kg upper boundary that arbitrarily is used to delimit microfauna from macrofauna, this group is characterized by a general suitability for fur exploitation. The represented species include weasel (*Mustela nivalis*), ermine (*M. erminea*), polecat (*M. putorius*), mink (*M. lutreola*), fox (*Vulpes vulpes*), lynx (*Felis lynx*), rabbit (*Oryctolagus cuniculus*) and hare (*Lepus* spp.). This group is therefore examined separately from the smaller microfauna (rodents and insectivores), as their introduction to the site in most cases was probably the result of direct human action. The pattern of these remains (overabundance of foot elements) suggests that these taxa were exploited for furs. These new data from El Juyo are discussed within the framework of small carnivoran/lagomorph recovery from other sites throughout the Cantabrian Paleolithic. An hypothesized specialized exploitation of these small mammals for their furs is related to the general pattern of faunal exploitation in Cantabrian Spain.

The changing human ecological interactions (faunal exploitation, landscape usage, site usage, technology and population movement) for the Cantabrian Paleolithic in general and the Lower Magdalenian period in particular are outlined in Chapter 10. These patterns are related to the paleoenvironmental reconstruction posited in the previous chapters. It will be shown that these parameters indicate an increasing intensity and organization of environmental exploitation with increasing separation from the constraints of the fluctuating Late Pleistocene climate and environment through the mediation of cultural mechanisms. By the time of the Lower Magdalenian occupation of El Juyo and the subsequent Upper Magdalenian and Azilian periods, resources were being "wild harvested" (Freeman 1973, 1975) with an efficiency and regularity that would presage the later changes in environmental exploitation of the Neolithic period.

Final conclusions are presented in Chapter 11. These include a summary of the presented research along with suggestions for future Upper Paleolithic research from a paleoenvironmental perspective.

NOTES ON TERMINOLOGY

For the sake of clarity, all teeth will be designated by the following shorthand system. Teeth listed with a single superscript (upper) or subscript (lower) number will refer to that side alone, as in "M_3" referring to the lower right third molar. The designation "${}^3M^3$" refers to the upper third molar in general, both left and right. This system is at odds with other authors (Hillson 1986), who use "M^3" to refer to either left/right upper third molar or specifically the right upper third molar and thereby may introduce confusion. The conventions of letter abbreviations still hold: "M" refers to molar, "P" refers to premolar, "C" refers to canine, and "I" refers to incisor. Uppercase letters will always refer to permanent adult dentition, while lowercase letters refer to juvenile deciduous dentition.

CHAPTER 2
TAPHONOMY

A series of factors influences the passage of microfaunal remains from living creatures to their final publication in archaeological monographs and articles. The taphonomic status and history of these remains are presented here. The intermediate stages proceed from living animals that are part of the local population, to the dead individuals culled from that population, to defleshed/digested skeletons, to their introduction into a cave and burial in sediments, to the preservation of a subset of these elements, and to their final recovery by fine screening and sorting. Each of these stages involves some loss of elements and hence potential loss of information from the preceding stage (Davis 1987; Klein and Cruz-Uribe 1984; Shipman 1981). Significantly, movement between these stages also may add information in terms of marks upon the bones themselves, their pattern and location, predatory agency of accumulation, and data inherent to the species represented such as mortality profile (Stiner 1994). Factors affecting the final state of microfaunal remains also leave their marks upon them, frequently allowing the determination of the stages in the winnowing process. The goal of taphonomic research therefore is to backtrack up this trail of intermediate stages in order to derive as nearly as possible a description of the original living populations of animals and their interactions (or lack thereof) with the human population of a site. Even a partial reconstruction of this passage allows a finer assessment of the biases affecting the proportions of a microfaunal assemblage and its relationship to the environment from which it was derived.

The conclusions of the analysis presented here are that owls were overwhelmingly responsible for the introduction of microfaunal remains into the site of El Juyo and are the most likely vector of introduction for most Cantabrian cave sites. More specifically, barn owls (*Tyto alba*) are by far the most likely source among the possible species of nocturnal raptor. The evidence for this conclusion comes from multiple sources, which include the site setting, the amount and distribution of microfauna in the cave, their age profiles, size distribution, habitat preferences, activity patterns, and the degree of breakage (light) and digestive corrosion (almost none) upon the individual remains. An exception to this pattern is found with the remains of hedgehog (*Erinaceus europaeus*), a larger species of microfauna which probably found its way into El Juyo as a commensal. Also exceptional are the remains of small fur-bearing mammals such as weasel (*Mustela nivalis*), which may have entered El Juyo from human trapping activity (Chapter 9). The majority of other microfaunal remains consists of rodents and insectivores and were introduced as the products of barn owl pellet regurgitation.

The identification of the principal microfaunal taphonomic agency at El Juyo as barn owls is significant. As discussed below, this particular predator is a well-researched dietary generalist whose prey accurately reflects the environment within an individual's hunting range. As such, corrections for any biases incurred when using its prey to interpret paleoenvironment are relatively straightforward (Chapter 4). Other predatory agencies of microfaunal accumulation have differing and usually more involved biases for which to account (greater differential destruction of prey bone, dietary specialization on a few species, or feeding habits that differentially concentrate the remains of prey taxa within an archaeological site). These other possible agencies are eliminated (below) on multiple grounds as the primary forces responsible for the El Juyo microfaunal sample. The present investigation therefore focuses upon barn owls as taphonomic agencies.

Microvertebrate taphonomy has only a brief history, since the field was formally defined five decades ago as "the science of the laws of embedding" (Efremov 1940:93). In broader terms taphonomy is the study of the processes affecting the remains of living creatures from the point of death through their excavation or discovery and subsequent manipulation. Microvertebrates (defined as species no larger than 5 kg) often make a large portion of this journey--from death until deposition in the place where they will be incorporated into sediments--inside an owl. This abbreviated journey is at the opposite end of a spectrum from ungulates that are often slain by one predator, stolen and consumed other predators or scavengers, disarticulated and differentially transported in pieces to provision young, fed and gnawed upon and dispersed by a succession of scavenger species (Blumenschine 1986; Haynes 1980, 1982), sometimes regurgitated as masses of bone by hyaenids (Kerbis Peterhans 1990; Kerbis Peterhans and Horwitz 1992; Sutcliffe 1970), and left exposed to the elements for extended periods of time before concentration or burial of their remains by geomorphic agencies (Behrensmeyer 1978, 1991). Owls thus encompass multiple, successive taphonomic processes in one organism: selection and mortality of prey, prey dismemberment, digestion of soft tissues, initial bone breakage and digestion, and the transport and concentration of vertebrate remains to the site of their incorporation into a paleontological or archaeological sequence.

Barn owls (*Tyto alba*) are among the most important accumulators of microvertebrate assemblages. The genus *Tyto* had evolved by the Miocene Period about 25 million years ago (Burton 1973), though the evolutionary history of birds is less well known due to the general fragility of their remains. Assuming continuity of modern behavior into the past, the taphonomic processes encompassed by barn owls or their near relatives have operated contemporaneously with hominid evolution for millions of years. Within the scope of the present analysis (less than 20,000 years), barn owl behavior is assumed not to have undergone any significant change.

Interpretation of microfaunal assemblages is crucial for the reconstruction of the environmental setting of human evolution. Barn owls tend to regurgitate largely intact bones of their small prey, giving a very complete record of what they have eaten and making identification of skeletal elements relatively easy. In contrast with other accumulators of microfaunal remains, barn owls readily roost and nest in abandoned or human-occupied structures and living sites, sometimes in close proximity to ongoing human activities. This near-commensalism means that remains accumulated by barn owls will often be intermixed with archaeological materials (Brain 1974, 1980, 1981; Brain and Brain 1977). One can therefore correlate evidence of human activity with an independent source of paleoenvironmental information. Research into the taphonomic signature of barn owls--the type of bone assemblage that they create while sampling small animal populations--is of importance regarding the paleoecological significance of those remains.

Most previous barn owl pellet research has focused upon dietary diversity and ecological interactions between predator and prey. Barn owl prey samples have been used to identify and quantify the small mammal populations of regions such as Britain (Glue 1967, 1970, 1974, 1975), Colorado (Marti 1973a) and Spain (Brunet-Lecomte and Delibes 1984; Delibes *et al.* 1984). These studies are valuable to the taphonomist in their analyses of the predation habits, hunting territories, and size range of prey taken by owls.

Taphonomic studies of microvertebrate assemblages have allowed a greater understanding of the relative depositional biases affecting the paleoecological information that they may yield. Many studies have focused upon the varying taphonomic signatures that a particular predator type may leave as it concentrates the bones of microvertebrate prey in or near a den, roost, or nest site. Concentrators of microvertebrate bones include nocturnal raptors (Andrews 1990; Avery 1982a, 1982b, 1987, 1992, 1993, 1995a, 1995b; Brain 1980, 1981; Denys and Mahboudi 1992; Dodson and Wexlar 1979; Hoffman 1988; Kusmer 1992; Levinson 1982; Pokines and Kerbis Peterhans 1998; Raczynski and Ruprecht 1974; Thackeray 1987; Thackeray and Avery 1990), diurnal raptors (Mayhew 1977), aquatic birds (Wilson 1987), small mammalian carnivores (Andrews and Nesbit Evans 1983; Mellett 1974; Oksanen 1983), and crocodilians (Fisher 1981a, 1981b, 1981c). Other taphonomic agencies that may concentrate microvertebrate bones include packrats (Elias 1987), harvester ants (Shipman and Walker 1980), natural sinkhole traps (Andrews 1990; Brain 1980, 1981) and moving water (Dodson 1973; Korth 1979; Voorhies 1969; Wolff 1973). Signatures of accumulating agencies include assemblage location, degree and type of skeletal breakage, digestive corrosion, skeletal part representation, size distribution among prey species, and the sex and age distribution within a prey species. Identification of the principal agency of accumulation of a microvertebrate assemblage will aid in the assessment of the relative biases which occurred in the sampling of the living populations.

OWL DIGESTION

Many bird species produce pellets (Bunn *et al.* 1982), the undigested remains of meals. Among avian predator species, those known to egest pellets are distributed among 60 families and 330 species according to a survey by Glue (1970:54). The members of both owl families (Tytonidae and Strigidae, order Strigiformes) egest pellets. These are primarily composed of bone, teeth, fur, feathers and arthropod exoskeleton. Diurnal raptors (families Accipitridae and Pandionidae of the order Accipitriformes; and Falconidae, order Falconiformes) produce pellets, although their osseous content tends to be much lower than those of owls. Members of other bird families such as Ardeidae (herons, bitterns) produce pellets, but these tend to leave few recognizable bones and consist mostly of fur (Glue 1970:54).

The mechanism of pellet formation in owls is itself an important taphonomic process. In owl stomachs recently dispatched prey are converted into compact lumps of soon to be egested bone and fur. While barn owls tend to swallow prey intact, other species such as great horned owl (*Bubo virginianus*) often take prey too large to consume whole (Jones 1952). These prey items must either be torn into pieces before consumption or their flesh eaten off their bones (Einarsen 1956; Howard 1958). This behavior naturally alters relative skeletal representation and breakage percentages in pellets among predator species. Some owl species about the size of barn owls, such as short-eared owls (*Asio flammeus*) habitually dismember prey before consumption even if these items are relatively small (Short and Drew 1962).

After swallowing, food passes to the glandular portion of the stomach (proventriculus), then through a sphincter to the muscular portion (ventriculus or gizzard). This latter stomach component tends to be bag-like and extensible in owls and other avian carnivores (Farner 1960:428-432). It is not heavily muscled in owls, thus contributing to bone preservation. The pyloric outlet is very small (1 mm or less) and located on the superior surface of the ventriculus at the same level as the esophageal opening (Reed and Reed 1928), thus preventing most outflow of solid indigestible elements. In a roentgenographic study of pellet formation in a great horned owl, Grimm and Whitehouse (1963) found that following a meal the ventriculus remains extended for about one hour. The pellet condenses over the next 2-8 hours and is complete by 10 hours after eating. Pellets were typically expelled about 16 hours after eating. Ventricular contraction and accessory contractions of the abdominal wall and proventriculus force the pellet upwards in small steps, the entire final egestion phase lasting about 4 minutes.

Low stomach acidity among owls relative to other carnivores contributes to bone preservation (Duke *et al.* 1976; Duke and Rhoades 1977). Basal stomach pH for Falconiformes is about 1.6, while for owls this figure is about 2.35 (Duke *et al.* 1975). Percentage weight of bone per pellet is therefore much lower among diurnal raptors. Averages around 6.5% of dry pellet mass were obtained for some species of hawk (*Buteo*), falcon (*Falco*) and eagle (*Haliaetus*) as compared to an average of 45.8% among two Strigidae: great horned owl (*Bubo virginianus*) and snowy owl (*Nyctea scandiaca*)

(Duke *et al.* 1975:653). Since the investigators found that average meal to pellet intervals among all the species tested (except for *Haliaetus*) fell between 18 to 22 hours, differences in pellet bone content were not a result of differences in digestion time (Duke *et al.* 1975:654). Clark (1972) found that prey bone made up only 17% of pellet weight for marsh hawks (*Circus cyaneus*) and 44% for short-eared owls (*Asio flammeus*) feeding upon the same prey species.

Stomach pH appears to be the main indicator of prey bone survivability between diurnal and nocturnal raptor pellets, although food habits play an obvious role. Cummings *et al.* (1976) measured loss from cleaned bones of house mouse (*Mus musculus*) in solutions of HCl in concentrations simulating falconiform and strigiform gastric juice (pH 1.66 and 2.35, respectively). Bovid pepsin was also added in varying concentrations. Bone loss by weight was significantly higher throughout the lower pH trials. Losses averaged over 70% at the falconiform concentration, while bones treated with the strigiform concentration averaged below 35% loss by weight (Cummings *et al.* 1976:57). Increasing the supplemental bovine pepsin concentration tended to increase bone losses slightly at each HCl concentration.

Other factors help to explain how most owls egest intact bones of prey. Low mixing of stomach contents (Reed and Reed 1928) contributes to low levels of bone breakage. The high pepsin activity in the stomachs of Strigiformes contributes to effective digestion of prey. Reed and Reed (1928) found the peptic activity of *Bubo virginianus* to be about three times that of a dog's gastric juice at comparable stages of digestion. In addition, gastric pH does not remain at a constant low level. Smith and Richmond (1972) found that barn owl stomach pH rises after feeding until pellet egestion to a maximum of around 6.0. This acidity level increases precipitously within one hour to around 2.0. Smith and Richmond (1972:183-184) also found that a captive barn owl could be induced to egest a pellet by letting it see live prey after a sufficient time lapse from its last feeding. The minimum time for induced egestion was 6.5 hours. Prey bones therefore can spend relatively short periods of time in an owl stomach under certain circumstances.

BARN OWL BEHAVIOR

Tyto alba is a particularly prolific accumulator of small animal remains (Davis 1959; Taylor 1994). Its near-cosmopolitan distribution (Burton 1973) spans a range of habitats, from temperate forests to tropical deserts and rain forest. Throughout this range, barn owl diet is generalist within the size range from large arthropods to subadult *Rattus* and *Arvicola* (about 100 g) (Avery 1993; Campos 1978; Glue 1967, 1970, 1974, 1975; Herrera and Jaksic 1980; Webster 1973). Therefore, barn owl diet tends to indicate local availability of prey species. Barn owls tend not to be selective in their prey choices, although the terrain occupied by a prey species (the owls prefer to hunt in more open habitats) and the activity patterns of prey (a tendency to freeze when startled or to spend more time on open ground, making capture easier) will influence a given species' chances of being taken (Derting and Cranford 1989). While insects, other arthropods such as scorpions, birds, reptiles, amphibians and bats are taken, the bulk of the diet is made up of nocturnal terrestrial mammals. Barn owls excel at sampling local small mammal fauna, such that some species have been found in owl pellets before their presence in a given range was discovered through trapping (Anderson and Long 1961; Hanney 1963; Heim de Balsac and de Beaufort 1969). Some taxa have been first discovered through the analysis of barn owl pellets (e.g. Anderson and Long 1961). The known range of small mammal species is also often extended through this analysis (Bunn *et al.* 1982).

Although occasionally observed to hunt in daylight just before sunset or just after dawn (Bunn 1972), barn owls are primarily nocturnal predators (Burton 1973; Gilliard 1958). Special adaptations include the ability to locate prey in very low light levels (Dice 1945) or in total darkness through hearing alone (Payne and Drury 1958; Payne 1971). Most hunting is done on the wing within a radius of a few kilometers from a daytime roost or nesting site. Evans and Emlen (1947:3) found that the hunting range of a barn owl studied in northern California (U.S.A.) comprised roughly 165 acres, or less than a 0.5 km radius in average distance from the roost. Webster (1973:186) found the hunting range of a nesting pair of barn owls in South Westmoorland (England) to cover 147 acres, again less than a 0.5 km radius. Most researchers estimate a larger radius, such as Lovari *et al.* (1976:173) in choosing 0.8 km to 5 km for hunting radii for the barn owl collections they studied in central Italy. Logically, such variables as prey density and competition with other raptors may effect the hunting range for any individual barn owl. Even these larger estimates of hunting radius, however, indicate extremely localized concentration of prey remains at the central roost.

Prey is consumed whole, which contributes to the completeness of dietary representation in pellets. Glue (1967:170) notes that barn owls are very rarely observed to decapitate small mammal prey before consuming it. Bird decapitation is more frequent, especially before feeding to nestlings. Since avian MNI's are usually based upon postcranial elements, avian prey counts should not be significantly affected. Most nights, a pellet containing the remains of the early evening's hunting is egested away from the day roost. Late night/early morning prey typically compose the pellet that a barn owl egests at its day roost before leaving for the following night's hunting (Smith and Richmond 1972). Pellets at a nesting site will also include those egested by the chicks. About half of the diet will therefore be egested at a day roost site.

In order to test whether barn owl pellets are an unbiased reflection of diet, Glading *et al.* (1943) fed captive members of this and other raptor species a natural diet of small mammals and birds. The pellets egested were then blind tested for the contained species. The hawk species fared poorly: their pellets contained almost no identifiable bone. The owls generally and barn owls in particular egested pellets which contained a highly accurate record of their diets. Young bird prey did tend towards underrepresentation in pellets of barn owls, indicating that their bones tend to be completely digested more often or rendered otherwise unidentifiable. This increased avian bone loss is most likely due to the fragility of the bones themselves. In a parallel study, Selleck and Glading (1943) enclosed barn owl nest sites with wire in order to enumerate the prey carried to the nestlings by their parents. They found that birds made up only 6.4% by bulk of the diet, not different significantly from the amount of bird prey in the diet (6.7%) indicated by pellets from the same site. Sparse avian prey in pellets therefore seems to indicate the actual hunting practices of the owls and not differential survivability of bird remains.

A final observation on the relationship between barn owl diet and local prey availability is the occasional hyperabundance of one species. An example of this situation is a pellet sample from Iraq that yielded a large proportion of bats (n = 53, 27.0% of total prey items). These species roosted in the same large structure as the owls and were therefore hyperabundant locally and correspondingly easy to prey upon (Nader 1969). Taylor (1994) also notes instances of barn owl predation upon locally abundant fish (beached grunion) or birds on an island with few terrestrial small prey species.

Barn owls and other owl species with similar diets and roosting habits can rapidly accumulate large numbers of pellets within a cave or human structure. Barn owl pellets are resistant to decay and can remain intact for many years in a dry depositional setting. Those deposited in moister settings are broken down rapidly by bacteria, fungi, and various insect larvae feeding on the fur matrix of the pellets (Levinson 1982), thus dispersing their bone contents. The spherical shape of pellets allows them to tumble into and be dispersed around a cave floor with slight hindrance. Their contents can then be dispersed into cave deposits by the decay of their fur matrix and the physical trampling by any subsequent or simultaneous human occupants.

TAPHONOMIC ANALYSIS OF EL JUYO

The physical layout and setting of the cave of El Juyo narrows the possible venues for entry of microfauna in large numbers. The former entrance (refer back to Figure 1.3) was wide and opened directly onto the main inhabited vestibule area before the partial collapse and sealing of the cave. This entrance was situated at the base of a very steep sinkhole leading directly into the cave. The entrance opened onto the bottom of a large, broad sinkhole in the limestone bedrock. Some small animals could be expected to die or be killed in this funnel-shaped area around the mouth of the cave. Their bones could have been tumbled, washed or tracked into the cave habitation area and incorporated into the sediments. Bones derived from carnivore scats or owl pellets dropped into this area could also have contributed to this source of bones.

Several factors argue against this source of microfauna remains being a major contributor to the assemblage from the cave. Primary is the large size of the assemblage itself, currently comprised of 7,707 identified teeth, 17,770 postcranial and fragmentary teeth remains identified to size class, and 19,976 unidentified microfaunal fragments. An additional 155 teeth were identified to larger species of microfauna (hedgehogs, etc.) discussed separately. While even very slow water flow may transport microfaunal remains (Korth 1979), it is difficult to envision more than a small proportion of these remains being washed into El Juyo from its immediate neighborhood. Andrews (1990) concludes that water is more likely to disperse small bones than to concentrate them. While the processes of microfaunal skeletal exposure degradation are less well researched than those of larger fauna, the two processes must be at least in part comparable due to their identical bone chemistry. Korth (1979) found through experimentation that aerially exposed small mammal carcasses must be buried quickly or bones lying on the surface will be destroyed. None of the El Juyo remains exhibits signs of exposure typically caused by extended aerial exposure on large faunal remains (Andrews 1990; Behrensmeyer 1978), such as cracking of exposed surfaces, flaking into pieces, or bleaching. The robust state of this assemblage (especially when compared with the fish remains recovered in an identical manner from El Juyo) also argues against washing into the cave from outside. Many delicate postcranial remains are found in abundance, including rodent and insectivore phalanges, metapodials, and vertebrae. Such tiny remains could not have been transported in bulk without a significant, clear flow of water, since their small size precludes tumbling action. No significant flow of water was ever entrant into the mouth of El Juyo. It is possible that some remains making this destructive journey from outside the cave did enter in previously greater numbers, only to be destroyed later by postdepositional forces or the excavation and recovery processes. Introduction in this manner does not characterize the El Juyo assemblage.

Limestone caves also can act as natural sinkhole traps for small animals (Andrews 1990; Brain 1980, 1981). Small mammals falling into a pit quickly die of starvation due to their very high metabolic requirements, with their largely intact remains becoming incorporated into site sediments. The physical layout of El Juyo precludes this scenario, since the wide former entrance allowed easy egress, nor does any natural chimney feature extend to the surface above the inhabited vestibule area. The potential for this process to accumulate microfauna must be examined in the case of each cave site setting. The current nearly sealed entrance of El Juyo does allow the occasional trapping of modern microfauna which cannot find an exit from the back galleries of the cave. Some recent remains of rat (*Rattus rattus*) and red fox (*Vulpes vulpes*) have been accumulated in this manner.

With the exception of snow vole (*Microtus nivalis*) the behavior of these small mammal species (Chapter 3) precludes their using the cave as a denning site and thereby adding their remains through natural mortality (Avery 1982a). *M. nivalis*, however, prefers rocky crevices among talus slopes and other areas of exposed rock over natural caverns. While many of the local rodent and insectivore species are naturally fossorial, their burrow systems are constructed in well-drained, easily tunneled topsoil. These tunnels in some cases also supply food: earthworms and arthropods in the case of moles (*Talpa* spp.) and plant roots in the case of arctic vole (*Microtus oeconomus*). None of these criteria are met by the densely trampled, object-dense, poorly-drained and vegetation-free sediments of an inhabited limestone cave vestibule.

Almost no burrowing disturbance has been detected in the El Juyo deposits. No Chiroptera (bat) remains have yet been recovered, though these appear in low amounts in other Cantabrian sites due to their natural cave-roosting tendencies. The remains of the larger commensal species hedgehog (*Erinaceus europaeus*) may have entered the El Juyo deposits through their natural in situ mortality and subsequent burial of skeletal elements. With the exception of this latter species, denning of microfauna species within the cave in between episodes of human occupation is not indicated.

Human utilization and disposal of these species, with the important exception of small fur-bearing carnivores (Chapter 9), is not a significant factor in their accumulation. Given their low yield of meat relative to the effort necessary for their capture, small rodents and insectivores (most under 40 g in total adult mass) have rarely been a significant human food resource. Their use as a food source is especially unlikely in a rich subsistence environment like Cantabrian Spain, where a wide range of larger or more efficiently harvested species existed (see Chapter 7). More directly, the pattern of bone recovery and breakage argues against their use by humans. Such complete bones are an unlikely result of any cooking process, which would weaken bones and cause greater postdepositional destruction than has been the case.

In addition, the spatial distribution of microfauna in the site indicates their introduction by non-human means. Though not all areas of the upper habitation levels have been completely excavated (leaving gaps in distribution), preliminary analysis indicates that microfaunal remains cluster towards the northeast corner of the cave vestibule, directly under a natural large fissure in the cave ceiling or in natural lowspots downslope from this location. This ceiling chamber is out of unaided human reach, and is by far the most suitable roosting place within this chamber for a raptor. It is also within the twilight zone of the cave, which is preferred by raptors. The roost's shape also allows easy spillage of egested pellets down into the main floor of the cave. Other locations within the vestibule offer no adequate platform space or are within easy access to terrestrial predators of raptors: humans, canids, felids and others. Since human structural foundations have been discovered, it remains slightly possible that owls roosted on abandoned huts or similar constructions within the vestibule before these collapsed or were displaced by subsequent human occupations. Unfortunately for modern investigation, the layout of the cave would also have spilled much owl egestation into an interior, uninhabited and unexcavated gallery of El Juyo, where they are lost to recovery among stratified human occupation remains. The El Juyo sample is therefore a random subset of the complete microfauna accumulated at the site.

The spatial arrangement of the cave also suggests that humans and birds would not have been occupants at the same time. The previous entrance was small enough to place any humans or human structures (such as in Level 6) in the direct flight path of the birds. The frequent use of fire in the cave might also deter roosting. The complete absence to date of any identifiable chiropteran remains within the Lower Magdalenian deposits also suggests that human presence was enough to deter simultaneous site occupation by these species and perhaps others. These species, however, are more likely to have roosted and died *in situ* in the (unexcavated) interior galleries of the cave. Pellets accumulated in the cave during episodes of owl occupation may then be incorporated into subsequent human occupation debris, intermixing the two even though each species occupies the cave at different times. Distribution patterns of microfauna may also illuminate the alteration of the site surface and structure building by humans, since round objects randomly trampled over the surface of the site will naturally tend to settle into open, low-lying spaces. Open pit structures, such as the stone room of Level 6 and especially the sanctuary structure of Level 4S would concentrate microfauna egested by the owls. These patterns and their implications for human site usage will be discussed in Chapter 10.

EL JUYO SKELETAL RECOVERY AND BREAKAGE

Table 2.1 lists microfaunal remains from El Juyo including identified teeth (described in Chapter 5) that may have been introduced by barn owls. These have been grouped by general size class, since for the most part these remains cannot be ascribed to species or genus. The main exception to this rule are the postcranial remains of *Talpa*, which are highly identifiable (discussed below). The main size classes (Small, Medium and Large) generally conform to the following taxa whose presence has been determined by their dental remains:

Small: This class includes the postcranial remains of shrews. Those identified from El Juyo are *Neomys fodiens*, *N. anomalus*, *Sorex minutus* and *S. coronatus*. Each of these species has an average adult mass of less than 15 g (Chapter 3). These have been identified in low numbers from the site, which is reflected by the relatively low total of non-identified tooth remains assignable to this general class (n = 261). These totals include incisor and fragmentary molar remains, since (unlike microtines) these were identifiable to species. Maxillary and mandibular counts include those with identifiable teeth that were assignable to species, as well as unidentifiable fragments assignable only to size class.

Medium: This class includes the remains other than identified teeth of small rodents. These have an average adult body mass of 20-50 g. The taxa whose presence has been determined from their molars and that conform to this class are *Apodemus* (Muridae) and *Pliomys lenki*, *Pitymys pyrenaicus*, *P. lusitanicus*, *Microtus nivalis*, *M. oeconomus*, *M. agrestis* and *M. arvalis* (Microtidae). The possible presence of the postcranial remains of other taxa conforming to this general size class (such as *Clethrionomys glareolus*) cannot be discounted. These remains also include rodent incisors, which were not assignable to species. Mandibular and maxillary remains are summed together even where these were identifiable to species, since this was possible only in rare cases. Microtine molar fragments are also included. The large total of these non-identified tooth remains (n = 11,287) corresponds with the large numbers of these taxa identified from their teeth.

Table 2.1: Microfaunal Remains from El Juyo by Size Class and Level

SMALL	11	10	9	8	8/9	7	6/9	6	5	4R	4S	4	2/4	3	2	1	Total	% Rep
Maxilla	0	0	0	0	0	0	0	4	0	0	3	2	0	0	0	0	9	13 6
Mandible	1	0	1	5	0	0	0	25	0	0	21	13	0	0	0	0	66	100 0
Teeth	2	0	2	17	0	1	0	94	0	0	65	45	0	0	0	0	226	*
Scapula	0	0	0	0	0	0	0	1	0	0	0	0	0	0	0	0	1	1 5
Humerus	0	1	1	5	0	3	0	14	0	2	23	3	0	0	0	0	52	78 8
Radius	0	0	0	0	0	0	0	1	0	1	2	0	0	0	0	0	4	6 1
Ulna	0	0	0	1	0	0	0	2	0	1	5	3	0	0	0	0	12	18 2
Pelvis	1	0	1	0	0	1	0	3	0	0	2	0	0	0	0	0	8	12 1
Femur	0	0	3	6	0	1	0	7	0	0	18	9	0	0	0	0	44	66 7
Tibio-fibula	1	0	1	3	0	0	0	8	0	0	7	7	0	0	0	0	27	40 9
Calcaneus	0	0	1	0	0	0	0	3	0	1	13	1	0	0	0	0	19	28 8
Talus	0	0	0	1	0	2	0	5	0	1	9	1	0	0	0	0	19	28 8
Total	5	1	10	38	0	8	0	167	0	6	168	84	0	0	0	0	487	
MEDIUM	**11**	**10**	**9**	**8**	**8/9**	**7**	**6/9**	**6**	**5**	**4R**	**4S**	**4**	**2/4**	**3**	**2**	**1**	**Total**	**% Rep**
Maxilla	0	0	2	12	0	0	0	40	0	8	80	48	0	0	1	0	191	15 0
Mandible	0	0	0	24	0	6	0	85	0	15	322	78	0	1	0	0	531	41 6
Incisor	13	4	33	105	2	22	2	410	0	76	1536	341	1	1	2	2	2550	100 0
Molar	12	3	41	230	0	64	7	728	1	138	2540	741	7	14	7	1	4534	67 6
Molar frag	4	1	10	34	0	9	1	60	0	18	373	121	6	3	0	0	640	**
Scapula	0	0	0	2	0	1	0	72	0	10	115	49	0	1	0	0	250	19 6
Humerus	2	0	12	26	0	9	1	157	1	31	311	169	3	0	0	1	723	56 7
Radius	0	0	0	13	0	1	1	42	3	7	90	65	1	0	0	0	223	17 5
Ulna	0	0	4	11	0	6	1	97	0	11	199	106	0	0	0	0	435	34 1
Pelvis	0	1	3	5	0	2	0	35	0	0	47	44	0	0	0	0	137	10 7
Femur	1	0	6	23	0	4	0	93	1	22	199	107	2	0	1	0	459	36 0
Tibio-fibula	1	1	7	14	0	3	0	92	0	13	178	97	2	0	0	0	408	32 0
Calcaneus	2	0	3	4	0	4	0	20	0	4	45	40	0	0	0	0	122	9 6
Talus	0	0	2	6	0	2	0	16	0	4	31	23	0	0	0	0	84	6 6
Total	35	10	123	509	2	133	13	1947	6	357	6066	2029	22	20	11	4	11287	
LARGE	**11**	**10**	**9**	**8**	**8/9**	**7**	**6/9**	**6**	**5**	**4R**	**4S**	**4**	**2/4**	**3**	**2**	**1**	**Total**	**% Rep**
Maxilla	0	0	1	2	0	0	0	17	0	0	38	11	1	0	0	0	70	14 5
Mandible	0	0	3	5	0	3	0	42	0	9	107	39	0	0	0	0	208	43 0
Incisor	10	2	12	34	0	6	0	166	0	45	531	89	4	0	1	2	902	93 3
Molar	3	3	39	112	0	46	1	459	0	89	1583	301	1	6	2	2	2647	100 0
Molar frag	0	2	3	12	0	1	0	14	0	5	200	16	0	1	0	0	254	**
Scapula	1	0	2	5	0	1	0	15	0	5	40	25	0	0	0	0	94	19 4
Humerus	0	3	6	9	0	3	0	52	0	10	123	36	1	0	0	0	243	50 3
Radius	0	1	1	2	0	0	0	30	0	8	74	24	1	0	0	0	141	29 2
Ulna	0	0	4	2	0	3	0	30	0	4	89	21	0	0	0	0	153	31 6
Pelvis	0	0	2	2	0	0	0	25	0	1	32	17	0	0	0	0	79	16 3
Femur	0	0	2	8	0	1	0	56	2	7	62	39	1	0	0	0	178	36 8
Tibio-fibula	2	0	2	12	0	3	1	36	0	8	55	37	0	0	0	0	156	32 3
Calcaneus	0	0	2	5	0	0	0	13	0	7	56	16	0	0	1	0	100	20 7
Talus	0	0	2	3	0	1	0	10	0	3	13	8	0	0	0	0	40	8 3
Total	16	11	81	213	0	68	2	965	2	201	3003	679	9	7	4	4	5265	

Table 2.1: (Cont.)

Talpa	**11**	**10**	**9**	**8**	**8/9**	**7**	**6/9**	**6**	**5**	**4R**	**4S**	**4**	**2/4**	**3**	**2**	**1**	**Total**
Teeth	3	0	11	11	0	22	1	107	0	11	101	32	0	1	0	0	300
Maxilla	0	0	1	0	0	1	0	3	0	0	2	3	0	0	0	0	10
Mandible	3	0	1	0	0	3	1	10	0	2	26	6	0	0	0	0	52
Scapula	0	0	0	0	0	0	0	1	0	0	1	1	0	0	0	0	3
Humerus	0	0	0	2	0	2	0	9	0	0	10	10	0	0	0	0	33
Radius	0	0	0	2	0	1	0	5	0	1	5	3	0	0	1	0	18
Ulna	0	0	0	0	0	3	0	4	0	0	7	4	0	1	0	0	19
Pelvis	0	0	0	0	0	0	0	0	0	0	0	0	0	0	0	0	0
Femur	0	0	1	0	0	0	0	6	0	0	5	5	0	1	0	0	18
Tibio-Fibula	0	0	2	0	0	1	0	8	0	1	19	6	0	0	0	0	37
Calcaneus	1	0	0	0	0	0	0	3	0	0	5	5	0	0	0	0	14
Talus	1	0	0	0	0	0	0	2	0	0	2	0	0	0	0	0	5
Phal/Carp/Meta	4	1	8	42	1	69	1	306	1	32	266	120	4	1	0	0	856
Total	12	1	24	57	1	102	3	464	1	47	449	195	4	4	1	0	1365
OTHER	**11**	**10**	**9**	**8**	**8/9**	**7**	**6/9**	**6**	**5**	**4R**	**4S**	**4**	**2/4**	**3**	**2**	**1**	**Total**
Rib	4	0	7	42	0	15	0	228	0	53	293	172	0	0	0	0	814
Vert./Stern.	4	3	61	131	1	26	4	534	3	83	1037	610	8	1	0	0	2506
Metapodial	5	1	24	38	0	26	3	285	0	57	438	124	4		1	1	1007
Phalanges	11	3	35	145	0	34	2	476	0	88	1153	511	6	4	4	1	2473
Total Ident.	92	30	365	1173	4	412	27	5066	12	892	12607	4404	53	36	21	10	25204
Unident. Frag	109	38	463	1163	14	246	6	3806	12	630	8691	4660	65	38	30	5	19976
Total All	212	78	837	2344	18	665	33	8868	29	1522	21298	9068	118	77	53	16	45180

* Not calculated due to differences in species total teeth.

** Included with molar calculations

Table 2.2: Comparison of Barn Owl Prey Skeletal Representation (%).

	Negev[1]	Kusmer (1990)		Dodson & Wexlar (1979)	Korth (1979)	Hoffman (1988)	Andrews (1990)		MEAN
MNI=	239	94	34	17	20	50	60	71	(585)[2]
Maxilla	92.2	94	74	100.0	87.5	86	54.2	60.5	83.2[3]
Mandible	88.4	94	90	94.1	92.5	98	71.6	100.0	90.2
Scapula	82.9	61	65	88.2	62.5	88	58.3	52.8	72.1
Humerus	91.1	82	87	82.4	82.5	99	68.3	72.5	84.9
Radius	78.5	68	76	85.3	62.5	75	50.0	80.9	73.4
Ulna	91.8	79	81	79.4	82.5	92	56.7	57.7	80.7
Pelvis	88.0	65	88	76.5	72.5	92	90.8	56.3	80.2
Femur	87.8	81	91	70.6	85.0	94	100.0	71.8	86.1
Tibio-Fibula	91.1	83	97	67.7	87.5	101	100.0	100.0	92.1
MEAN	88.0	79	83	82.7	79.4	92	72.2	72.5	82.5

[1] (From Pokines and Kerbis Peterhans 1998).

[2] Total sample MNI.

[3] Per part means are weighted according to sample size.

Table 2.3: El Juyo Skeletal Element Breakage--Small.

		11	**10**	**9**	**8**	**8/9**	**7**	**6/9**	**6**	**5**	**4R**	**4S**	**4**	**2/4**	**3**	**2**	**1**	**Total**
Maxilla	Total	0	0	0	0	0	0	0	4	0	0	3	2	0	0	0	0	9
	Comp	0	0	0	0	0	0	0	1	0	0	0	0	0	0	0	0	1
	Frag	0	0	0	0	0	0	0	3	0	0	3	2	0	0	0	0	8
Mandible	Total	1	0	1	5	0	0	0	25	0	0	21	13	0	0	0	0	66
	Comp	0	0	0	1	0	0	0	3	0	0	5	3	0	0	0	0	12
	Frag	1	0	1	4	0	0	0	22	0	0	16	10	0	0	0	0	54
Scapula	Total	0	0	0	0	0	0	0	1	0	0	0	0	0	0	0	0	1
	Comp	0	0	0	0	0	0	0	0	0	0	0	0	0	0	0	0	0
	Prox	0	0	0	0	0	0	0	0	0	0	0	0	0	0	0	0	0
	Dist	0	0	0	0	0	0	0	1	0	0	0	0	0	0	0	0	1
Humerus	Total	0	1	1	5	0	3	0	14	0	2	23	3	0	0	0	0	52
	Comp	0	1	1	2	0	1	0	4	0	0	9	3	0	0	0	0	21
	Prox	0	0	0	0	0	0	0	0	0	0	2	0	0	0	0	0	2
	Prox Epi	0	0	0	0	0	1	0	0	0	0	0	0	0	0	0	0	1
	Shaft	0	0	0	2	0	1	0	2	0	0	2	0	0	0	0	0	7
	Dist	0	0	0	1	0	0	0	8	0	2	10	0	0	0	0	0	21
Radius	Total	0	0	0	0	0	0	0	1	0	1	2	0	0	0	0	0	4
	Comp	0	0	0	0	0	0	0	1	0	0	1	0	0	0	0	0	2
	Prox	0	0	0	0	0	0	0	0	0	1	1	0	0	0	0	0	2
	Dist	0	0	0	0	0	0	0	0	0	0	0	0	0	0	0	0	0
Ulna	Total	0	0	0	1	0	0	0	2	0	1	5	3	0	0	0	0	12
	Comp	0	0	0	1	0	0	0	1	0	0	0	2	0	0	0	0	4
	Prox	0	0	0	0	0	0	0	1	0	1	5	1	0	0	0	0	8
	Shaft	0	0	0	0	0	0	0	0	0	0	0	0	0	0	0	0	0
Pelvis	Total	1	0	1	0	0	1	0	3	0	0	2	0	0	0	0	0	8
	Comp	1	0	0	0	0	0	0	0	0	0	0	0	0	0	0	0	1
	Frag	0	0	1	0	0	1	0	3	0	0	2	0	0	0	0	0	7
Femur	Total	0	0	3	6	0	1	0	7	0	0	18	9	0	0	0	0	44
	Comp	0	0	1	1	0	0	0	1	0	0	4	4	0	0	0	0	11
	Prox	0	0	0	4	0	1	0	6	0	0	12	4	0	0	0	0	27
	Prox epi	0	0	0	0	0	0	0	0	0	0	0	0	0	0	0	0	0
	Shaft	0	0	1	1	0	0	0	0	0	0	1	0	0	0	0	0	3
	Dist	0	0	1	0	0	0	0	0	0	0	1	1	0	0	0	0	3
	Dist Epi	0	0	0	0	0	0	0	0	0	0	0	0	0	0	0	0	0
Tibio-fibula	Total	1	0	1	3	0	0	0	8	0	0	7	7	0	0	0	0	27
	Comp	0	0	0	0	0	0	0	1	0	0	0	1	0	0	0	0	2
	Prox	1	0	1	1	0	0	0	1	0	0	4	4	0	0	0	0	12
	Prox Epi	0	0	0	0	0	0	0	0	0	0	0	0	0	0	0	0	0
	Dist/Shaft	0	0	0	2	0	0	0	6	0	0	3	2	0	0	0	0	13

Table 2.4: El Juyo Skeletal Element Breakage--Medium.

		11	10	9	8	8/9	7	6/9	6	5	4R	4S	4	2/4	3	2	1	Total
Maxilla	Total	0	0	2	12	0	0	0	40	0	8	80	48	0	0	1	0	191
	Comp	0	0	0	2	0	0	0	3	0	1	0	0	0	0	0	0	6
	Frag	0	0	2	10	0	0	0	37	0	7	80	48	0	0	1	0	185
Mandible	Total	0	0	0	24	0	6	0	85	0	15	322	78	0	1	0	0	531
	Comp	0	0	0	3	0	0	0	4	0	0	5	0	0	0	0	0	12
	Frag	0	0	0	21	0	6	0	81	0	15	317	78	0	1	0	0	519
Incisor	Total	13	4	33	105	2	22	2	410	0	76	1536	341	1	1	2	2	2550
	Comp	5	0	8	38	0	9	2	116	0	28	357	89	1	0	0	2	655
	Frag	8	4	25	67	2	13	0	294	0	48	1179	252	0	1	2	0	1895
Molar	Total	16	4	51	264	0	73	8	788	1	156	2913	862	13	17	7	1	5174
	Identified	12	3	41	230	0	64	7	728	1	138	2540	741	7	14	7	1	4534
	Unident Frag	4	1	10	34	0	9	1	60	0	18	373	121	6	3	0	0	640
Scapula	Total	0	0	0	2	0	1	0	72	0	10	115	49	0	1	0	0	250
	Comp	0	0	0	0	0	0	0	0	0	0	0	0	0	0	0	0	0
	Prox	0	0	0	0	0	0	0	0	0	0	0	0	0	0	0	0	0
	Dist	0	0	0	2	0	1	0	72	0	10	115	49	0	1	0	0	250
Humerus	Total	2	0	12	26	0	9	1	157	1	31	311	169	3	0	0	1	723
	Comp	1	0	0	3	0	2	1	24	1	3	25	31	0	0	0	0	91
	Prox	0	0	3	7	0	2	0	51	0	8	84	36	1	0	0	0	192
	Prox Epi	0	0	2	3	0	2	0	17	0	2	26	26	1	0	0	0	79
	Shaft	0	0	2	1	0	0	0	8	0	4	20	10	0	0	0	0	45
	Dist	1	0	5	12	0	3	0	57	0	14	156	66	1	0	0	1	316
Radius	Total	0	0	0	13	0	1	1	42	3	7	90	65	1	0	0	0	223
	Comp	0	0	0	0	0	1	1	10	2	1	11	20	0	0	0	0	46
	Prox	0	0	0	10	0	0	0	31	1	6	79	45	1	0	0	0	173
	Dist	0	0	0	3	0	0	0	1	0	0	0	0	0	0	0	0	4
Ulna	Total	0	0	4	11	0	6	1	97	0	11	199	106	0	0	0	0	435
	Comp	0	0	0	2	0	1	1	5	0	1	8	12	0	0	0	0	30
	Prox	0	0	4	9	0	5	0	89	0	10	177	88	0	0	0	0	382
	Shaft	0	0	0	0	0	0	0	3	0	0	14	6	0	0	0	0	23
Pelvis	Total	0	1	3	5	0	2	0	35	0	3	47	44	0	0	0	0	140
	Comp	0	0	0	1	0	0	0	1	0	0	0	0	0	0	0	0	2
	Frag	0	1	3	4	0	2	0	34	0	3	47	44	0	0	0	0	138
Femur	Total	1	0	6	23	0	4	0	93	1	22	199	107	2	0	1	0	459
	Comp	0	0	1	2	0	1	0	9	1	2	12	15	0	0	0	0	43
	Prox	1	0	3	15	0	1	0	67	0	13	146	61	2	0	1	0	310
	Prox Epi	0	0	0	0	0	0	0	0	0	0	1	0	0	0	0	0	1
	Shaft	0	0	0	0	0	0	0	2	0	1	5	2	0	0	0	0	10
	Dist	0	0	2	5	0	1	0	8	0	5	19	15	0	0	0	0	55
	Dist Epi	0	0	0	1	0	1	0	7	0	1	16	14	0	0	0	0	40
Tibio-fibula	Total	1	1	7	14	0	3	0	92	0	13	178	97	2	0	0	0	408
	Comp	0	0	0	2	0	0	0	1	0	0	5	7	0	0	0	0	15
	Prox	0	1	3	3	0	1	0	27	0	4	57	35	0	0	0	0	131
	Prox Epi	0	0	2	1	0	0	0	4	0		4	4	0	0	0	0	15
	Shaft/Dist	1	0	2	8	0	2	0	60	0	9	112	51	2	0	0	0	247

Table 2.5: El Juyo Skeletal Element Breakage--Large.																		
		11	**10**	**9**	**8**	**8/9**	**7**	**6/9**	**6**	**5**	**4R**	**4S**	**4**	**2/4**	**3**	**2**	**1**	**Total**
Maxilla	Total	0	0	1	2	0	0	0	17	0	0	38	11	1	0	0	0	70
	Comp	0	0	1	1	0	0	0	2	0	0	0	0	0	0	0	0	4
	Frag	0	0	0	1	0	0	0	15	0	0	38	11	1	0	0	0	66
Mandible	Total	0	0	3	5	0	3	0	42	0	9	107	39	0	0	0	0	208
	Comp	0	0	1	0	0	0	0	3	0	1	6	2	0	0	0	0	13
	Frag	0	0	2	5	0	3	0	39	0	8	101	37	0	0	0	0	195
Incisor	Total	10	2	12	34	0	6	0	166	0	45	531	89	4	0	1	2	902
	Comp	0	0	2	19	0	2	0	30	0	6	84	30	1	0	0	2	176
	Frag	10	2	10	15	0	4	0	136	0	39	447	59	3	0	1	0	726
Molar	Total	3	5	42	124	0	47	1	473	0	94	1783	317	1	7	2	2	2901
	Identified	3	3	39	112	0	46	1	459	0	89	1583	301	1	6	2	2	2647
	Unident Frag	0	2	3	12	0	1	0	14	0	5	200	16	0	1	0	0	254
Scapula	Total	1	0	2	5	0	1	0	15	0	5	40	25	0	0	0	0	94
	Comp	0	0	0	0	0	0	0	0	0	0	0	0	0	0	0	0	0
	Prox	1	0	0	1	0	0	0	0	0	0	1	0	0	0	0	0	3
	Dist	0	0	2	4	0	1	0	15	0	5	39	25	0	0	0	0	91
Humerus	Total	0	3	6	9	0	3	0	52	0	10	123	36	1	0	0	0	243
	Comp	0	0	0	1	0	0	0	8	0	4	12	5	0	0	0	0	30
	Prox	0	1	1	4	0	0	0	13	0	1	32	6	0	0	0	0	58
	Prox Epi	0	0	3	3	0	1	0	12	0	2	29	11	0	0	0	0	61
	Shaft	0	0	1	0	0	0	0	2	0	0	4	4	0	0	0	0	11
	Dist	0	2	1	1	0	2	0	17	0	3	46	10	1	0	0	0	83
Radius	Total	0	1	1	2	0	0	0	30	0	8	74	24	1	0	0	0	141
	Comp	0	0	0	0	0	0	0	1	0	1	5	6	0	0	0	0	13
	Prox	0	1	1	2	0	0	0	27	0	7	69	16	0	0	0	0	123
	Dist	0	0	0	0	0	0	0	2	0	0	0	2	1	0	0	0	5
Ulna	Total	0	0	4	2	0	3	0	30	0	4	89	21	0	0	0	0	153
	Comp	0	0	0	0	0	0	0	1	0	0	1	2	0	0	0	0	4
	Prox	0	0	4	2	0	3	0	27	0	4	84	18	0	0	2	0	144
	Shaft	0	0	0	0	0	0	0	2	0	0	4	1	0	0	0	0	7
Pelvis	Total	0	0	2	2	0	0	0	25	0	1	32	17	0	0	0	0	79
	Comp	0	0	0	0	0	0	0	0	0	0	0	0	0	0	0	0	0
	Frag	0	0	2	2	0	0	0	25	0	1	32	17	0	0	0	0	79
Femur	Total	0	0	2	8	0	1	0	56	2	7	62	39	1	0	0	0	178
	Comp	0	0	0	0	0	0	0	1	1	0	5	2	0	0	0	0	9
	Prox	0	0	2	3	0	1	0	34	0	3	32	24	1	0	0	0	100
	Prox Epi	0	0	0	0	0	0	0	0	0	1	1	2	0	0	0	0	4
	Shaft	0	0	0	0	0	0	0	0	0	0	4	0	0	0	0	0	4
	Dist	0	0	0	3	0	0	0	15	0	2	13	7	0	0	0	0	40
	Dist Epi	0	0	0	2	0	0	0	6	1	1	7	4	0	0	0	0	21
Tibio-fibula	Total	2	0	2	12	0	3	1	36	0	8	55	37	0	0	0	0	156
	Comp	0	0	0	1	0	0	0	0	0	0	5	1	0	0	0	0	7
	Prox	1	0	1	5	0	1	1	7	0	3	10	12	0	0	0	0	41
	Prox Epi	0	0	1	0	0	1	0	4	0	3	6	9	0	0	0	0	24
	Shaft/Dist	1	0	0	6	0	1	0	25	0	2	34	15	0	0	0	0	84

Table 2.6: El Juyo Skeletal Element Breakage--*Talpa*.

		11	10	9	8	8/9	7	6/9	6	5	4R	4S	4	2/4	3	2	1	Total
Maxilla	Total	0	0	1	0	0	1	0	3	0	0	2	3	0	0	0	0	10
	Comp	0	0	0	0	0	0	0	0	0	0	0	0	0	0	0	0	0
	Frag	0	0	1	0	0	1	0	3	0	0	2	3	0	0	0	0	10
Mandible	Total	3	0	1	0	0	3	1	10	0	2	26	6	0	0	0	0	52
	Comp	0	0	0	0	0	0	0	0	0	0	1	0	0	0	0	0	1
	Frag	3	0	1	0	0	3	1	10	0	2	25	6	0	0	0	0	51
Scapula	Total	0	0	0	0	0	0	0	1	0	0	1	1	0	0	0	0	3
	Comp	0	0	0	0	0	0	0	0	0	0	0	0	0	0	0	0	0
	Frag	0	0	0	0	0	0	0	1	0	0	1	1	0	0	0	0	3
Humerus	Total	0	0	0	2	0	2	0	9	0	0	10	10	0	0	0	0	33
	Comp	0	0	0	0	0	1	0	3	0	0	4	3	0	0	0	0	11
	Frag	0	0	0	2	0	1	0	6	0	0	6	7	0	0	0	0	22
Radius	Total	0	0	0	2	0	1	0	5	0	1	5	3	0	0	1	0	18
	Comp	0	0	0	0	0	0	0	0	0	0	0	0	0	0	0	0	0
	Frag	0	0	0	2	0	1	0	5	0	1	5	3	0	0	1	0	18
Ulna	Total	0	0	0	0	0	3	0	4	0	0	7	4	0	1	0	0	19
	Comp	0	0	0	0	0	0	0	0	0	0	0	0	0	1	0	0	1
	Frag	0	0	0	0	0	3	0	4	0	0	7	4	0	0	0	0	18
Pelvis	Total	0	0	0	0	0	0	0	0	0	0	0	0	0	0	0	0	0
	Comp	0	0	0	0	0	0	0	0	0	0	0	0	0	0	0	0	0
	Frag	0	0	0	0	0	0	0	0	0	0	0	0	0	0	0	0	0
Femur	Total	0	0	1	0	0	0	0	6	0	0	5	5	0	1	0	0	18
	Comp	0	0	0	0	0	0	0	1	0	0	2	0	0	0	0	0	3
	Frag	0	0	1	0	0	0	0	5	0	0	3	5	0	1	0	0	15
Tibio-Fibula	Total	0	0	2	0	0	1	0	8	0	1	19	6	0	0	0	0	37
	Comp	0	0	0	0	0	1	0	1	0	0	2	0	0	0	0	0	4
	Frag	0	0	2	0	0	0	0	7	0	1	17	6	0	0	0	0	33

Large: The remains in this class conform in almost all cases to *Arvicola*, a large microtine genus with a maximum adult mass of 200-300 g. Confusion with remains of smaller taxa, even other microtines, is therefore unlikely. No other prey in this size class has similar postcranial remains. Rejected possibilities include *Mustela nivalis* (weasel). While this species falls in this size class, it is a small carnivoran and has distinguishable skeletal characteristics. While adult *Talpa* can exceed 100 g, their remains are highly distinctive due to their specialized fossorial lifestyle. It is, however, possible that some fragmentary *Talpa* pelvic remains may have been included in the totals for the Large size class since none were directly assignable to this genus. All totals for molar fragments in the Large class are directly attributable to *Arvicola*, since it is the only microtine genus within this class in Europe. The large incisor remains may similarly be attributed to *Arvicola*, since there is generally no other rodent of this size possible. The large number of non-identified tooth remains in this class (n = 2618) corresponds with the large numbers of molars identified to this genus.

***Talpa*:** As mentioned above, the remains attributable to this highly distinctive genus are listed separately from the other large microfauna. These are also not included in calculations for breakage, since the postcranial elements of *Talpa* are robust, highly specialized and do not conform in general to the breakage and part representation of the rodent and small insectivore taxa discussed here. A total of 1065 non-identified tooth remains were identified to *Talpa*. These belong to the species *T. occidentalis* and *T. europaea*.

Other: Total numbers of ribs, vertebrae/sternebrae, metapodials and phalanges were tallied regardless of size class, since this could be determined with certainty only rarely. Some of these elements were assignable to *Talpa* and are included in that category's totals, since the phalanges, carpals and metacarpals of the front limbs of this genus are highly distinctive. Despite their small size, these elements make up the bulk of the identified *Talpa* remains. For non-*Talpa* categories, these elements total 6800.

Not Listed: Other non-dental remains are not included in this tabulation. For larger taxa such as lagomorphs, preliminary analysis indicates that these occur in extremely small numbers in El Juyo. This includes analysis of remains from small screens, large screens and mapped items. This pattern of recovery is discussed in greater depth elsewhere (Chapter 9). Non-dental remains of one medium microfauna species, *Erinaceus europaeus* (hedgehog) are discussed separately at the end of this chapter.

Percentage representation: This calculation (following Andrews 1990) allows comparison among the various skeletal elements. The MNI per size class is derived from the total number (fragmentary or whole) of each element is divided by the number of that element expected per individual. This amount is 12 for molars, 4 for incisors, and 2 for all remaining major appendicular elements (humeri, scapulae, etc.). The largest total among all these elements is the MNI. Each element is then scaled against the MNI based upon the most abundant element:

$$\text{\% Representation} = \frac{\text{Maximum Element MNI}}{\text{Element MNI}} \times 100$$

Percentage representation for the relatively most abundant element will therefore equal 100%, with other more fragile elements falling between 100% and 0%. Differential breakage of elements into multiple pieces and differential identifiability of these fragments makes this calculation only a rough guide to relative abundance. Since these biases apply to all assemblages, interassemblage comparison on these criteria remains valid. The gross characteristics of relative element representation are sufficient to eliminate most possible accumulating predators as discussed below.

Table 2.2 compiles percentage representation for key microfaunal skeletal elements derived from studies of barn owl pellets. These figures must be considered as maximal amounts for the survivability of these elements. In each case they are derived from undisturbed pellet collections, and two of the studies (Dodson and Wexlar 1979; Hoffman 1988) utilized pellets egested by captive owls fed a controlled diet. Breakage in an archaeological or paleontological setting will naturally be much greater, and further increased by the mechanical forces of excavation and recovery. These values do reflect the exceedingly light damage incurred by barn owls upon microfaunal bone, and the relative durability among these elements.

Not surprisingly, the more durable elements are the tibiofibula (mean representation of 92.1%), mandible (90.2%), femur (86.1%) and humerus (84.9%). These elements are more robust and identifiable when in a damaged state. The second survivorship tier is occupied by the maxilla (83.2%), ulna (80.75) and pelvis (80.2%). The lower representation of the maxilla is explicable by the frequency at which the cranial elements of young microfauna fall apart along their suture lines when consumed by owls. The ulna is both more gracile than the femur or humerus and is less recognizable. The juvenile pelvis, while superficially robust, often falls apart into relatively unidentifiable ischial, pubic and iliac portions when an individual is consumed before these elements have fully fused. The radius (73.4%) and scapula (72.1%) are the least recovered elements. The radius is difficult to separate from small mammal metapodials, which are similarly long and thin. Low percentage representation of this element therefore may be due less to gracility and more to a lack of distinguishing identifying features. The low percentage representation of the scapula is attributable to its fragility.

Application of these breakage guidelines to an archaeological sample cannot be direct. Specifically, breakage patterns among rodent incisors are likely to bias relative part representation percentages (Tables 2.3-2.6). These elements are highly durable and identifiable, yet minor breakage will simply produce more fragments that are durable and identifiable. Other tooth fragments also fit this pattern, in that they make up a very small portion of unidentified microfaunal fragments: a tiny fragment can at least be identified as a tooth.

Many tiny non-dental fragments remain unidentified. Elements with large, durable and identifiable unfused epiphyses or a tendency to fragment into more than one (proximal and distal) identifiable piece (humeri, femora and tibio-fibulae) also increase their relative part representation. Elements lacking diagnostic characters (ulnae, radii) bias their part representation negatively. These differences are considered in interpretation of relative part representation.

The skeletal element representation pattern among the Small, Medium and Large classes indicates high rates of survival among even the most fragile elements. The humerus is highly represented (78.8%, 56.7% and 50.3%, respectively) among all three classes. The extremely fragile and less identifiable scapula was found in significant amounts in the Medium and Large classes (19.6% and 19.4%). This relatively high representation is indicative of very light damage to scapulae. In almost all cases the more durable distal extremity of the scapula was identified, compared with the proximal epiphysis, proximal end, shaft, and distal end of fragmented humeri. The humerus therefore is relatively overrepresented as compared with the scapula, since the former can separate/fragment into multiple component pieces, all coming from a single original element.

Comparison of how frequently the more delicate scapula survived with the robust humerus is more illuminating when considering these factors of identifiability, recoverability and survivability of fragments. When comparing the total number of distal scapulae (n = 250) with proximal humeri (n = 283, disregarding unfused epiphyses) in the Medium class, this delicate element survives at 88.3% the rate of the latter. This pattern is not repeated in the Small class, perhaps because even the distal extremities of shrew scapulae were too delicate to survive in an identifiable state. These might also be small enough when fragmented to pass through a 1 mm mesh screen and be lost due to the analytical procedure used. This pattern is, however, repeated with the distal scapulae and proximal humeri of the Large size class, which correspond to the taxon *Arvicola*. In this case, the number of distal scapulae (n = 91) exceeds the number of proximal humeri (n = 88), or 103.4%. Great destructive forces upon this microfaunal assemblage would have caused a much greater discrepancy between inherently more durable and less durable elements. The taphonomic agency responsible for their accumulation therefore caused very little damage during this process.

The general breakage character is very light damage, even when considering postdepositional trampling and the effects of excavation and wet screening. Proximal and distal humeri and femora and proximal tibio-fibulae survived well. Other elements survived less well. Mandibles, though numerous, were rarely complete. This breakage of such normally durable elements is explained by the propensity of rootless microtine molars to fall from their sockets. Microtine tooth loss is frequent even among modern owl pellet assemblages. This loss leaves a large gap along the midline of the mandible and makes it more susceptible to postdepositional crushing. The same is not true for murid or soricid mandibles, which (though few in number) were less often crushed than microtine mandibles. Microtine tooth loss similarly makes maxillae more susceptible to crushing. In the Medium size class (the remains of which are dominated by microtine species), mandibles are represented by 531 total pieces. Only 12 (2.3% of the total) were complete and unbroken. In the Large size class (corresponding to the microtine genus *Arvicola*), 13 of 208 (6.3%) were complete. Mandibles among the Small species (corresponding to shrews, which have rooted teeth) 12 of 66 (18.2%) were complete. Only 1 of 52 (1.9%) of *Talpa* mandibles were complete. While this genus has rooted teeth, its mandibles are relatively long, gracile and hence easily fragmented.

Pelves did not fare well generally. A lack of fusion combined with lesser identifiability of fragments made this element less visible. Breakage at the acetabulum was common, though if this area had fused sometimes the acetabular portion alone survived. This element was rarely recovered in a complete state: 1 of 8 (12.5%) in the Small class, 2 of 140 (1.4%) in the Medium class, and 0 of 79 in the Large class were recovered complete.

Small elements (ribs, vertebrae/sternebrae, metapodials and phalanges) were recovered in relatively large amounts. This may be attributable to the size of the screens employed in laboratory processing (0.25 mm). These elements are typically recovered in large amounts in the assemblages produced by many species of owl and are rarely preserved in the assemblages produced by diurnal raptors and carnivores (discussed below).

DIGESTIVE CORROSION

Evidence of digestive corrosion was examined for on all elements identified by the author. Surface corrosion may be attributed almost entirely to predator action, since the pH of each sediment lot in this limestone cave vestibule was above 7.0. Among all teeth identified in this study (n = 7486), only 277 (3.7%) had traces of digestive corrosion, typically rounding the occlusal surface and the sides exposed above the sockets. This type of damage is more likely to occur on teeth, since these have the highest mineral content among skeletal elements (Hillson 1986). They are also the most exposed skeletal elements: they enter a predator's stomach exposed to digestive acid, while only after initial digestion do other elements (other than those exposed through breakage) become exposed to acid corrosion. Among the identified large non-dental remains (thereby excluding foot elements, vertebrae and ribs), only 14 of 9841 (0.14%) have visible acid corrosion. This apparent lack of corrosion of course may have been negatively biased, since corroded elements are more likely to be destroyed or rendered unidentifiable than non-corroded elements. Clearly, however, digestive corrosion is not a significant feature of this microfaunal assemblage.

PREY ACTIVITY PATTERNS

The activity patterns of the rodents and insectivores recovered (Chapter 3) are overwhelmingly nocturnal, coinciding with most owl hunting activity. This pattern is typical among small mammal accumulations, since nocturnal behavior among small mammals is a general evolutionary adaptation for predator avoidance. Many of these species fall prey to diurnal predators as well when their activity patterns overlap during the crepuscular hours. These crepuscular species include *Crocidura russula*, *Sorex coronatus*, *Microtus agrestis* and *Pitymys lusitanicus* all of which are common prey of buzzards (*Buteo buteo*) and barn owls in the modern environs of El Juyo (Chapter 4). Patterns of bone recovery and damage are much more compelling evidence of owl accumulation than nocturnality of prey.

An apparent exception to the correspondence of owl and prey activity is the presence of small amounts of moles (*Talpa* spp.) in this assemblage. Moles, being fossorial, tend to exclude themselves from predation by avian predators. The modern barn owl pellet samples examined from the immediate environs of El Juyo (Chapter 4) yielded only a *Talpa europaea* MNI of 3 out of a total MNI of 934 vertebrate prey items. Glue (1974) found that *Talpa europaea* made up less than 1% of the diet (26 of 47,865 vertebrate prey items) of the barn owl taken from 188 pellet samples from Britain and Ireland. In contrast, a total of 300 teeth of *Talpa* spp. were included in the total sample of 7,862 identified teeth (3.8% of the total). Part of this apparent overabundance is explicable in terms of the greater number of teeth per individual in this genus of mole. *Talpa* has the full placental mammalian complement of 11 teeth per demi-mandible/maxilla (3/1/4/3). All these teeth are generally identifiable, even the tiny incisors if found intact in their sockets. This contrasts with the teeth of all the rodents recovered at El Juyo, which have only the evolutionarily reduced pattern of 1/0/0/3, or 16 teeth per individual. This amount is winnowed to 12 molars per individual when one considers the general unidentifiability of rodent incisors beyond size class. *Talpa* therefore has nearly 3 times as many potentially identifiable teeth when compared with a member of the genus *Microtus*, whose teeth are recovered in great numbers from El Juyo and other Cantabrian sites.

An additional 1,065 *Talpa* remains besides identified teeth were recovered from a total of 25,204 from all levels of El Juyo. An overrepresentation of mole postcranial elements is easily explicable: these elements are highly distinctive, identifiable, and in the case of the forelimb elements (phalanges, carpals, metacarpals, radius, ulna and humerus) extremely robust due to their evolution as specialized excavation appendages. Postcranial elements were identified only to size class for the remainder of the small insectivore and rodent species recovered, with the exception of water voles (*Arvicola terrestris*) which are mostly distinguishable by their larger size.

Moles do fall to owls in significant numbers during one phase of their yearly cycle: the late summer dispersal of juveniles who must leave their natal burrows to establish their own. At this time moles traveling on the surface are easy prey for owls. The presence of a slight overabundance of moles may indicate that the season of owl occupation of the cave tended to include late summer.

COMPARISON OF PREDATOR TYPES

The damage pattern of these remains precludes their concentration by terrestrial carnivores including canids, mustelids or felids. Many species of mammalian carnivore habitually accumulate scats in latrine areas, sometimes near their dens or to mark territories (see Andrews and Evans 1983:290; Stuart 1977). Mellet (1974) coined the term "coprocoenosis" to indicate a fossil assemblage derived from an accumulation of mammalian carnivore scats. Andrews and Evans (1983) examined the scat contents of a battery of species from three families of mammalian carnivore. These were the Viverridae (represented by mongoose *Ichneumia albicauda* and genet *Genetta genetta*), Canidae (red fox *Vulpes vulpes*, bat-eared fox *Otocyon megalotis*, arctic fox *Alopex lagopus* and coyote *Canis latrans*) and Mustelidae (pine marten *Martes martes*; weasel, stoat and polecat *Mustela* spp.). Overall bone breakage was much greater for all these predator species as compared to owl-produced assemblages. Proportions of complete humeri relative to the total recovered among these predators ranged from 0% to 43%, ulnae ranged 0% to 57%, femora 0% to 25%, and tibiae 0% to 57% (Andrews and Evans 1983:295). Overall bone breakage ranged from moderate (genet) to very great (canids), and rounding or polishing of elements ranged from little (bat-eared fox) to great (remaining canids).

Mammalian predators also damage and digest small prey bones much more than do owls (Day 1966; Greenwood 1979; Lockie 1959; Robertson and Whelan 1987; Stuart 1977; Weaver and Hoffman 1979). Korth (1979) compared small mammal remains from coyote scats to those extracted from samples of great horned and barn owl pellets. The coyote-derived sample included many sharply broken bone fragments, although major limb bones often survived whole: 62% of all recognizable humeri and femora were unbroken (Korth 1979:248). Smaller mammal species skeletal elements generally were more likely to escape damage, probably because less mastication was necessary for their consumption. Mandibles were noted rarely to retain their original complement of teeth, in contrast to mandibles derived from owl pellets.

Degree of digestive corrosion on skeletal remains is great among Carnivora-produced assemblages. Andrews and Evans (1983) found corrosion of bones and teeth by digestive acid was common among the mammalian carnivore species examined. While mongoose-consumed prey bones and teeth showed no such alteration, the remaining species ranged from little to severe damage. Andrews and Evans (1983:301) also summarize the characteristics of felid-altered prey bones, which were found to incur extreme breakage and corrosion and great rounding and polishing. A lesser degree of bone breakage and corrosion among smaller mammalian prey was noted.

The damage pattern of the El Juyo remains also precludes their accumulation by diurnal raptors such as hawks, falcons or eagles. Diurnal raptors leave far fewer recognizable bones per consumed prey item in their pellets than do owls. Eating habits are partially responsible for this low bone representation. Hawks and other diurnal raptors tend to be far more fastidious eaters than most owls, often carefully flensing prey of their edible portion or tearing it into small chunks (Clark 1972; Einarsen 1956). These methods of feeding tend to leave prey bones behind at the site of eating (thus not leading to accumulation in a single locality) or to break bones to a much greater degree than would swallowing prey whole. The lower diurnal raptor gastric pH leads to increased digestion of bone (Duke *et al.* 1975). Hoffman (1988) fed captive species of owl and diurnal raptor a controlled diet of house mouse (*Mus musculus*). Pellets of the hawk species tested (red-tailed hawk *Buteo jamaicensis*, rough-legged hawk *Buteo lagopus*, and sparrow hawk *Falco sparverius*) averaged less than one identifiable bone for each mouse eaten. Not a single complete skeletal element was recovered from the pellets of any of these species, and most of the fragments were in very poor condition. Andrews (1990:38) examined the pellets of several common diurnal raptors and found skeletal part relative abundance to fall to levels much lower than among barn owls, generally less than half.

The recovery and damage patterns of the El Juyo microfaunal assemblage are in stark contrast to assemblages derived from terrestrial carnivores or diurnal raptors. The large amount of identifiable postcranial remains, many of them largely intact, is in complete contrast to the characteristics of these predator types discussed above. Gastric corrosion was also very low among these remains. This pattern of damage and recovery is entirely consistent with many owl-produced microfaunal assemblages.

Some owl species may be quickly eliminated as potential accumulators of the El Juyo assemblage. Raczynski and Ruprecht (1974) found that long-eared owls (*Asio otus*) destroyed prey bone to a much greater extent than did the barn owls studied. Bone destruction was much higher among the juveniles of this species. Long-eared owls, like tawny owls, hunt and roost in trees (Burton 1973), making a cave at the base of a sink hole a very unlikely roosting site. Little owls (*Athene noctua*) do not create large bone accumulations due to their dietary focus upon invertebrates. Snowy owl (*Nyctea scandiaca*), great grey owl (*Strix nebulosa*), Ural owl (*Strix uralensis*), Tengmalm's owl (*Aegolius funereus*), Scops owl (*Otus scops*), hawk owl (*Surnia ulula*) and pygmy owl (*Glaucidium passerinum*) may be eliminated by multiple factors such as specialization of diet, range, habitat requirements, and bone accumulation habits.

Certain characteristics of the El Juyo assemblage point to barn owls as its principal accumulators. Some owl species with European ranges produce assemblages with comparable amounts of prey bone breakage: barn owls (*Tyto alba*), eagle owls (*Bubo bubo*) and short-eared owls (*Asio flammeus*). While some differences in breakage and corrosion pattern are evident among modern pellet collections from these predator species, these nuances are likely submerged in an archaeological microfaunal assemblage that has undergone postdepositional breakage. These three species, however, are distinguishable by other aspects of behavior. Kusmer (1990) found that short-eared owl (*Asio flammeus*) pellets (MNI = 19) had a slightly greater part representation mean (91%) and completeness mean (75%) than the barn owls she studied. These results may indicate that separation of short-eared from barn owl assemblages based only upon skeletal part representation and breakage may not be possible. Feeding behavior is similar to barn owls, with small mammal prey usually swallowed whole and larger prey torn into pieces first (Chitty 1938:270). Short-eared owls are open-country dwelling, however, and they invariably roost in grass nests on the ground near marshy areas (Burton 1973). They are also primarily diurnal in their activity.

Eagle owls can leave behind bones with similar degrees of breakage, though this species tends to tear its prey into smaller pieces before consumption (Einarsen 1956; Howard 1958). The pellets of the closely related and ecologically equivalent New World great horned owl (*Bubo virginianus*) have been studied in more detail. Kusmer (1990) found that great horned owl pellets yield prey bones with very little breakage and high representation of elements. Mean part representation for nine skeletal elements was 82% and 87% for two samples of great horned owl pellets. Overall barn owl pellet part representation for a sample from the Negev Desert (Israel) was 87.9% (Pokines and Kerbis Peterhans 1998), with a grand mean of 82.5% for the barn owl pellet studies summarized below. Korth's (1979) results indicate an overall part representation mean of 84.4% for the same nine elements in his study of great horned owl pellets, and Hoffman's (1988) results indicate a mean of 87.3% for his sample from this species. Dodson and Wexlar (1979) noted a similarly high part representation mean (85.2%) for the great horned owl pellets that they examined, but found a much lower overall mean of complete elements (34.0%). Moon (1940:461) found similar bone weight percentages for pellets of barn owls and great horned owls: 38% and 40%, respectively.

Being much larger than barn owls (3,000-4,000 g vs. 350-600 g, respectively), eagle owls have a diet averaging a much higher body mass per individual taken. Hedgehogs, lagomorphs and large birds are taken frequently (Burton 1973; Donázar and Ceballos 1989; Donazár *et al.* 1989; Einarsen 1956; Frey and Walter 1986; Goodman 1990; Orsini 1985). Barn owls do not focus upon prey greater than 100 g in size and concentrate their attention upon prey in the 11-50 g range (Goszczynski 1977:409). An assemblage dominated by small rodents and insectivores is therefore unlikely to have been accumulated by such a large species of owl.

In addition, the setting of El Juyo argues against its use as an eagle owl roost. This (largest) species of owl would have found great difficulty fitting its large wingspan into the small confines of the available roosting space: its wingspan is 1.5 m, with a body length of 0.5 m. In comparison, European barn owls (*Tyto alba alba*) have a wingspan about 760 mm and a body length under 400 mm (Burton 1973). While they do roost in caves and fissures, such low places as El Juyo (located at the base of a sinkhole) are avoided. Eagle owls choose roosts for a clear field of view and high location. This species also shuns human contact and avoids roosting in human habitations (Mikkola 1983).

Table 2.7: Summary of Microvertebrate Taphonomy.

	Accumulating Predator							
Assemblage Character	Small Carnivore	Diurnal Raptor	Tawny Owl	Long-eared Owl	Eagle Owl	Short-eared Owl	Barn Owl	**El Juyo**
Degree of Breakage	Very High	Very High	High	High	Very Low	Very Low	Very Low	Very Low
Completeness of Skeletal Elements	Very Low	Very Low	Low	Low	Very High	Very High	Very High	Very High
Degree of Digestive Corrosion	Very High	Very High	Low	Low	Very Low	Very Low	Very Low	Very Low
Density of Microfaunal Accumulation	Low to High	Moderate	Low	Moderate	Very High	Very High	Very High	High
Size Class of Most Prey	Variable	Variable	5-100 g	5-100 g	30 to 3000+ g	5-100 g	5-100 g	5-100 g
Activity Pattern of Prey	Variable	Diurn.	Noct.	Noct.	Noct.	Diurn.	Noct.	Noct.
Habitat of Predator	Variable	Variable	Forest	Forest	Most	Open	Most	Open
Prey Specialized	Variable	Variable	No	No	No	No	No	No
Dens/Roosts in Caves	Some	Some	No	No	Yes	No	Yes	Yes

Other species of owl found in Europe can be eliminated as accumulators of microfauna in El Juyo on multiple grounds. Tawny owls (*Strix aluco*) tend to egest their pellets away from their roosts (Southern 1954), thus greatly diminishing their potential to accumulate large numbers of microfaunal remains beneath a roost. Tawny owls destroy prey bone to a much greater extent than do barn owls (Raczynski and Ruprecht 1974). They are also typically associated with and hunt while perched among the branches in a forest, either coniferous or deciduous (Burton 1973). All environmental reconstructions of Cantabrian Spain during the last glacial period based upon non-microfaunal evidence (Chapter 7) indicate sparse forest cover. Tawny owls also roost in trees, which makes the consistent selection of a low-lying cave such as El Juyo an unlikely choice for individuals of this species. The concentration and distribution of microfauna within the cave rules out accumulation by owls roosting in a tree just outside the cave mouth. Other aspects of tawny owl behavior do coincide with those of barn owls (non-specialized diet, range including Europe, size range among prey species), so one must consider the general possibility that tawny owls may have contributed to the El Juyo microfaunal assemblage. However, the overall evidence indicates that they are not significant accumulators of the El Juyo remains.

Barn owls leave a high proportion of complete elements, much higher than great horned owls do. Delicate longbones are rarely damaged. The highest degree of damage is to scapulae, but even these are normally recovered in an identifiable state. Only a few instances of slight gastric corrosion are typical. Groups of articulated elements (vertebrae or metapodials) are occasionally recovered from pellets. Teeth often remain in their sockets, but could be easily removed. Barn owl damage to bone assemblages is minimal, in keeping with their swallowing of whole prey, relatively high gastric pH, slight stomach musculature, and frequent pellet egestion.

PREDATOR SUMMARY

Table 2.7 (above) summarizes the characteristics of those predators that must be seriously considered as among the possible accumulators of the El Juyo assemblage. Non-predatory agencies have been ruled out on the grounds of the size and state of the microfaunal accumulation and the physical layout of the cave vestibule. Moving water could not have produced the pattern of remains found and would have left traces of damage on the bones. Natural sinkhole trapping is not possible, since the vestibule of the cave is not a suitable design. Natural denning deaths of microfauna would not have occurred in this depositional setting nor would it have produced the spectrum of species accumulated. Some animal agency therefore must have been the primary accumulator these remains.

Accumulation in a small carnivore (mustelid, canid or felid) latrine area is eliminated by the very high degree of prey skeletal breakage, low degree of skeletal completeness (more durable elements greatly outnumber less durable elements) and very high degree of gastric corrosion caused by predators in this class. Some small carnivores such as weasels and stoats (*Mustela* spp.) do sometimes cache uneaten small vertebrate prey in a den for later consumption (Oksanen 1983). If not recovered by the accumulating predator such collections could leave a concentration of highly intact bones within the area of an archaeological site. Such behavior is highly unlikely to have contributed significantly to the El Juyo assemblage due to the lack of suitable denning areas within the vestibule of the cave, and the lack of any finds of highly complete small vertebrate skeletons concentrated in a very small area. The large microfaunal concentration at the base of Tumulus 4 in the Sanctuary feature of Level 4 was accumulated at the base of a large open pit. This location would not have been a suitable small carnivore denning area, nor are very large concentrations produced in this manner.

Diurnal raptors (eagles, hawks, or falcons) can be eliminated as significant contributors by many factors. These predators cause much gastric corrosion on prey bones and typically break them much more greatly than is the case for the El Juyo assemblage. This breakage discrepancy is true even though postdepositional and excavation/screening forces have caused further damage to this sample. While many species of raptor will roost in rock fissures, a deeper cave setting like El Juyo is less likely.

All aspects of the El Juyo assemblage indicate its accumulation by barn owls. Other species of owl common to Western Europe can be eliminated by gross characteristics. Long-eared owls and tawny owls destroy bones of their prey to a much greater extent than do eagle owls, short-eared owls and barn owls. Long-eared owls and tawny owls are also forest-dwelling and roost in trees. Tawny owls also have a much lower potential to concentrate prey bones at a single point in their range, since pellets are egested most often away from their roosts. Short-eared owls and eagle owls can be eliminated by aspects of their behavior: choice of roosting site and in the case of the latter size range of prey.

POSTCRANIAL FUSION

The relative extent of postcranial epiphyseal fusion was recorded among the three main size classes of elements recovered (small, medium and large). These data are informative due to the difficulty of age determination using teeth of most microtines, which grow continuously throughout the life of these species. Microtines make up the majority of small mammal remains from the site. These results are listed in Table 2.8. The recorded amounts indicate the total number (right and left) of unfused proximal or distal elements for three most visible and robust postcranial elements: proximal humerus, proximal femur, distal femur and proximal tibiofibula. While many of the unfused epiphyses themselves were recovered, they were not included in these calculations since they would inflate the proportions of unfused elements: these would in some cases be represented twice. The fused/unfused diaphyses were tabulated, which in the case of some femora included both ends. These were tabulated separately for proximal and distal fusion. These data are presented cumulatively for all levels of El Juyo, since in general single level totals were too small and in no case was a significant difference noted between levels for any of the size classes. This data does not include the remains of

Table 2.8: Postcranial Epiphyseal Fusion for All Levels of El Juyo.

	Small		Medium		Large	
	x/y	%	x/y	%	x/y	%
Prox Humerus	9/23	39.1%	234/283	82.7%	72/88	81.8%
Prox Femur	2/38	5.3%	45/353	12.7%	18/109	16.5%
Dist Femur	1/14	7.1%	63/98	64.3%	33/49	67.3%
Prox Tibia	9/14	64.3%	92/146	63.0%	33/48	68.8%

x = # of unfused elements (does not include epiphyses)
y = total # of elements
%= % unfused

Talpa, which has very different postcranial elements from the other small mammal species tabulated.

Small postcranial elements generally must conform to the smaller taxa identified from teeth. These are *Neomys* and *Sorex* spp., though the presence of some remains of species present in such low amounts that they could not be detected from dental remains cannot be discounted.

The medium remains correspond to the majority of taxa identified from teeth. These include *Apodemus* (Muridae) and *Microtus*, *Pliomys* and *Pitymys* (Microtidae). Most of these taxa cannot be aged by dental attrition, since *Microtus* and *Pitymys* have continuously growing molars. Methods of determining these growth rates (Von Koenigswald and Golenishev 1979) are not applicable to archaeological materials. Length of mandibular diastema (Saint Girons 1973) could not be used since so few mandibles were recovered in a measurable state. Assignment to a species within these two genera for toothless mandibles was generally not possible. Species differences would alter relative diastema length, further complicating results. The molars of *Pliomys* grow continuously but develop roots and can be aged using this criterion. This species is represented, however, by a single tooth in Level 4S. The molars of *Apodemus* have been assigned relative ages based upon tooth wear stages. These results are included in Chapter 6. This genus, based upon identified tooth counts, constitutes a small proportion of medium rodent remains. Hence, the majority of individuals taken in the medium size class must be given relative age based upon postcranial fusion at large.

Relatively little is known about the times of epiphyseal fusion for small mammals (even in otherwise extensively studied mice and rats), so age classes beyond fully grown/subadult cannot be assigned (Pucek and Lowe 1975). Postcranial fusion indicates that the majority of individual prey items taken in the Medium size class were not fully grown adults. While the femoral head was unfused only in 45 of 353 cases (12.7%), the remaining elements were unfused in 63.0% to 82.7% of all cases.

These data correspond with reported age profiles of small mammal prey taken by owls (Brown and Twigg 1971; Buckley 1977; Dean 1973; Derting and Cranford 1989; Dickman *et al.* 1991; Saint Girons 1973). Most individuals taken fall into the subadult and juvenile classes. This age preference is attributable to the restriction of smaller individuals (both within and among species) to foraging areas with poorer cover (Kotler *et al.* 1991; Kotler *et al.* 1988; Longland and Jenkins 1987). This competitive exclusion results from dominance hierarchies typical to small rodents (Crowcroft 1966; Noyes *et al.* 1982; Lloyd 1975). Adult males tend to exclude non-mated females and all smaller males from home territories. Smaller individuals of other species are similarly excluded. This exclusion from the best areas of cover or densest food forces these individuals into more open microhabitats and/or situations where more time is required away from burrows to acquire food. Owls take these individuals in greater numbers due to their greater vulnerability. Age and sex effects may be less pronounced among terrestrial carnivores better able to pursue small mammals into dense cover or burrows.

The remains in the Large size class correspond almost entirely to *Arvicola*. Aging this large microtine species by dental attrition again is impossible due to its continually growing molars. Most postcranial remains of the other larger small mammal taxon detected from teeth (*Talpa*) are distinguishable and have been tallied separately. Age classes of the *Arvicola* individuals taken are particularly important, since these species (both *A. terrestris* and *A. sapidus*) fall at the upper range of prey normally taken by a barn owl or other medium owl species. Adult mass in these species approaches 300 g (Chapter 3). While *Arvicola* is regular barn owl prey where their ranges overlap, it is usually preyed upon in smaller numbers (Chapter 4).

Barn owl predation upon small mammals exceeding 100 g results mainly from the relative availability of small prey in a given environment. Over this predator's cosmopolitan range, larger species can make up a significant proportion of diet (Avery 1993; Davis 1958; Evans and Emlen 1947; Herrera and Jaksic 1980; Marti 1969; Perrin 1982). Most small mammals (primarily rodents and insectivores) tend to have average adult mass well below 100 g. This propensity by the barn owls to consume the smaller taxa reflects their greater general availability. In situations where larger taxa are more abundant, they tend to be taken in greater numbers. These larger taxa, however, are preferentially preyed upon in their pre-adult age classes.

The epiphyseal data supports the interpretation that mostly juvenile *Arvicola* were taken. With the exception of the proximal femur which fuses at a younger age, nonfusion among the elements ranged from 67.3% to 81.8%. Age classes for the majority of *Arvicola* taken must have fallen in the juvenile and subadult classes. These individuals are much smaller than adults and conform more closely to an upper limit of 100 g per individual prey item for barn owls. They were also likely excluded from areas of prime forage and cover by larger, more dominant individuals and therefore were available in greater relative numbers.

The relative state of epiphyseal fusion per size class was compared. No statistical difference was found comparing Small, Medium and Large. Increasing sample size among the Small size class might change these results, since these remains show a higher degree of fusion. Relatively more adults among this size class (shrews) may fall prey, since the subadults and juveniles of these species may be so small as to be a less desirable food choice among the owls. Since barn owls will also consume arthropods in small but regular amounts, this interpretation is less tenable. The striking agreement between individuals of medium and large size class may reflect similar growth patterns among microtines, since effectively all of the large remains and almost all of the medium remains are assignable to this rodent family. Also indicated are similar patterns of barn owl predation, where younger individuals are preyed upon preferentially to adults.

TOOTH ERUPTION

Tooth eruption for *Arvicola* indicates that some very young individuals were taken. Among *Arvicola* molars identified by the author (n = 2486), 73 (2.9%) were not fully erupted (the occlusal surface had not come into complete wear). In the case of third molars only ($_3M_3$ and $^3M^3$), 27 of 833 (3.2%) had not fully erupted at the time of death. Final tooth eruption occurs developmentally very early in rodents, generally within the first few weeks of life and corresponds with weaning and emergence of new litters from maternal care. These individuals are very vulnerable to predation at the time of initial dispersal. Individuals only slightly older would not be detectable by tooth eruption, since their occlusal surfaces would have come into wear along the entire surface. Continuous molar growth would replace subsequent dental attrition.

The presence of these very young individuals indicates occupation of the cave by barn owls at least partly during late spring, summer, or early autumn, since this is the breeding season for *Arvicola* (Chapter 3). Individuals with unerupted molars must have entered the population during this period, though older individuals could have been taken at any time of the year.

The molars of *Apodemus* (Muridae) do not grow continuously and allow assignment to general age class. This genus belongs to the Medium size class, which is dominated by microtines that cannot be aged by this criterion. General wear stage assignment for *Apodemus* is presented in Table 2.9 and follows the general criteria of Adamczewska-Andrzejewska (1967), Bishop and Delany (1963), and Delany and Davis (1961). Both upper and lower molars were aged following these general stages based upon *A. sylvaticus* and *A. flavicollis*, both of which may be present in the El Juyo sample. Since these age classes are only approximate, variations by species are unimportant.

Table 2.9: El Juyo *Apodemus* Age Classes Based upon Tooth Wear.

Wear Stage	Age Class	Approximate Age	Number of Teeth	%
I	Neonate	4 weeks	0	0%
II	Juvenile	2 months	27	39.7%
III	Subadult	5 months	27	39.7%
IV	Adult	9 months	14	20.6%
Total			68	100.0%

(Following: Adamczewska-Andrzejewska 1967; Bishop and Delany 1963; Delany and Davis 1961)

The majority of individuals taken (79.4%) fall into the juvenile and subadult categories, placing them within the first half-year of life. As discussed above, the majority of small rodents are taken by predators before reaching adulthood. In the case of *Apodemus* and barn owls, this preference represents greater vulnerability of younger individuals to predation and not selection by the owls for smaller prey: adult *Apodemus* are easily within the prey size limits of this predator. No neonates were taken, unlike among *Arvicola*. This age class among *Apodemus* is normally unavailable to avian predators, since the young have not yet dispersed from their natal nest. Individuals this young may fall to terrestrial predators, especially small mustelids that can hunt prey in small burrows.

Season of occupation by the owls is suggested by these remains. Late spring to autumn must have been at least part of the occupation span due to the presence of juveniles. The breeding season for *Apodemus* ends in early autumn, so juveniles could not be present in a population past the end of this season. The presence of some adult molars (20.6%) indicates occupation may have continued during the winter months, though this age class is present throughout the year.

TAPHONOMY OF HEDGEHOG REMAINS

The skeletal element recovery pattern for the remains of hedgehog (*Erinaceus europaeus*) does not conform to the larger, fur-bearing taxa (Chapter 9), nor does it conform to the pattern found for smaller insectivores and rodents discussed above. This species (Chapter 3) is much larger (about 1 kg) than the typical microfauna otherwise recovered (most below 40 g, with some reaching up to 300 g). It is rarely prey of species such as barn owls. Hedgehogs were entirely absent from a total sample of 47,865 vertebrate prey items taken from 188 barn owl pellet samples from Britain and Ireland (Glue 1974). While it is possible that juvenile hedgehog could be taken by a barn owl (as are the occasional juvenile rabbit), this taxa is generally outside the size of this predator. Hypothetical juvenile prey, being too large to swallow whole, are likely to have their edible portions flensed from their bones and spiny hides. This feeding pattern would leave most bones behind at the kill site and prevent their return to a roosting site. Eagle owls will take hedgehogs, but this prey was recovered in very low numbers. Prey in this size range is typically a much larger proportion of eagle owl diet. The preponderance of cranial elements also argues against owl predation upon this species, since these bones are difficult to swallow and are the most likely to be left behind at a kill site. Low meat utility elements such as foot bones are also less likely to be swallowed, yet these appear in high relative abundance at El Juyo.

The recovery pattern of these remains (Table 2.10) is consistent with natural in-site mortality of this species. Hedgehogs do not shun human habitation areas (Chapter 3) and may have entered the cave even while occupied. With the exception of a single fragmentary tibio-fibula diaphysis, all elements recovered from this species were only the most durable: teeth, mandible/maxilla fragments or whole elements, and foot elements. This pattern indicates a high degree of postmortem damage. This could have occurred as a result of trampling upon previously intact skeletons left on the site surface, with the more delicate bones generally being destroyed. No gastric corrosion was noted on any of these elements. Since they are generally of low meat utility, they may represent the remains left behind by a larger predator inhabiting the area of the cave that had consumed the missing elements.

SUMMARY

Taphonomic analysis of the microfauna from El Juyo is a necessary precursor to paleoenvironmental analysis, since in this manner the primary accumulator of the assemblage can be determined. Analysis indicates that barn owls (*Tyto alba*) are responsible for nearly all the microfaunal remains in the site. This conclusion takes into consideration the spatial layout of the cave: its available denning/roosting places, external access, relative location and utilization by humans. The size range, activity patterns and age profiles of the prey were also considered, as were the per element recovery pattern, digestive corrosion, and breakage of the skeletal remains. This data was compared with the known ecological activity and taphonomic signatures of the possible accumulating agencies. All of these except barn owls were eliminated on multiple grounds. The relationship between barn owls and the ecological signal that is recorded by their small mammal prey is discussed in Chapters 4 and 5, where in the latter chapter paleoenvironmental reconstruction for El Juyo is presented.

Table 2.10: *Erinaceus europaeus* Remains from El Juyo.

	Level							
Element	**9**	**8**	**7**	**6**	**4R**	**4S**	**4**	**Total**
teeth	10	33	20	11	8	23	17	122
1st phal		1		3				4
2nd phal				5				5
3rd phal				1		2		3
metapodial		2		1				3
carpal		2						2
astragalus				1				1
tibia frag.						1		1
mandible/frag.		1		3	1		1	6
maxilla/frag.	1							1
Total	12	38	20	25	9	26	18	148

CHAPTER 3
SPECIES CHARACTERISTICS AND ECOLOGIES

This chapter describes the behavioral, morphological, and ecological characteristics of the small mammal species present in Cantabrian Spain during the Late Pleistocene and Holocene. The author assumes that the behavior of these species has not significantly changed in the few thousand intervening years. In almost all cases, this assumption is not radical since these species have undergone no major evolutionary changes. One possible exception is the evolution of commensal behavior on the part of some of these taxa, which necessarily cannot predate the rise of major human-induced impacts upon the landscape (forestry, agriculture, purposefully set fires) and the extensive construction of habitations. Most of this coevolutionary process therefore must have occurred beginning with the Neolithic Period and its resultant changes in landscape usage. One may infer that this process was in an incipient stage during the Upper Paleolithic, particularly with the barn owls responsible for the introduction of small mammal remains to the site.

The metrical characteristics of these species have varied somewhat. This variation is described in Chapter 6. Following Bergmann's Rule, some species may have had a larger mean size during the generally colder Pleistocene. The possibility of such microevolutionary change was taken into consideration in the identification of these taxa.

Ecological behavior presents a more difficult problem, since the Pleistocene ecology of Cantabrian Spain does not necessarily conform to any extant landscape. The modern climate of Europe owes its temperate nature to the warming influence of the Gulf Stream and general ocean proximity. The deflection of the Gulf Stream away from Europe by the intrusion of polar ice stretching southwards may have resulted in climatic change unpredictable by the simple glacial regime of rising/falling ocean level. In addition, simple analogy to more northerly regions of Europe fails to consider differences in solar influx, which was maintained even during glacial conditions. Topography for the northern coast of Spain is also a unique consideration, due to the proximity of the highland zone so close to the Atlantic coast. Species interactions that no longer occur, such as *Microtus oeconomus* (arctic vole) with endemic Iberian or Pyrenean vole species such as *Pitymys lusitanicus* and *P. pyrenaicus* may have affected their ecological behavior and habitat exploitation. Some ecological conditions of the Late Pleistocene are no longer directly observable even in modern northern climates, which reduces the usefulness of direct analogy to modern conditions (Andrews *et al.* 1979).

The large-scale human induced changes to the European landscape alluded to above had not yet occurred, so our understanding of the Pleistocene fauna is also partially limited by having an altered modern fauna for study and comparison. This alteration includes the introduction of new species such as gray squirrel (*Sciurus carolinensis*), house mouse (*Mus musculus*), rat (*Rattus rattus*) and (*R. norvegicus*), wallaby (*Macropus rufogriseus*), American mink (*Mustela vison*), Canadian beaver (*Castor canadensis*), muskrat (*Ondatra zibethicus*), racoon (*Procyon lotor*), racoon dogs (*Nyctereutes procyonoides*), and feral and domesticated forms of multiple species. Other alterations include the range extension of some species already a part of the European fauna such as harvest mice (*Micromys minitus*), rabbit (*Oryctolagus cuniculus*) and edible dormouse (*Glis glis*), and the range contraction/local extirpation of most large predators. This range of problems must be taken into consideration when using the contemporary fauna of Europe to describe extinct ecological conditions.

RODENTIA, FAMILY GLIRIDAE

This rodent family includes 6 genera in the Palearctic region, 5 of which are monospecific, and has remained taxonomically stable for some time (Corbet 1978, 1984). These species are characteristically arboreal and hibernating.

Glis glis (edible dormouse, fat dormouse)

(Figure 3.1, end of chapter) *G. glis* is found today in Europe, ranging from northern Spain to Ukraine. Southern range includes the Mediterranean region and Turkey, and isolated populations are found in the Caucusus region, southern England, and some Mediterranean islands. This species' English name derives from the fact that it was raised and eaten by Romans. Its range may therefore have been artificially extended to previously uninhabited regions (as in England: Corbet and Ovenden 1982:153). Mean adult mass is 140 g, though prior to hibernation this may extend upward to 250 g (Corbet and Harris 1991:264). *G. glis* is the largest member of the family Gliridae. Its dental pattern is 1.0.1.3/1.0.1.3, with rooted molars exhibiting distinctive transverse ridges. This species is thoroughly arboreal and prefers deciduous woodlands of varying ages, and will also occupy orchards and gardens. Some commensalism has been noted (Corbet and Harris 1991:265). Its altitude ranges up to the limits of the tree line, in the Pyrenees about 2,000 m (Corbet and Ovenden 1982:153). The diet of *G. glis.* is varied and includes fruits, nuts, buds, bark, invertebrates, fungi, eggs, and nestlings.

G. glis lives in groups with strong hierarchies. Both young and the male may remain with the adult females during winter hibernation. This occurs in nests both above and below ground, as this species can also burrow. Home ranges are tiny (ca. 100 m diameter) and dispersal is only 1-2 km. Activity is entirely nocturnal, with the daylight hours spent holed up in a tree crevice or a nest close to the trunk (Corbet and Harris 1991:265-266).

Sexual maturity typically is not reached until after the first winter. Litter sizes of 4-6 are common, through the known range is 1-11 (Corbet and Harris 1991:263-266). Gestation lasts 22-24 days, with weaning occurring at about 30 days, at which time the young leave the nest. The breeding season extends only from mid-June through August. Life expectancy in the wild is 3-4 years and can be higher in captivity.

Eliomys quercinus (garden dormouse)
(Figure 3.2) *E. quercinus* occupies temperate Europe from the Mediterranean to Finland, including Spain and European Russia. This species of dormouse is smaller than *G. glis*, but has the same dental formula (1.0.1.3/1.0.1.3) and lamellar pattern to molar ridges. Adult mass is extremely variable across its range. Means range from 60 g in central Europe to 170 g on Formentera Island, Spain. 100 g is more typical for this species (summarized by Moreno and Collado 1989). Although its primary habitat is coniferous and deciduous forest, *E. quercinus* is less arboreal than other species of dormouse and can be encountered in lower foliage zones, on the ground, and is occasionally commensal. This preference for trees tends to make this species less vulnerable to owl predation, though this may increase during times of juvenile dispersal. Inhabited altitudes range up to 2,000 m in the Alps and Pyrenees (Corbet and Ovenden 1982:152).

E. quercinus breeding season in Spain runs from February through October. Sexual maturity is reached at about 60 g adult body mass. Two litters per adult female each year are typical, with recently weaned young appearing in March-June and September-October. Gestation lasts for about 25 days. Weaning occurs at 40-45 days, at which point the juveniles have reached a body mass of about 32 g. Average litter size is 5.5, with a range of 4-8 (Moreno 1988).

RODENTIA, FAMILY MICROTIDAE
Members of this rodent family dominate the microfauna of El Juyo and other Cantabrian sites. This taxon is variously listed as the family Microtidae due to popular usage, but taxonomic rules give precedence to Arvicolidae. To avoid confusion, the author will retain the popular usage of the term microtines to describe this group. This family is also variably described as a subfamily of Cricetidae (with hamsters comprising Cricetinae--see for example Corbet 1978). The members of this rodent family are characterized by their prismatic, continuously growing molars which form roots only in one extant genus, *Clethrionomys*. Molar roots also formed in one species (*Pliomys lenki*), which went extinct at the end of the Pleistocene and is found in small numbers in Cantabrian sites. A short, abrupt face, small external pinnae, and relatively short tails also characterize this taxon. As is the case for all rodents, the upper and lower incisors grow continuously. The dental formula for all species of microtine is 1.0.0.3/1.0.0.3.

Microtine evolutionary history begins by the Pliocene, at which time all genera had closed roots. This shift towards hypsodonty occurred in multiple parallel lineages throughout their subsequent evolution (Brunet-Lecomte 1990; Kowalski 1966). The genus *Pitymys* evolved during the Middle Pleistocene, but was later supplanted by *Microtus*, which is still dominant today and has undergone radiation into dozens of Holarctic taxa (see Gromov and Polyakov 1992 for a recent summary). *Arvicola* evolved in the Middle Pleistocene from *Mimomys*, and has two extant forms that differ cytologically (*A. sapidus* and *A. terrestris*). Nomenclature is further confused in this clade, since some authors lump *Chionomys* and *Pitymys* into subgenera of *Microtus*. The author will in this case retain the usage of splitting into distinct, but closely related genera. Other members of this family include lemmings (*Lemmus lemmus* and *Dicrostonyx torquatus* in northern Europe), true tundra-dwelling species never yet recovered from any Spanish site. The steppe lemming (*Lagurus lagurus*), today endemic to the steppes of European Russia, has been identified very rarely from Cantabrian sites. The tundra vole (*Microtus gregralis*) has been found in one site only and in very small numbers. Members of the microtid family are typically open grassland or tundra dwellers, and burrow extensively but shallowly into suitable topsoil.

Microtus agrestis (field vole; short-tailed vole)
(Figure 3.3) *M. agrestis* ranges across temperate and boreal Europe to the Urals, with discontinuous distribution extending eastward to Lake Baikal in Siberia. Modern distribution does include northern Spain, but does not include the Mediterranean region. Corbet and Harris (1991:204) found an adult average of 39.7 g for males (n = 20) and 30.9 g for females (n = 25) for second year (fully grown) adults at Loch Tay, Scotland. Mean mass for a population including juveniles and subadults would therefore be somewhat lower. Preferred habitats are rough, ungrazed grasslands and marginally into woodland, hedgerows, bogs, dunes, scree, and moorlands (Corbet and Harris 1991:205). Feeding is herbivorous, primarily on green leaves and grass stems.

This species is primarily nocturnal in summer, becoming more diurnal in winter. Juveniles tend to be more nocturnal than adults (Corbet and Harris 1991), perhaps due to their competitive exclusion from favored microhabitats with denser cover by larger adults. Breeding occurs from March through October, with multiple litters of generally 2-7 each. A gestation period of 18-20 days is the norm. Weight at birth is 2 g, and weaning occurs by 28 days (Corbet and Harris 1991:206) by which point all teeth are fully erupted and have come into wear. Very few individuals survive through two winters in the wild. This species is common prey of diurnal raptors, nocturnal raptors, and mustelids and is a preferred prey item of foxes.

Microtus arvalis (common vole)
(Figure 3.4) Modern range extends across temperate Europe, from northern Spain and into Siberia nearly as far as Lake Baikal. Mediterranean and far northern regions are not included is this distribution. Corbet and Harris (1991:209) indicate a mean of 42 g for males (n = 39, range 29-67 g) and 36 g for females (n = 38, range 22-55 g) on Orkney (the Channel Islands). *M. arvalis* is a habitat generalist and can be found in deciduous and coniferous forest, marsh, heather moorland, hay meadows, and gardens but is mostly absent from arable land and short pastures (Corbet and Harris 1991:210). It does prefer meadow over these other habitats (Corbet and Ovenden 1982:160), and can be found in its highest densities there. Most feeding is herbivorous: leaves and stem of grasses and a variety of dicotyledons (Corbet and Harris 1991:210). This species is active both night and day.

Breeding occurs somewhat earlier than other species of microtid, starting in February and running through September, with multiple litters in a season being the norm. Females born early enough in a season may even reproduce before its end. Two to five are typical for one litter, with 3.3 being average. Gestation and lactation both last about 20 days (Corbet and Harris 1991:211).

Microtus nivalis (alpine vole, snow vole)
(Figure 3.5) *M. nivalis* has a relict distribution across the alpine regions of Europe, including Spain, France, Italy, Greece, Switzerland, and the former Yugoslavia. Niethammer and Krapp (1982:264) found a mean size of 41.3 g (n = 20, range 18-59) for European museum specimens of this species. Its preferred habitat is loose rock and grassland above the tree line and more open montane woodland. It is found up to snowline, and in Switzerland is most populous at 1600-2600 m (Corbet 1966:247). *M. nivalis* lives up to 4,000 m on Mount Blanc (Corbet and Ovenden 1982:161), but can be found as low as 30 m a.s.l. in the former Yugoslavia (Krystufek 1990; Krystufek and Kovacic 1989). Its distribution therefore seems more dependent upon the availability of rocky/fissure habitats than altitude itself. This species also will enter underground caverns, a potentially important consideration when dealing with archaeological cave deposits.

Breeding begins at the end of April, with litters produced from May through August. This season can extend slightly at the lower altitudinal extremes of *Microtus nivalis* range, but it is relatively short when compared with those of other species of European microtines. Females born early enough in the year are able to reproduce before the season's end. Adult mass is obtained at about 3 months of age. Litter size ranges from 2-4, with a mean of 3, and an adult female normally produces two litters per year. Gestation lasts for less than one month (Le Louarn and Janeau 1975).

For simplicity, the author has retained the genus designation *Microtus* for this species. It is sometimes classified in a separate genus *Chionomys*, or in a subgenus *Microtus* (*Chionomys*) *nivalis*.

Microtus oeconomus (arctic vole; root vole)
(Figure 3.6) This abundant species today has a Holarctic distribution, from Europe into Canada. Its southward range does not include Spain, with its presence indicating a definite cold climate. This medium vole species averaged 31.8 g (n = 306, range 10-65 g) for a population including juveniles in the Netherlands (Ligtvoet and Wijngaarden 1994). It prefers wet, marshy areas and humid grassland (Corbet 1966:246; Corbet and Ovenden 1982:161). Tundra and wooded steppe are also occupied. It also prefers moist grassland microhabitats such as reedbeds, but can be found in dryer dike slopes and among dunes, hedgerows and fields during peak years and in the absence of direct competition from species such as *Microtus arvalis* (Ligtvoet and Wijngaarden 1994). Although slightly larger on average than *M. arvalis*, *M. oeconomus* tends to be outcompeted due to its more solitary lifestyle: *M. arvalis* tends to live in small groups and thus dominates agonistic encounters between the two species. Where modern ranges overlap (such as in the Netherlands) *M. oeconomus* has the advantage in humid microhabitats, which are significantly cooler than the dry grassland (including pasture and cropland) favored by other species of *Microtus*.

Breeding for *M. oeconomus* in Europe extends from early May through late September. Gestation lasts for about 20 days (Nadeau 1985). Litter size ranges from 2-8, with an average of 5 in Europe, with higher values in the Nearctic (Keller 1985; Ligtvoet and Van Wijngaarden 1994).

Microtus gregalis (tundra vole, narrow-skulled vole)
(Figure 3.7) *M. gregalis* has a discontinuous range across the tundra and forested steppe zones of Asia. Its penetration into northern Spain during a cold episode therefore indicates significant alpine tundra/steppe in the Cantabrian Cordillera. Typical adult mass is about 40 g (Ognev 1964:404). This species has been identified from only one Spanish Paleolithic site: Erralla Lower Magdalenian Level V (Pemán 1985). Since this site is located well away from the modern coastline and is within the hunting radius of upland habitats, there is no need to postulate that this habitat penetrated to lowland coastal Cantabrian Spain even during glacial maxima.

Complex burrow colonies with numerous exits (30-40) are excavated and inhabited perenially. Activity is nocturnal and peaks after sunset. The breeding season is relatively short for *M. gregalis* in the northern part of its range, commencing in late June/early July and lasting at least through September. Two litters are normal per adult female each season. Litter size is very high, averaging 8-9. Farther south, litters are produced as early as May, with 5-12 neonates typical. Members of the first litter each breeding season typically reach sexual maturity and reproduce that same year (Ognev 1964).

Clethrionomys glareolus (red-backed vole; bank vole)
(Figure 3.8) *C. glareolus* ranges across temperate Europe into central Asia as far as Lake Baikal. Northern Spain, Britain and Sweden are included in this range, but the Mediterranean zone generally is not. This vole genus differs from the others described here in the presence of molar roots, which develop after birth and grow throughout the individual's lifetime. The molars have the typical microtine morphology of division into prisms which are worn down in cross-section. Their roots become visible by age 3 months (Alibai 1980; Bishop and Delany 1963). Differences in prismatic morphology also make the molars of this species distinguishable from those of the other microtines present.

Body size of this species is smaller among the microtines relevant here. Niethammer and Krapp (1982:110) report a mean mass of 16.4 g (n = 30, range: 13.2-23.2g) for museum specimens from Finland. This sample includes males, females and juveniles and thus reflects more closely the average mass of those individuals likely to fall prey to a small predator.

While *C. glareolus* prefers mature deciduous forests with thick ground cover, its habitats also extend marginally into grassland, hedgerows and coniferous forests. This species is normally crepuscular, so predation by nocturnal and diurnal predators is common. The breeding season extends from April through October. Gestation periods range from 18 to

27 days, with a positive correlation between length and litter size. This size ranges from 1-7 young, with 3-4 average. Weaning occurs at 17-18 days, and all molars have erupted by 28-30 days Corbet and Harris 1991:195-199).

Pitymys pyrenaicus (Pyrenean pine vole)
(Figure 3.9) *P. pyrenaicus* has a relatively restricted range and today is found only in northeastern Spain and southwestern France. Niethammer and Krapp (1982:443) indicate a mean size of 21 g (n = 8, range 17-25 g) for specimens from Spain. This species is found in the upper reaches of beech forest to 1500 m in the French Pyrenees and also in Spanish parkland (Niethammer and Krapp 1982:446). Borghi *et al.* (1994) report this species at over 2,000 m in the Spanish Pyrenees.

P. pyrenaicus is now considered to be a distinct species from *P. savii* (Savi's pine vole), which is distributed discontinuously from the former species' range in Italy and Sicily. *P. pyrenaicus* is fossorial and herbivorous, preferring deep soils for tunneling and dense vegetation cover. In areas of overlap with *P. lusitanicus*, the latter species is competitively excluded to areas of sparser vegetative cover (Borghi *et al.* 1994).

Breeding occurs from April through November. Litter size is somewhat small for a rodent of this size, with a range of 1-3 and a mean of 2.5 (Castién and Gosálbez 1995b).

Pitymys lusitanicus (Lusitanian pine vole)
(Figure 3.10) *P. lusitanicus* is found only in the north and west Iberian peninsula, including northern Portugal and Spain as far as the western Pyrenees. Niethammer and Krapp (1982:478) report a mean size of 16.3 g (n = 20, range: 10.0-23.0 g) for specimens from both Spain and Portugal. Preferred habitats are open meadows, agricultural land and shrubland (Niethammer and Krapp 1982:482). *P. lusitanicus* is fossorial, preferring shallow soils with low proportions of rocky outcrops (Borghi *et al.* 1994). Recent owl pellet analysis from Spain (see Chapter 4) corroborates these results, with *P. lusitanicus* comprising a significant percentage of prey in a coastal lowland zone of mixed pasture, meadow, and small patches of deciduous forest.

The reproductive biology of this species is less well reported than for other microtine species (which generally have larger ranges). The breeding season extends from spring through fall, with the highest reproductivity in spring (Veiga 1986). Gestation lasts for about 21 days, with multiple litters in a season being typical.

Arvicola terrestris (northern water vole)
(Figure 3.11) *A. terrestris* today occupies a large range, including northern temperate Europe and stretching into Siberia east of Lake Baikal and south to Iraq. Its discontinuous range includes populations in Britain and Cantabrian Spain, but excludes the Mediterranean and western France. Corbet and Harris (1991:214) indicate a mean size of 311 g for males (n = 18) and 272 g for females (n = 17), in both cases for overwintered adults from Oxfordshire, England. These sizes may therefore be taken as maximal adult weights, with a lower population mean expected if juveniles and subadults are included. It is found most frequently in lowlands, along the banks of slow water available year round. *A. terrestris* sometimes lives away from water and forms extensive burrows (Corbet and Harris 1991:214). This bimodal lifestyle led to original designation of water-dwelling (*A. amphibius*) and burrowing (*A. terrestris*) as separate species. The name *A. amphibius* has technical precedence, but *A. terrestris* has been retained due to longtime usage.

A. terrestris pursues a mostly herbivorous lifestyle. Invertebrates and fish are rare dietary inclusions (Corbet and Harris 1991:217). Activity is both at night and day. Breeding normally occurs from April through September, with an average of about 6 young each. Gestation ranges from 20-30 days. Birth weight is from 3.5 to 7.5 g and increases to about 22 g by 14 days. Weaning occurs at around 22 days (Corbet and Harris 1991:217; Ventura and Gosálbez 1990), at which time the third molar has fully erupted and started to come into wear. Males disperse by the age of 4 months from their natal territory. Dominant females settle in their natal territories, with the remainder dispersing (Corbet and Harris 1991:215). These dispersing juveniles are the most frequent prey of raptors and carnivores, as they tend to be excluded to more open microhabitats and must forage over greater distances for food.

Arvicola sapidus (southwestern water vole)
(Figure 3.12) This species is found only in the Iberian peninsula and southwestern and central France. The range of *A. sapidus* overlaps with that of *A. terrestris* only in central France, Cantabrian Spain and the Pyrenees. Niethammer and Krapp (1982:212) indicate a mean size of 201 g (n = 28, range: 165-275 g) for specimens from France, which includes not fully grown individuals. Their habitat is always along water such as rivers, canals and lakes where the cover vegetation is sufficient (Corbet and Ovenden 1982:166). The ecology of this species is very similar to that of *A. terrestris* discussed above.

Pregnant females appear in the population from February through October. These reach sexual maturity at about 170 g body mass. Litter size range is 2-5, with an average of 3.3 (Ventura and Gosálbez 1987).

RODENTIA, FAMILY MURIDAE
The rodent family Muridae is comprised of Old World mice and rats. Some authors (Corbet and Ovenden 1982) place the subfamilies Microtinae (voles and lemmings, above), Spalacinae (mole rats), and Cricetinae (hamsters) under the family Muridae with the Murinae. For the sake of clarity, each of these groupings will be treated as separate families until their evolutionary relationships can be better sorted out. The family Muridae is notable for its commensalism and resultant cosmopolitan range spread of house mouse (*Mus musculus*), black rats (*Rattus rattus*) and brown rats (*R. norvegicus*). The early spread of these species may be linked to the spread of naturally and artificially created grasslands (agricultural fields and other alteration of landscape). For example, *M. musculus* shows up in Israel for the first time around 120,000 B.P. (Tchernov 1975) and *R. rattus* during the late Mousterian (Tchernov 1979). Arrival in Spain of these species probably did not occur before Neolithic times

and hence their appearance in paleolithic levels there would be strong indication of mixing of deposits.

Physical characteristics include large external pinnae, a relatively long tail, and a longer, thinner body than voles. Dental formula remains 1.0.0.3/1.0.0.3, with continuously growing incisors and rooted molars with many small cusps arranged in staggered rows.

Apodemus sylvaticus (wood mouse)
(Figure 3.13) *A. sylvaticus* today has an extensive range that has perhaps been extended by commensal contact with humans, as evidenced by the presence of this species on Iceland. Its range also includes all of temperate and Mediterranean western Europe, as well as part of North Africa, Turkey, and eastward through Russia past the Urals. Corbet and Harris (1991:222) report a mean size of 19.1 g for males (n = 20) and 17.8 g for females (n = 13) from Perthshire, England. Trapping in Cantabrian Spain (see Chapter 4) resulted in an average size of 21.3 g for males (n = 26, range: 14-32 g) and 19.1 g for females (n = 21, range: 12-25 g) for a mostly adult sample. *A. sylvaticus* is a habitat generalist and can be found in most types if they are not too wet, including woodland, cropland, grassland, heather, blanket bog and sand dunes (Corbet and Harris 1991:224). Deciduous forest, however, is their favored habitat. This species is also commensal and will follow human shelters to higher altitudes than their normal treeline limit (Corbet and Ovenden 1982:170).

Activity is mainly nocturnal, though this tends to become biphasic during long nights. *A. sylvaticus* forms extensive burrow systems and usually nests below ground. Diet is opportunistic: seeds, vegetation, fruit, and even animal foods are all consumed (Corbet and Harris 1991:226-227). Litters of 4-7 are common and are produced from March through October. Gestation lasts 19-20 days, with weaning occurring at 18-22 days. Few adults survive for more than one winter. This species is common owl prey, especially due to the overlap of activity.

Apodemus flavicollis (yellow-necked mouse)
(Figure 3.14) *A. flavicollis* has a discontinuous range throughout temperate Europe, including isolated populations in northern Spain and the Pyrenees, southern England, southern Norway and Sweden, and the Levant. The Mediterranean area is populated in Italy and Greece. Size ranges up to 1.5 times that of *A. sylvaticus* (Corbet and Harris 1991:229). Trapping data from Cantabrian Spain (see Chapter 4) yielded masses of 25.1 g for males (n = 16, range: 17-34 g) and 23.4 g for females (n = 15, range: 16-32 g) for a population including juveniles. These means are only slightly above those of *A. sylvaticus* trapped in the same locations in Cantabrian Spain (see above). *A. flavicollis* inhabits mainly deciduous woodlands and marginally extends into hedges, rural gardens and buildings, but is not a true commensal. It is more common than *A. sylvaticus* in alpine and coniferous forests and lives up to the treeline, and is less common in scrub and fields (Corbet and Harris 1991:230-231).

This species is active mostly at night. *A. flavicollis* forms extensive burrow systems, in which it may store food. In winter, nesting in small groups may occur. Its diet is similar to *A. sylvaticus* and includes a variety of plant and some animal materials. Litter size ranges from 2 to 11, and breeding occurs from February through October. Gestation and weaning are similar to those of *A. sylvaticus*, with both around 20 days (Corbet and Harris 1991:232).

Micromys minutus (harvest mouse)
(Figure 3.15) *M. minutus* ranges across the temperate Palearctic zone from Spain to Korea. This range may have been extended in some areas with the spread of agriculture, since this species inhabits the stalk zone in grasslands and has readily adapted to croplands (Trout 1978). Tall, dense grasslands are preferred, though some spillover into mature forests has been reported. The ground level runways of other rodents are inhabited in winter (Corbet and Harris 1991:236). Predation therefore seems to increase in autumn when less cover is available and juveniles disperse to their new territories. Activity is normally nocturnal, with peaks after sunset and before dawn. This elusive species is a regular but numerically low prey item, never reaching >1% of the diet of British raptors (Glue 1975). Its importance to raptor diet is further reduced by its small average body mass of only 4-10 g.

The breeding season of harvest mice extends from May through October, with most litters produced at the end of this period. Litters range from 4 to 8, with an average of 5.4. Gestation last 17-19 days, and the young are weaned at 15-16 days. Litters are raised in nests in the stalk zone, which further reduces their susceptibility to predation by raptors until dispersal occurs (Corbet and Harris 1991:237).

INSECTIVORA, FAMILY SORICIDAE
The teeth of these species are distinguishable on multiple grounds. The teeth of *Sorex* have a red color on their cusps which is highly distinctive unless severe attrition has occurred. Dental formula follows 3.1.3.3/ 1.1.1.3. *S. minutus* is much smaller than *S. coronatus*. *S. araneus* may be distinguished from the latter species by slight differences in mandibular morphology (Hausser and Jammot 1974; Niethammer and Krapp 1990; Turni and Müller 1996) where these features have been retained in archaeological and paleontological materials. The teeth of *Neomys* exhibit this red staining to a lesser degree and follow a 3.1.2.3/1.1.1.3 formula. Teeth of *N. fodiens* and *N. anomalus* are distinguishable from each other primarily upon metrical grounds (Ruprecht 1971). The teeth of *Crocidura* follow a 3.1.1.3/1.1.1.3 formula and lack red staining of their cusps.

Crocidura russula (greater white-toothed shrew; house-shrew)
(Figure 3.16) This species is found today only in Western Europe and North Africa, including the countries of Spain, Portugal, France, and Morroco. Body mass for a sample of male and female adults from western Switzerland averaged 11.4 g (n = 21, range: 9.0-16.1 g) (Niethammer and Krapp 1990:432). Seven individuals trapped in Cantabria (see Chapter 4) yielded a mean mass of 7.9 g (6 males, 1 female, range: 6.5-10 g). *C. russula* is separable from *C. suaveolens*

(lesser white-toothed shrew) only by dentition and is otherwise slightly larger in most dimensions. *C. russula* is a habitat generalist and can be found in forests, hedgerows, grassland, agricultural lands, maquis scrubland and amongst human habitations (Corbet and Harris 1991:74).

This essentially solitary species is active both day and night and pursues the typical shrew diet of primarily invertebrates and a few small vertebrates. Average litter size is 4, with a range of 2-11, and breeding occurs from as early as February through October. Gestation lasts from 28-33 days. Birth weight is 0.8-0.9 g, with weaning occurring at 28 days. Sexual maturity is reached at 58-71 days, so breeding during the season of birth is common (Corbet and Harris 1991:74). Densities have been recorded to peak at 77-100/ha in Europe during the summer breeding season (Corbet and Harris 1991:75). This species is common raptor prey in Spain, but was a much rarer component of Pleistocene faunas.

Sorex minutus (pygmy shrew)
(Figure 3.17) *S. minutus* today ranges throughout temperate Europe, from northern Portugal to Lake Baikal in Siberia, and extends from southern Greece to northern Scandinavia. Its average size is one of the smallest among mammals, with a mean of 3.6 g (n = 20, range: 2.6-5.2 g) for specimens from Germany and France (Niethammer and Krapp 1990:193). It is found in all habitat types, though primarily in those with abundant ground cover. *S. minutus* prefers grassland over woodland, and while largely ground-dwelling, it will climb foliage (Corbet and Harris 1991:62). It tolerates lower ground cover than *S. coronatus*, and its range extends even to montane heather (Corbet and Ovenden 1982:125).

This species is active both day and night, with peaks around 0800-1000 and 2100-2300 (Corbet and Harris 1991:62). *S. minutus* is both solitary and aggressive, and thus reaches typical densities of only 12/ha, with a mean of 6/ha found in England (Corbet and Harris 1991:63). The diet consists primarily of invertebrates. This species' energetic needs are such that it must consume 1.25 times its own body weight each day. Breeding occurs from March through October and peaks in June/July. Four to seven per litter are typical, after a gestation period of 22-25 days. These newborns weigh only 0.25 g and are weaned at about 22 days. They generally do not obtain sexual maturity until after their first winter. Lifespan in the wild is approximately 13 months. *S. minutus* is common prey of barn owls and tawny owls, though in low numbers reflective of its low overall density.

Sorex coronatus (Millet's shrew)
(Figure 3.18) *S. coronatus* is found only in northern Spain and throughout France, the Low Countries, and into Switzerland. It is a sibling species to *S. araneus*, whose nonsympatric range extends eastward through Europe. A sample of mostly adult *S. coronatus* from Switzerland (Niethammer and Krapp 1990:238) averaged 7.8 g (n = 19, range: 6.5-11.75 g). Five specimens trapped in northern Spain (see Chapter 4) had an average mass of 7.4 g (2 males, 3 females). *S. coronatus* can be found in all habitats with sufficient cover, including forests, hedgerows, dunes, fields and high, barren plains. Its montane habitats extend as far as the summer snowline (Corbet and Ovenden 1982:124). The primary association of *S. coronatus* is with humid meadows.

The breeding season for this species extends from late January until mid-November (López-Fuster 1989; López-Fuster *et al.* 1988), so individuals born early in the year are likely to reach sexual maturity and reproduce before the year ends. Litter size ranges from 2-6 with a mean of 3.7. Gestation lasts for 24-25 days. Juveniles leave the nest after 20-25 days and have a total life expectancy of 14-21 months.

Neomys anomalus (Miller's water shrew)
(Figure 3.19) *N. anomalus* has a discontinuous distribution through Europe, from Spain into Russia. Its northern range extends as far as Germany, and its southern range extends into Greece. Niethammer and Krapp (1990:318) report a mean size of 10.9 g (n = 9, range: 10.1-12.3 g) for mostly juvenile museum specimens from Germany. The primary habitat of this species is along water courses, but it also can be found away from them in humid forests and meadows (Corbet and Ovenden 1982:127). *N. anomalus* is found between 180-1,850 m a.s.l. and normally below 1,000 m today in northeastern Spain (Torre and Tella 1994).

The breeding season of *N. anomalus* extends from March through September, with the first litters produced by the end of May. Litter size range is from 5-13 (Niethammer and Krapp 1990:332). Weaning occurs at 29-30 days (Michalak 1987).

Neomys fodiens (water shrew)
(Figure 3.20) *N. fodiens* ranges from northern Spain to Lake Baikal in Siberia. Its range includes England and extends to northern Scandinavia and into the Mediterranean region in Italy. Isolated populations can also be found in the far East, including the Korean peninsula and Sakhalin Island. Corbet and Harris (1991:65) indicate a typical size range of 12-18 g for adults of this species. Four specimens trapped in Cantabria (see Chapter 4) had an average mass of 12.5 g (1 male, 3 females, range: 10-17 g). Its habitat preferences are mainly the banks of clear, fast-flowing waters and also by ponds. It is especially numerous in watercress beds. *N. fodiens* is also found among boulders on rocky beaches and sporadically away from water in deciduous woodland, hedgerows, grassland, and occasionally in coniferous woodland (Corbet and Harris 1991:65). It is the most aquatic of the European shrew species, and it can be found up to 2500 m in the Swiss Alps (Corbet and Ovenden 1982:126).

N. fodiens is active both by day and night, though night activity predominates with a peak just before dawn. Individuals are solitary, but live in close proximity. Their diet consists primarily of invertebrates and also includes small aquatic vertebrates. Foraging occurs in the water, and this species will also form extensive burrows (Corbet and Harris 1991:65-67). The breeding pattern is somewhat different than other species of this size discussed above. Breeding occurs from April to September and peaks in May/June. Average litter size is 6, ranging from 3-15. Gestation is typical for a mammal species this size and lasts for 19-21 days, but weaning time is longer, at 38-40 days (López-Fuster and Ventura 1995). A maximum of 3 litters are therefore possible during a breeding season, and females normally breed only after overwintering. Lifespan is typically no more than 14-19 months in the wild (Corbet and Harris 1991:67). This species does not normally obtain large densities, and is regular (but not abundant) prey of barn owls and tawny owls.

INSECTIVORA, FAMILY TALPIDAE

Talpa europaea (northern mole)

(Figure 3.21) *T. europaea* can be found from northern Spain to European Russia, but its range does not include northern Scandinavia or the Mediterranean region, except for southern France. Britain is included in this range. Size is somewhat variable. Corbet and Harris (1991:46) report a mean size of 110 g for males (n = 42) and 85 g for females (n = 57) from Suffolk, England. Maurizio and Hausser (1990:225) indicate a mean size of 63 g (n = 36, 18 males and 18 females, range: 51-88 g) for a population from Val Bregaglia, Italy. *T. europaea* is present in most habitats where the soil permits tunneling. It was originally a deciduous woodland species but is now found in pastures and cropland and is uncommon in coniferous forests, moorland, and dunes.

Most of the lifespan of this species is spent within its tunnels, so predation by raptors must occur during those rare times when this species is above ground. Predation risk thus increases dramatically during the time of juvenile dispersal and also during mating season when males leave their territorial burrows to encounter females with overlapping ranges. Diet consists almost entirely of invertebrates obtained by the pitfall trap effect of tunnel entrances or those obtained while tunneling, especially earthworms.

Breeding is slower in this species than in other insectivores. A litter of 3-4 is typically produced in the spring, with one additional litter produced in the more southerly portions of this species' range. Birth weight runs around 3.5 g, and the nest is left after one month. Dispersal of juveniles then occurs at age 5-6 weeks, with nearly full growth obtained by age 3 months. Sexual maturity is reached the following spring. Lifespan in the wild is up to 6 years, but less than 3 years is more usual (Corbet and Harris 1991:47-48).

Talpa occidentalis (Iberian mole)

(Figure 3.22) This species' distribution is limited to all of the Iberian Peninsula and the southern edge of France. The range of *T. occidentalis* therefore overlaps with that of *T. europaea* throughout most of the former's range, especially in Spain. Niethammer and Krapp (1990:158) report a mean mass of 48.9 g (n = 20, range: 34-63g) for a population of males and females from Spain. Its habitat preferences are similar to those of *T. europaea*: fields and deciduous forests where suitable soils for tunneling are found, and rarely into coniferous forests and heath. *T. occidentalis* has demonstrated karyotypical differences from *T. caeca* (Jiménez *et al.* 1984), with which it was previously considered to be conspecific. *T. caeca* has a discontinuous range from that of *T. occidentalis* and is found from southern France to Italy and Greece. The range of both of these species overlaps with that of (the generally larger) *T. europaea.*

Adult male and female *T. occidentalis* are sexually active from September through May. Pregnant females are found in the population from October through April, while nursing females are found from November through May (Jiménez *et al.* 1990). Sexual maturity in both sexes is reached after one year. The long breeding season permits multiple litters per year for some females. The duration of gestation and lactation in this species is probably similar to those of *T. europaea*, though these remain unreported.

Galemys pyrenaicus (Pyrenean desman)

(Figure 3.23) *G. pyrenaicus* is endemic to the Pyrenees, their westward extension (the Cantabrian Cordillera), and highlands further south in northern Portugal and the Spanish meseta. This species belongs to the family Talpidae within the subfamily Desmaninae. This subfamily contains only two widely separate taxa, *G. pyrenaicus* and *Desmana moschata* found only in the area north of the Black and Caspian seas in the former European U.S.S.R. This large insectivore has a cranium generally similar to that of *Talpa*, with a long rostrum containing the maximum placental mammalian tooth pattern (3.1.4.3/3.1.4.3). These teeth are morphologically and metrically distinguishable from those of *Talpa.* Body mass ranges from 35.5 g to 80 g (summarized by Palmeirim and Hoffman 1983:1). This species is therefore in the proper size range to be taken by many species of European owls, but is usually a numerically small prey item.

Breeding last from January through July, with a gestation period of about 30 days. Litter size ranges from 1-5, with a mean of 3.6 (Palmeirim and Hoffman 1983:3).

G. pyrenaicus is an alpine and riverine species (Castién and Gosálbez 1992b; Palmeirim and Hoffman 1983). Its altitude zone is from 300 to 2200 m, though the lower end of this range is more typical. Its presence in near sea level in a coastal lowland site therefore indicates significant downward altitudinal zone shift during cold phases, at least placing its habitat within the hunting radius of a predator inhabiting the lowland zone. This species is well adapted for a semi-aquatic lifestyle, possessing four webbed feet, a keeled tail, a mobile sensory snout for probing stream bottoms, and insulating pelage. It is associated with cold, fast-running oxygenated (trout-abundant) streams from which it derives its main diet of crustaceans and other arthropods. Activity is typically nocturnal, though this may extend past dawn sometimes (personal observation).

INSECTIVORA, FAMILY ERINACEIDAE

Erinaceus europaeus (hedgehog, western hedgehog)

(Figure 3.24) This species is distributed throughout Western Europe, including southern Scandinavia, but it is replaced in Eastern Europe by the eastern hedgehog (*E. concolor*). *E. europaeus* weighs ca. 120 g at weaning and increases to over 1.1 kg as an adult (Corbet and Harris 1991:38). Habitat preferences are grassland/woodland edges of all types, including hedgerows. They are found rarely in coniferous woodland, marshes or moorland (Corbet and Harris 1991:40). This species is partially commensal, readily entering campgrounds and other open human habitations and rapidly losing its fear of humans (personal observation). Such behavior is an important consideration in the interpretation of archaeological cave faunas.

This species is almost entirely nocturnal, with some activity at dawn and dusk. Winter hibernation is also common, though it is frequently interrupted and sometimes burrows are changed. Diet is mostly invertebrate and also includes some carrion and bird eggs. One litter of 4-6 is normally born in June, but a second one may be born later in the season. Birth weight is 10-25 g. Deciduous dentition replacement begins by 24 days and weaning occurs at 4-6 weeks, with dispersal happening thereafter. Young do not breed in the season of their birth. Corbet and Harris (1991:42) also report that densities are about 1/ha, while noting that no conclusive data exists. The author has noted greater densities than this in northern France, where many individuals may gather around a human encampment to scavenge food. Lifespan in the wild does not usually exceed 7 years, with 60-70% of mortality occurring during the first year of life (Corbet and Harris 1991:43). Due to their size and protective spines, *E. europaeus* is very rare tawny owl prey and even rarer barn owl prey. Despite their protective spines, they are often taken by eagle owls.

FUR-BEARING SPECIES

These species are grouped separately not only for taxonomic differences, but also for differences in their probable mode of procurement and presumed deposition within Cantabrian Upper Paleolithic sites: through human hunting. This acquisition may represent fur trapping by humans (Chapter 9). The majority of these species are too large to be common barn owl prey in most instances, though eagle owls commonly feed upon the smaller taxa or juveniles of the larger. An exception is the weasel (*Mustela nivalis*) which is small enough to be preyed upon by barn owls, yet still may offer a usable pelt to human trappers for the larger members of this species. Some of these species may also have been acquired for their meat, though among the carnivores this is a less desired resource for humans (due to the generally poorer tasting meat). These taxa generally occur in low numbers among Cantabrian Upper Paleolithic sites.

FAMILY SCIURIDAE

Marmota marmota (alpine marmot)

This large rodent species is found today in the alpine region of Europe (Figure 3.25) and has been reintroduced to the Pyrenees and the Balkans (Herrero *et al.* 1992). The mass of this species is extremely variable throughout the year due to seasonal fat storage for hibernation. *M. marmota* reaches 300-400 g by six weeks of age, over 2000 g by 2 years of age and over 3000 g by 3 years of age when full size is reached. With seasonal fat gain, adult mass can reach 4500 g in autumn (Niethammer and Krapp 1978:174).

Marmots inhabit the alpine zone generally above the treeline in open pastures. Altitudes from 1800 to 2400 m are most typical, but this species can be found down to 1450 m in the Pyrenees (Herrero *et al.* 1992). Soil depth sufficient for the construction of their extensive burrow systems is another requirement (Corbet and Ovenden 1982:148). Large family groups occupy these systems. Group cooperation for predator avoidance makes this species difficult to hunt, but they may be trapped or snared by devices placed near their burrow entrances. Since this species hibernates each winter in its burrow system, its acquisition by humans must occur during the warmer part of the year.

Breeding occurs shortly after the adult females emerge from hibernation in March. Gestation lasts for 30 days, and weaning occurs after an additional 30 days. The juveniles emerge from their natal burrows in May. Mean litter size is about 2.4, but this species reproduces every year, unlike some species of *Marmota* that reproduce biennially (Barash 1989).

LAGOMORPHA, FAMILY LEPORIDAE

The three species discussed below all have a dental formula following 2.0.3.3/1.0.2.3. These dental remains are distinguishable on slight morphological grounds, especially with the ${}^{2}P^{2}$, ${}_{3}P_{3}$ and ${}_{3}M_{3}$ (Hillson 1986:80). *Lepus* is also significantly larger than *Oryctolagus*.

Oryctolagus cuniculus (rabbit)

(Figure 3.26) The original range of this species is thought to have been confined to Northwest Africa and the Iberian peninsula (Corbet 1978). Human habitat alteration and deliberate introductions/escapes from farms has led to huge range expansion to the British Isles, France, Germany, Italy, some Mediterranean islands, and notably to Australia and New Zealand. This range expansion includes northern Spain, where this species had not been previously numerous. Preferred habitats are natural or artificial areas of short grasses providing good cover on or in close proximity to burrowable topsoil (Corbet and Harris 1991:149-150).

Breeding begins as early as January and extends through August, though the onset of the breeding season can be delayed by severe weather. Litter sizes of 3-7 are typical. Gestation lasts about 30 days, after which the female is often pregnant again within one day. Weight at birth is 30-35 g, which increases to 150-200 g at weaning (ca. 21 days), then 1200-2000 g at adulthood (Corbet and Harris 1991:149-152).

Lepus capensis (brown hare; common hare)

(Figure 3.27) *L. capensis* is found in the temperate and Mediterranean zone of the Palearctic, into the Arabian peninsula and all of Africa. This wide-ranging species is divided by some authors into *L. capensis* for the southerly portion of this range and *L. europaeus* for the northerly portion, including Europe. This separation is primarily based upon size (Corbet 1978:72), variation of which across such a large and varied range is typical within many species. The author therefore retains for the present the monospecific designation, although this is not always the case for European Paleolithic researchers.

Corbet and Harris (1991:155-157) report mean adult masses of 3.23 kg for males (n = 723) and 3.43 kg for females (n = 689) for a population from East Anglia (England). Their size makes them unlikely prey except as juveniles for smaller species of owl (barn, tawny or long-eared), since weight at birth is normally over 100 g. They are common prey of the much larger eagle owl and particularly that of red foxes. Breeding continues from February through October and expands in mild climate, with litter sizes of 3 the most common (range: 1-4). Gestation lasts for 41-42 days, with weaning occurring after 23 days.

L. capensis is originally a species of open grasslands, though as is typically the case they have adapted to and spread their original range with the expansion of agriculture and the opening of formerly forested habitats. This species may leave visible trails on open meadows and characteristic gaps in denser vegetation, which makes the placing of snares or other trap types a practical method of harvest by humans. Tracking through snow is also possible. While winter pelage does not turn white in this species, their fur is still prized. They are also hunted for their meat, though most modern hunting is for purposes of agricultural pest control.

Lepus timidus (arctic hare; mountain hare)
(Figure 3.28) *L. timidus* inhabits the tundra and coniferous forest zone of the Holarctic, from the northern British islands to Greenland and Labrador. An isolated population also exists in and around Switzerland and may be a Pleistocene relict. This species is replaced ecologically to the south of its range by *L. capensis* in Europe. *L. timidus* tends to be the smaller of the two. Adults averaged 2.8 kg (n = 45) in a Scottish population and 3.2 kg (n = 27) in an Irish one (Corbet and Harris 1991:164). This species seems to follow Bergmann's rule in Europe, as northern individuals tend to be larger than southern individuals (Angerbjörn and Flux 1995). As with *L. capensis*, juveniles could be taken occasionally by smaller owl species, with adult predation falling mostly to larger predators like red fox.

This species is associated with open habitats, particularly heather moorland, and has not surprisingly colonized farmland as well. Highest densities are reached in mixed open/forest habitats along hedgerows and clearings (Angerbjörn and Flux 1995:5). Most activity is nocturnal and in groups which measuring up to 100 individuals. Simple, short burrows 1-2 m long are excavated for daytime shelter, though less permanent surface scrapes are also common. Burrows could have drawn the trapping attention of Paleolithic hunters. The pelage of *L. timidus* turns from dusky brown to mostly white in winter, perhaps increasing its value (*i.e.* supplying natural camouflage to human hunters against a snow backdrop).

Breeding continues from January through October, with three litters per season per adult female typical. Litter size ranges from 1-8, with means for first litters (2.15) smaller than those for the second litters (3.24) in a season. Gestation averages about 50 days, with weaning occurring after 4 weeks (Angerbjörn and Flux 1995:4-5).

CARNIVORA, FAMILY MUSTELIDAE
This carnivoran family is abundant and diverse in Europe, with up to 8 species found naturally today in the Iberian peninsula (Corbet and Ovenden 1982; Santos Reis 1983). Among *Mustela* and *Gulo*, dental formula is 3.1.3.1/ 3.1.3.2, while *Martes*, *Meles* and *Lutra* have additional teeth.

Mustela nivalis (weasel, least weasel)
(Figure 3.29) This smallest member of the mustelids (and of the order Carnivora) has a widespread distribution, including nearly all of the Holarctic region and also parts of North Africa. This range may have been altered somewhat by human activity and for example spread by Romans to Egypt, but Dayan and Tchernov (1988) suggest that this Egyptian population is a Pleistocene relict. Size across this range is variable, with a very high degree of sexual dimorphism. Populations throughout Britain (summarized by Corbet and Harris 1991:391) had mean sizes ranging from 106-131 g for males and 55-63 g for females. Sheffield and King (1994:2) summarize mean adult male body masses, ranging: 30-100 g in North America, 55-202 g in the United Kingdom, 33-108 g in Sweden, 40-220 g in (the former) Czechoslovakia, 36-250 g in (the former) U.S.S.R., 72-185 g in (the introduced population of) New Zealand, and 125-250 g in Africa. In direct opposition to Bergmann's rule, size tends to increase significantly towards the southern end of this species' range (Corbet and Harris 1991:388). Reductions in size in some areas may also reflect sympatry with *M. erminea* and *M. frenata*: where these two species overlap with it, *M. nivalis* tends to be smaller. In North America, this relationship is less clear, and size variation does not seem to be related to sympatry or Bergmann's rule (Sheffield and King 1994:2).

Habitat preferences are extremely varied and have in common adequate cover to protect this species from the many that prey upon it (most larger carnivores, owls, hawks, humans, etc.). Hence, *M. nivalis* may be found where its (typically small rodent) prey may be found: open forests, agricultural fields, meadows, hedgerows, alpine meadows and woods, scrublands, steppe and semi-desert, prairies, and coastal dunes. Dense forest and true sandy deserts are avoided (Sheffield and King 1994:5). Water voles are a frequent component of the diet where the ranges of these two species overlap in the Palearctic (Delattre 1983; Jedrzejewski *et al.* 1995). Small prey is frequently cached (Oksanen 1983). *M. nivalis* is active both day and night, and is trapped more frequently during the day (Sheffield and King 1994:6). This broad activity pattern combined with small size makes this species vulnerable to most larger predators, including diurnal and nocturnal raptors, and most larger carnivores. This species is rare prey (well below 1% of diet) of smaller species of owl, such as barn or tawny (Glue 1974; Southern 1954).

One or two litters are produced each year, without delayed implantation of embryos (Corbet and Harris 1991:393-394). Pregnancy typically occurs during late February through July, with births beginning in April. Gestation lasts 34-37 days. Litter size varies (Sheffield and King 1994:3-4), but an average of about 5 is typical for many portions of this species' range. Newborns weigh less than 2 g. Their permanent dentition is completely in place by day 42, with full adult body mass being obtained by weeks 12-15. Life expectancy is little over a year in the wild, though up to three years is possible. Their pelage turns white in winter in areas of significant snow cover, which may increase their value to human trappers.

Mustela erminea (ermine, stoat)
(Figure 3.30) *M. erminea* also has a Holarctic distribution, though primarily in the tundra, boreal, and deciduous zones (Corbet 1978). Sexual dimorphism is less pronounced than

in *M. nivalis*, with males about 1.5 times the size of females. Adult male mean weight (summarized by Corbet and Harris 1991:378) for European countries ranges from 208-283 g, in Russia 134-191 g, in North America 56-206 g, and in Britain ca. 320 g. This distribution does not follow Bergmann's rule and may reflect variation in local prey size (Erlinge 1987; King 1991). The habitat requirements of this species are similar to those of *M. nivalis*: most that offer sufficient cover, either for itself or for its prey to exist (Sleeman 1993). Farmland, forests, wetlands, and alpine zones are all occupied, but mature forests lacking in cover and small mammal prey are avoided (Corbet and Harris 1991:379). Depredation upon poultry raising operations are also common, but no true commensal relationship with humans exists. This species also caches small prey in large numbers (Oksanen 1983), a behavior which may leave patterns detectable in the fossil record.

Mating in this species occurs in summer, but implantation of the embryos is delayed until the following spring when a single litter of 6-12 is produced. Active pregnancy therefore lasts about 4 weeks. The young are weaned at 7-12 weeks (Corbet and Harris 1991:384-385).

This species is generally too large to fall prey except as juveniles to smaller species of owl, though predation by eagle owls is not uncommon. Winter pelage of *M. erminea* is mostly white, which historically has increased its value to human hunters.

Mustela putorius (European polecat)
(Figure 3.31) The range of this species covers all of temperate and most of Mediterranean Europe, excluding Greece. This range, however, may have been altered due to local extirpation by humans in some areas and expansion into other areas where the landscape has been altered. Body size in Britain (Corbet and Harris 1991:399) averages 1111 g for males (n = 28, range: 800-1710 g) and 689 g for females (n = 14, range: 530-915 g). *M. putorius* is the probable wild progenitor of domesticated ferrets (*M. furo*). Occupied habitats are varied, again with those offering sufficient cover being favored. These include forests, hedgerows, agricultural lands, wetlands, and water margins. Upper altitudes tend to be more thinly populated (Corbet and Harris 1991:401).

Both males and females become sexually mature and breed for the first time in the year following their birth. Mating typically takes place in April, and a litter of usually 5-10 young is produced after a gestation period of 42 days. Most litters are thus produced in May or June, with newborns weighing 9-10 g. Permanent dentition is place by 11-13 weeks, and juveniles reach adult dimensions by the autumn following their birth. The young are fully independent by late summer and disperse in the early autumn (Corbet and Harris 1991:403-404).

Mustela lutreola (European mink)
(Figure 3.32) This species may be found currently in Eastern Europe and Russia, and in southwestern France. Due to the continuing value of mink pelts, this range in some populated areas must be a significant contraction from its former state (Youngman 1982:3). The range of this species is further confused by the past introduction of American mink (*M. vison*) to Europe in order to replenish local supplies of trappable animals. Further confusion results from the interfertility (in captivity) of these two species (Corbet and Harris 1991:408) and their very similar appearance. This range may also have expanded into some areas, as into France within the past few centuries and into Cantabrian Spain as recently as 1951 (Youngman 1982:12-13). A sample of 7 males from France averaged 871 g (range: 650-1005 g), with two females weighing 505 g and 580 g (Youngman 1990:1). The preferred habitats of this species are marshy areas and the borders of rivers and lakes (Corbet and Ovenden 1982:182; Youngman 1990:2). This preference is in keeping with its semi-aquatic lifestyle and diet.

Mating occurs in the early spring (Youngman 1990:1-2), with a litter of 2-7 born 40-43 days later. Implantation of embryos is slightly delayed, so active pregnancy lasts only about 28 days (Corbet and Harris 1991:413). Size at birth runs about 8 g. Full adult dentition is in place by 60-67 days, and the young begin their breakup from the litter and dispersal by age 4 months. Males and females can breed in the year following their birth.

The pelts of this species are still coveted for their beauty, warmth, and water repellence, though most today are obtained from commercial mink farms. Traps or snares may be placed at low points along river banks where these are likely to pass.

Martes martes (pine marten)
(Figure 3.33) This species is found throughout most of Europe and into European Russia. Northern Spain is included in this range, as are Scandinavia and the central Mediterranean region. As with most large predators, its range may have been significantly reduced from some areas of Europe. Since it is a valued fur-bearing animal, its range may have been artificially or accidentally extended into other areas. The genus *Martes* has more teeth than other mustelids, with a formula of 3.1.4.1/3.1.4.2. Corbet and Harris (1991:373) report both significant sexual dimorphism and variation between populations. In Ireland, males range from 1500-1950 g (n = 31), with females ranging 1110-1450 g (n = 38). While populations in Scotland are similar in size, specimens from a population in the (former) U.S.S.R. have a male range of 670-1050 g (n = 59) and a female range of 484-850 g (n = 49). Habitat preference is coniferous or mixed forest, but this species may also be found at times in deciduous forest or more open vegetation (Corbet and Ovenden 1982:184). It may thus be found up to the tree line, which in the Pyrenees and Alps is approximately 2,000 m. *M. martes* is arboreal and highly agile. Human habitations are not avoided and some scavenging or livestock raiding may occur.

Most activity is solitary and nocturnal, with daytimes spent in a den of various configurations (rockfall, tree crevice, human constructions, etc.). Diet is mostly carnivorous upon small mammals and birds, but also includes invertebrates and berries (Goszczynski 1976). Delayed embryo implantation increases the gestation time between the summer mating season and birth the following March to April. Active pregnancy therefore lasts only about 30 days. Litter size is from 1-5, with an average size of 30 g each at birth. Weaning occurs at 5-6 ½ weeks. Adult weight is reached by the summer following birth (Corbet and Harris 1991:374).

Martes foina (beech marten, stone marten)
(Figure 3.34) Current range of this species is primarily in western and southeastern Europe, including northern Spain and the central Mediterranean region, but excluding Scandinavia. Isolated populations are also found in Central Asia. As with *M. martes*, this range may have been altered by human actions. *M. foina* inhabits deciduous forests within this zone, and may therefore be found to the limits of the treeline. Human habitations are not avoided and food is often scavenged from these settings (Corbet and Ovenden 1982:185; Delibes 1983). Their lifestyle is mainly arboreal. Their diet is eclectic and opportunistic: small mammals, birds, eggs, reptiles, amphibians, fish, invertebrates, vegetable matter and carrion are all commonly consumed (Amores 1980; Goszczynski 1976). The combination of association with the biotopes around human dwellings and their attractability to a wide variety of possible baits would have made this species an attractive fur trapping target.

M. foina reaches reproductive maturity in the second year of life. Mating occurs between July and mid-August, with delay of embryo implantation until the following January/February. Birth takes place in March/April after an active pregnancy of just over one month. Litter size ranges from 1-5, with an average of around 3. Weaning occurs no later than July (Madsen and Rasmussen 1985).

Lutra lutra (otter)
(Figure 3.35) *L. lutra* range covers the expanse of the Palearctic region from the British to the Japanese islands. Spain and the Mediterranean region are included in this area, as are China and Southeast Asia. The tundra zone is not inhabited. This highly aquatic species is associated with river, lake, stream and marsh habitats. This species' adult mass averaged 10.1 kg for males (n = 433) and 7.0 kg for females (n = 220) as summarized for populations in the Great Britain (Corbet and Harris 1991: 427). Dental formula follows 3.1.4.1/3.1.3.2.

Breeding in this species is aseasonal, though both more births and more successful births are recorded for summer months, especially in those portions of its extensive range where winters are severe. Mating therefore is most likely to occur in spring. There is no delayed implantation of embryos, and gestation lasts about 62-63 days. Litters are typically 2-3 newborn, though the known range is 1-5. The young are weaned at 14 weeks, but stay with the mother for 7-12 months and do not reach sexual maturity until during their second year of life (Corbet and Harris 1991:429).

Otter fur is particularly prized for its insulating and waterproof characteristics, as well as large size making fewer furs necessary to complete a useful garment. Males and females inhabit solitary ranges, with much overlap between them. This species leaves visible trails and hauling-out places near water edges. During the daytime, shelter is found above ground under cover or below ground in natural fissures or abandoned burrows. Trapping at such junctures is the most viable method of capture by humans, since this species is solitary, nocturnal and very elusive (Santos Reis 1983).

Meles meles (badger)
(Figure 3.36) *M. meles* is found from Portugal to Japan and China, including the British Isles, Scandinavia, Asia Minor and the Mediterranean region in its range. Its dental formula is 3.1.3-4.1/3.1.3-4.1, with minute vestigial first premolars sometimes occurring, more often mandibularly (Corbet and Harris 1991:415). Mean mass for individuals at least one year old from southwest England (males) is 11.6 kg (n = 33, range: 9.1-16.7 kg) and (females) 10.1 kg (n = 84, range: 6.6-13.9 kg) (Corbet and Harris 1991:419).

M. meles prefers deciduous forest and open pastures, and preferentially habitats which offer a mixture of these two vegetations. This species is more gregarious than other mustelids, living in large burrows with multiple entrances occupied for multiple generations. An average group size of 6 is usually presided over by a dominant male (Corbet and Harris 1991:419). Adaptations for a significantly fossorial lifestyle include eyesight best in poor light and front limbs and claws adapted for tunneling. The diet is opportunistic and omnivorous, with earthworms as a primary food source and including other invertebrates, small mammals, fruits, tubers, birds, fungi and carrion. Activity is primarily nocturnal and crepuscular.

Litters of 1-5 are born around February, though mating is during the previous spring and delayed implantation of embryos occurs. Active pregnancy lasts about 7 weeks. Birth weight ranges from 75-132 g, and near adult weight is obtained by the end of the first calendar year of life. Permanent dentition is in place by age 4 months, which coincides with weaning. Lifespan in the wild is unlikely to exceed 6 years (Corbet and Harris 1991:422).

The denning behavior of *M. meles* makes its trapping or snaring less difficult, since these may be place so as to intersect entrances or other chokepoints.

Gulo gulo (wolverine, glutton)
(Figure 3.37) This largest member of the Mustelidae ranges throughout the forest tundra and boreal regions of the Holarctic, though competition with humans may have caused contraction from the more southerly parts of its range. Adult mass normally ranges from 10.9-18.1 kg, but may reach up to 32.0 kg (summarized by Pasitschniak-Arts and Larivière 1995:1-2). The favored habitat of this species is mature timber stands, especially conifers. Shrub, rock, tundra, and ice are also occupied (Pasitschniak-Arts and Larivière 1995:6). Variations in habitat usage primarily reflect changes, both seasonal and cyclical, in prey availability.

Their diet is carnivorous. Smaller prey are usually hunted, while larger animals such as caribou and elk are usually procured through scavenging. Deep snow to slow or trap larger mammals seems to be a requirement for their direct procurement (Pasitschniak-Arts and Larivière 1995:5). Individuals are solitary and have large home ranges, but these freely overlap with those of others. Activity is mostly nocturnal, but this may vary.

Since embryo implantation is delayed, mating occurs in summer and birth is not until the end of the following winter, usually by late March. Active pregnancy ranges from 30-50 days. Litter size varies by region, but 2-3 is average. Birth weight averages 84 g, weaning occurs at 7-8 weeks, and adult size is reached by the winter following birth when juvenile dispersal occurs. Sexual maturity is reached for both sexes at the age a year and a half (Pasitschniak-Arts and Larivière 1995:5). *G. gulo* fur is particularly long, lush and warm, and their propensity for scavenging makes them attractable to traps baited with carrion.

CARNIVORA, FAMILY CANIDAE

The teeth of the two species described below are similar in morphology, with those of *Alopex lagopus* being smaller and having reduced distal molars as compared with *Vulpes vulpes* (Hillson 1986). Their dental formula follows 3.1.4.2/3.1.4.3.

Vulpes vulpes (red fox)

(Figure 3.38) This ubiquitous Holarctic species has an expansive range extending also into North Africa and the Arabian peninsula. Corbet and Harris (1991:355) report body size means ranging from 6.4-7.3 kg for males and 5.4-6.2 kg for females from populations in the United Kingdom. Its habitat requirements are part of the reason for its adaptive success: it adapts readily to most habitats, even urban, within this range. Fragmented, highly variable habitats supplying an abundant array of foraging choices are preferred (Corbet and Harris 1991:357-358). Small mammals, especially rabbits, make up the majority of prey items, but additionally birds, amphibians, invertebrates, and plant materials are consumed (Goszczynski 1974, Howes 1978, Lockie 1959, Robertson and Whelan 1987, Scott 1941). Human-supplied resources such as farm poultry or garbage dumps are readily exploited. Such camp-follower behavior may have begun early in prehistory, and it may be reflected in the patterns of skeletal remains from prehistoric sites. Activity is mainly nocturnal and crepuscular, but even this is mutable. *V. vulpes* is territorial and lives in family groups.

Mating occurs mainly in January, with births following in March. Gestation lasts 52-53 days, and litters of 4-5 are typical. Birth weight ranges from 80-120 g and weaning occurs at about 7 weeks. Young are born in a burrow and emerge for the first time around 4 weeks, at which age they therefore are more susceptible to predation. Adult size and independence is reached by September of the birth year (Corbet and Harris 1991:363-364).

Alopex lagopus (arctic fox)

(Figure 3.39) This arctic species has a circumpolar distribution, extending in the Palearctic from Sweden to the Kamchatka peninsula. This range extended far southwards during glacial episodes and included intermittent penetration into northern Spain. Its numbers there were never numerous and probably reflect areas of alpine tundra expansion during the coldest periods. Abundance in France where true tundra conditions were obtained was more common and was accompanied by other extreme cold indicative species like reindeer and snowy owl.

Mating occurs about April, with a 6 week gestation period and birth of pups in May or June. Older females sometimes have a second litter in the same summer. Litter size is quite large: 5-8 and 4-10 are reported as typical, and some extreme cases of up to 23 in a single litter are also known. Both parents participate in raising the young, with the male often provisioning the lactating female (Ognev 1962).

A. lagopus fur is long, lush, warm and white, making it a valuable acquisition for hunters and trappers into this century (Ognev 1962) and ideally suited for winter clothing in glacial age Eurasia.

CARNIVORA, FAMILY FELIDAE

Both of the two species discussed below has the reduced dental formula characteristic of the Felidae: 3.1.2.1/ 3.1.2.1 and are clearly distinguishable from other small carnivores on these grounds. Size differences between these two species allows their separation by tooth size.

Felis silvestris (wildcat)

(Figure 3.40) *F. silvestris* can today be found from Spain, through the circum-Mediterranean region, Arabia, India, Africa, and to western China. Populations also exist in a few northern European areas, such as Scotland and northern France. Mediterranean forms of this species are believed to be the wild progenitors for domestic cats, *F. catus*. As with any domesticated species, one may expect that the range and characteristics of the wild form may have been greatly altered by human intervention. The existence of feral and hybrid forms further blurs understanding about *F. silvestris* in its predomesticated state.

Typically, *F. silvestris* is larger and less gracile than its domesticated counterpart. Corbet and Harris (1991:434) report mean weights of 5.2, 4.7 and 4.4 kg for males and 4.5, 3.9 and 4.0 kg for females of three Scottish (adult) populations. Diet is eclectic, reflecting the large and diverse range of this species with its many food options. Most small mammals and birds are taken and are supplemented with smaller amounts of amphibian, reptiles, and insects. Preferred habitats are forest and brush areas (Corbet and Ovenden 1982:188). Activity is primarily solitary and nocturnal, and territories are maintained.

One litter with an average size of 3.7 is produced (range: 2-6), normally in May after a gestation period of about 66 days (Corbet and Harris 1991:436). Birth weight is from 100-163 g, with permanent dentition arriving at about 6 months. Most adult mass is obtained by 10 months, but epiphyseal fusion is not complete until 18-19 months. Young leave their mothers in the autumn and disperse to form their own home territories, and therefore they may be more vulnerable to predation at that time.

Felis lynx (lynx)

(Figure 3.41) This species has a Holarctic, boreal forest distribution that stretches from the montane regions of the Iberian peninsula through Siberia. Its range is similarly extensive in this zone of North America. Its current range in Europe is severely reduced to isolated alpine regions. This species is sometimes placed into a separate genus (*Lynx*) along with *F. rufus* (the North American bobcat), though Corbet (1978) retains this distinction only at the subgenus level. The Iberian variety is also sometimes given separate species status (*F. pardina*), but the author does not recognize this separation as valid. Additionally, the Nearctic lynx is sometimes designated as a separate species (*Felis/Lynx canadensis*) from the Palearctic variety, but this distinction must await further investigation. Adult body mass for males in Sweden averages 17.9 kg and is generally larger in the Palearctic than in the Nearctic (summarized by Tumlison 1987). Sizes in Spain tend to be much smaller, averaging 5.7 kg (n = 10) for juveniles less than 1 year old and >9.3-12.8 kg (n = 13) for subadults/adults in Doñana National Park (Beltrán and Delibes 1993).

Mating occurs in February and March in the Palearctic region, and sometimes earlier in the more southerly portions of this range. Estimation of gestation time span 63-74 days, with births of litters in late May or early June. Litter size ranges from 1-6, with 2-3 typical in the Palearctic region. Deciduous dentition begins replacement at 4 months. Weaning is complete by six months of age, though some litters remain with the mother for a full year after birth. Complete epiphyseal fusion is reached by 32 months (Tumlison 1987:3-4).

Favored habitat is coniferous forest, but in Spain areas of thick brush are sometimes inhabited (Corbet and Ovenden 1982:189). Activity patterns are solitary and nocturnal, and principle prey includes small mammals such as lagomorphs and rodents (Saunders 1963).

CARNIVORA, FAMILY VIVERRIDAE

Genetta genetta (genet)

(Figure 3.42) The members of this genus (numbering about 10) all have a sub-Saharan African distribution, with the exception of *G. genetta* which extends from throughout this area into the Iberian peninsula, France, North Africa and Arabia. Corbet (1978:177) therefore suggests that the European range of genets resulted from human introduction. This species, however, has been identified from one northern Spanish paleolithic context: 2 remains from Azilian Level I at El Pendo (Fuentes Vidarte 1980). Significant northern spread of this tropical/subtropical species was unlikely to have occurred previous to near-complete glacial retreat.

G. genetta is a small carnivore ranging from 1360-2250 g adult body mass (Niethammer and Krapp 1993). Its teeth are identifiable morphologically from other small carnivores of similar size (such as *Martes* or *Mustela*) and follow a 3.1.4.2/3.1.4.2 formula. It is solitary, strictly nocturnal and survives on a generalist diet of most small animals, including birds, fruits, mammals, and invertebrates (Le Jacques and Lode 1994; Virgós *et al.* 1996). This species is associated with forest and shrub in its European range. Its behavior makes it easiest to acquire by trapping. Its appearance indicates significant warming and closing of environments during the Holocene in Cantabrian Spain.

Litters are produced during the summer months, roughly from April through August (though the breeding season is reversed in the Southern Hemisphere portion of its range). Litter size is from 2-4, with 2.6 average. Gestation lasts for 10-11 weeks, with weaning occurring at about 9 weeks. The young are born and weaned in a reused burrow or natural hollow (Skinner and Smithers 1990:477).

Table 3.1: Summary of Ecological Characteristics.

	Forest	Shrubs	Forest Edge/ Meadow	Alpine	Water Edge/ Wetlands	Humid Meadow	Dry Meadow	Boreal/ Tundra	General
Glis glis	X								
Eliomys quercinus	X								
Apodemus flavicollis	X								
Apodemus sylvaticus	X								
Clethrionomys glareolus	X								
Martes martes	X								
Martes foina	X								
Felis silvestris	X	x							
Felis lynx	X	x							
Genetta genetta	X	x							
Erinaceus europaeus			X						
Meles meles			X						
Microtus nivalis				X					
Galemys pyrenaicus				X	X				
Marmota marmota				X					
Arvicola terrestris					X	X			
Arvicola sapidus					X	X			
Microtus oeconomus					X	X			
Neomys fodiens					X				
Neomys anomalus					X				
Mustela lutreola					X				
Lutra lutra					X				
Pitymys lusitanicus						X			
Talpa europaea						X			
Talpa occidentalis						X			
Sorex minutus						X			
Sorex coronatus						X			
Microtus arvalis							X		X
Microtus agrestis							X		
Micromys minutus							X		
Crocidura russula							X		
Lepus capensis						X	X		
Lepus timidus						X	X		
Oryctolagus cuniculus						X	X		
Microtus gregalis								X	
Gulo gulo								X	
Alopex lagopus								X	
Pitymys pyrenaicus									X
Mustela nivalis									X
Mustela erminea									X
Mustela putorius									X
Vulpes vulpes									X

X= primary association x= secondary association

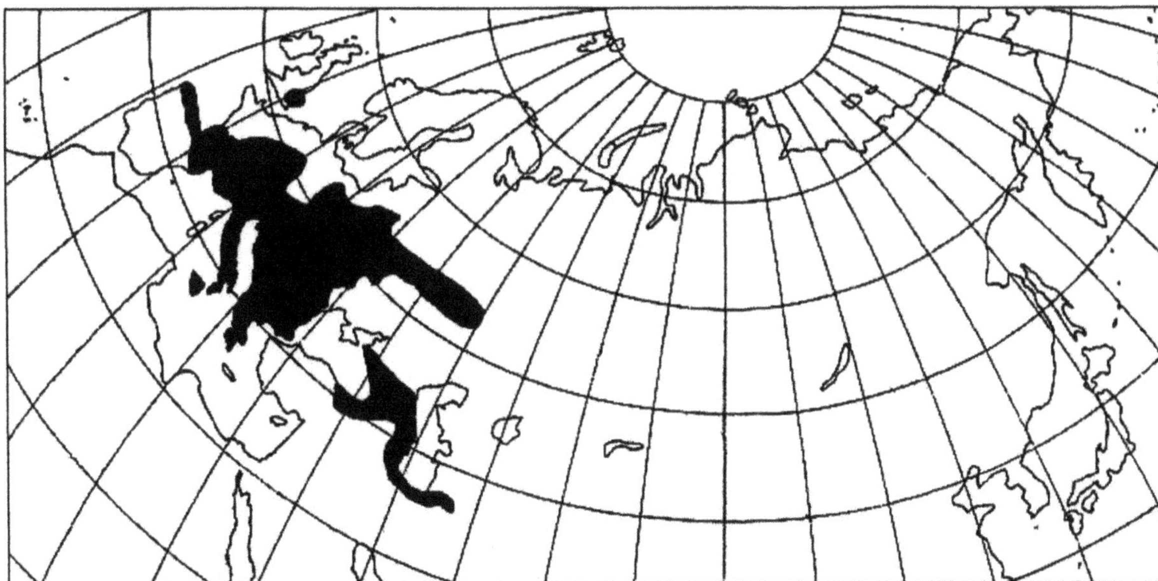

Figure 3.1: Modern Range of *Glis glis* (redrawn after Corbet 1978; Corbet and Harris 1991; Corbet and Ovenden 1982).

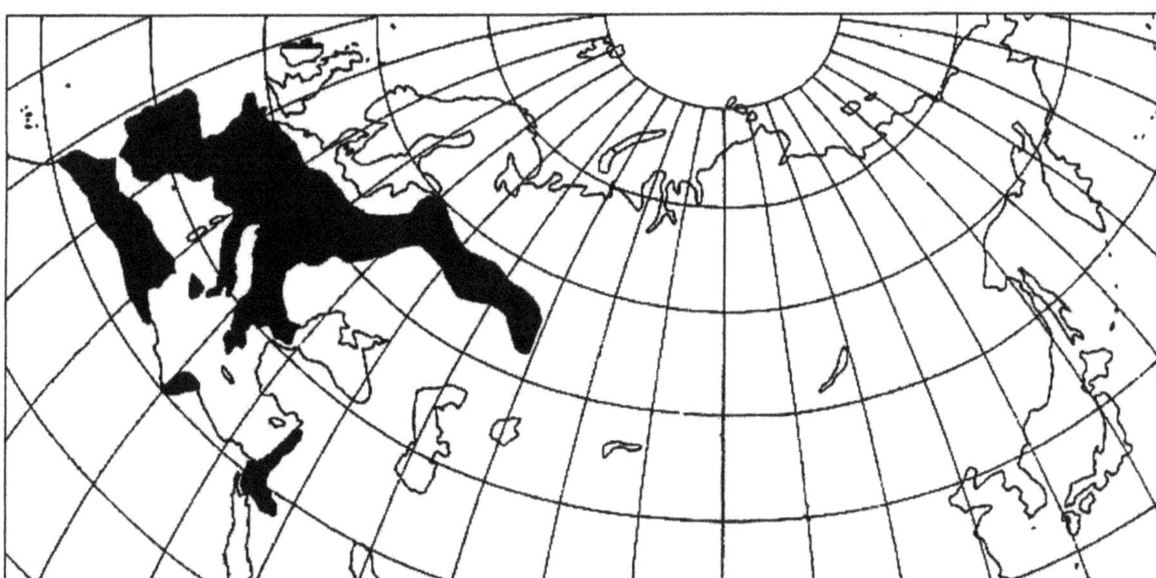

Figure 3.2: Modern Range of Eliomys *quercinus* (redrawn after Corbet 1978; Corbet and Harris 1991; Corbet and Ovenden 1982).

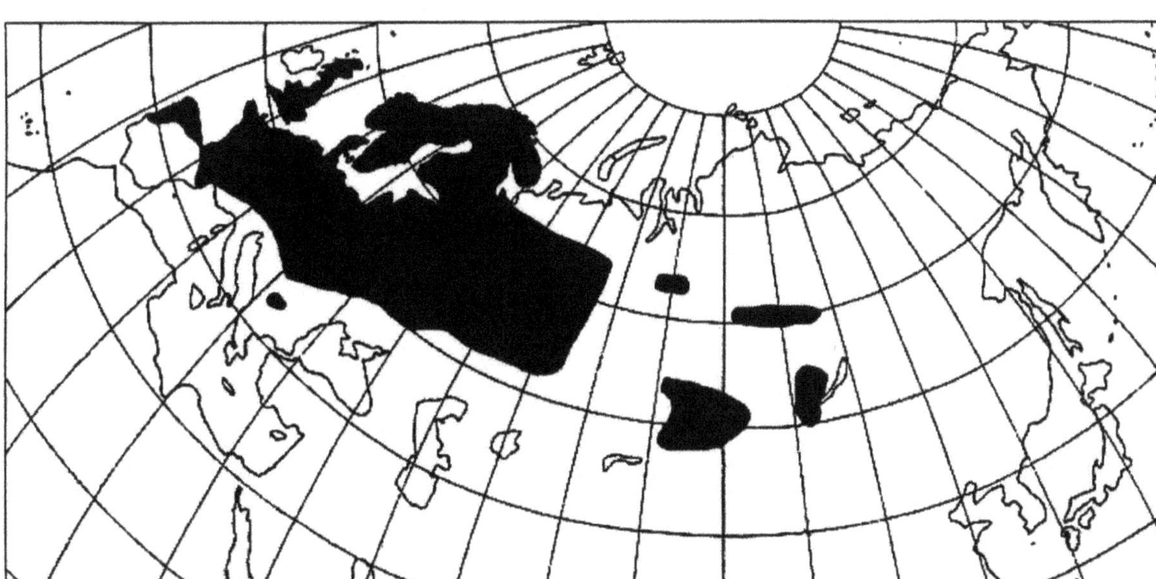

Figure 3.3: Modern Range of *Microtus agrestis* (redrawn after Corbet 1978; Corbet and Harris 1991; Corbet and Ovenden 1982).

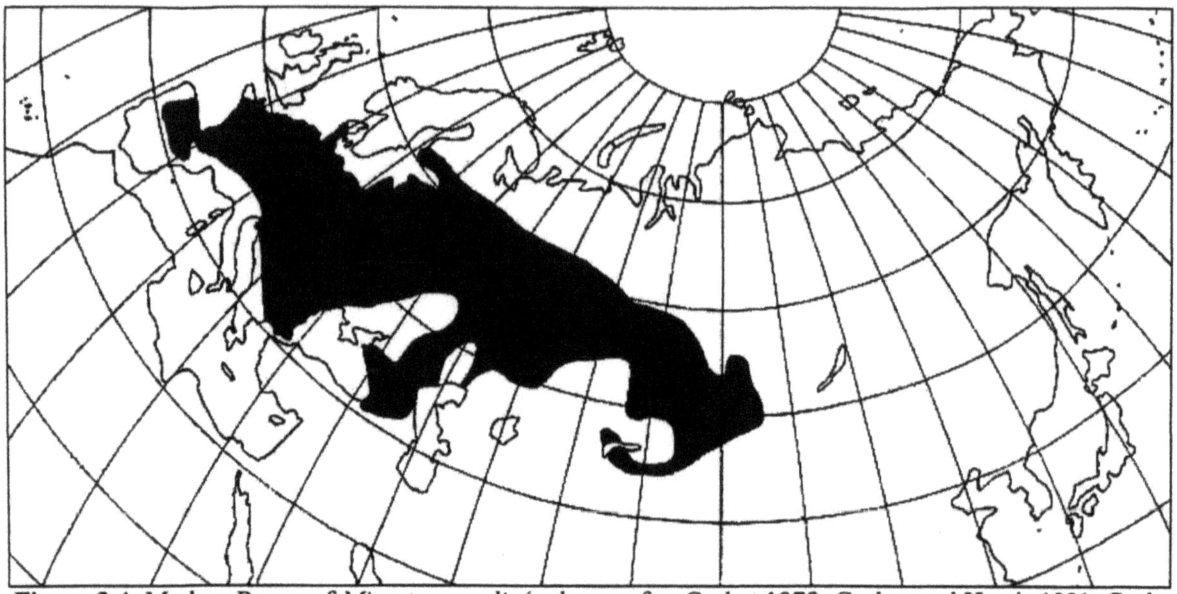

Figure 3.4: Modern Range of *Microtus arvalis* (redrawn after Corbet 1978; Corbet and Harris 1991; Corbet and Ovenden 1982).

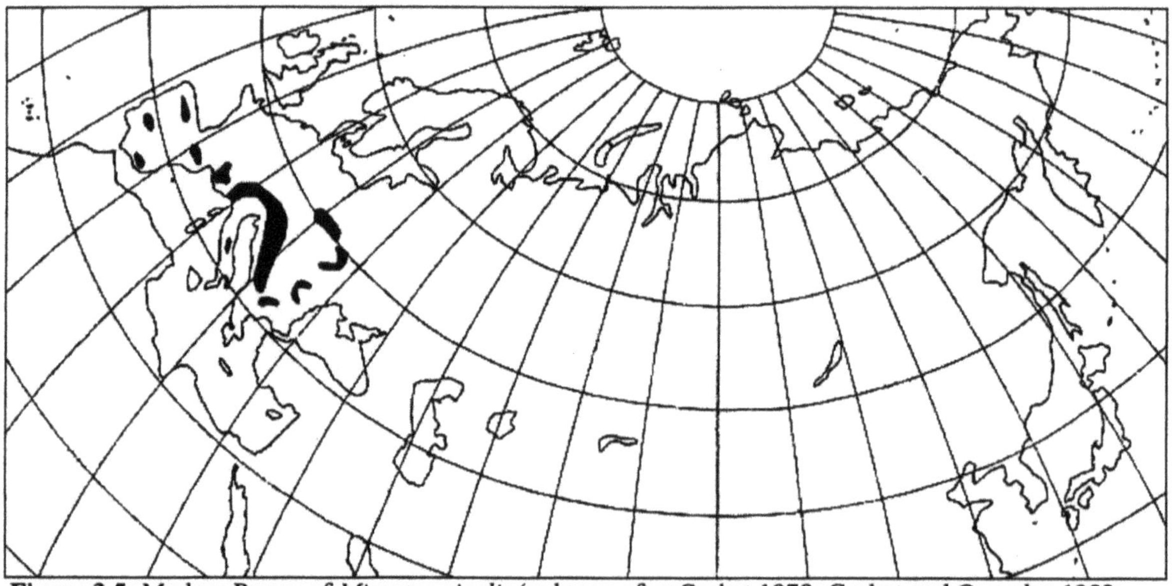

Figure 3.5: Modern Range of *Microtus nivalis* (redrawn after Corbet 1978; Corbet and Ovenden 1982; Niethammer and Krapp 1982).

Figure 3.6: Modern Palearctic Range of *Microtus oeconomus* (extending into the Nearctic as far as Canada) (redrawn after Corbet 1978; Corbet and Ovenden 1982).

Figure 3.7: Modern Range of *Microtus gregalis* (redrawn after Corbet 1978).

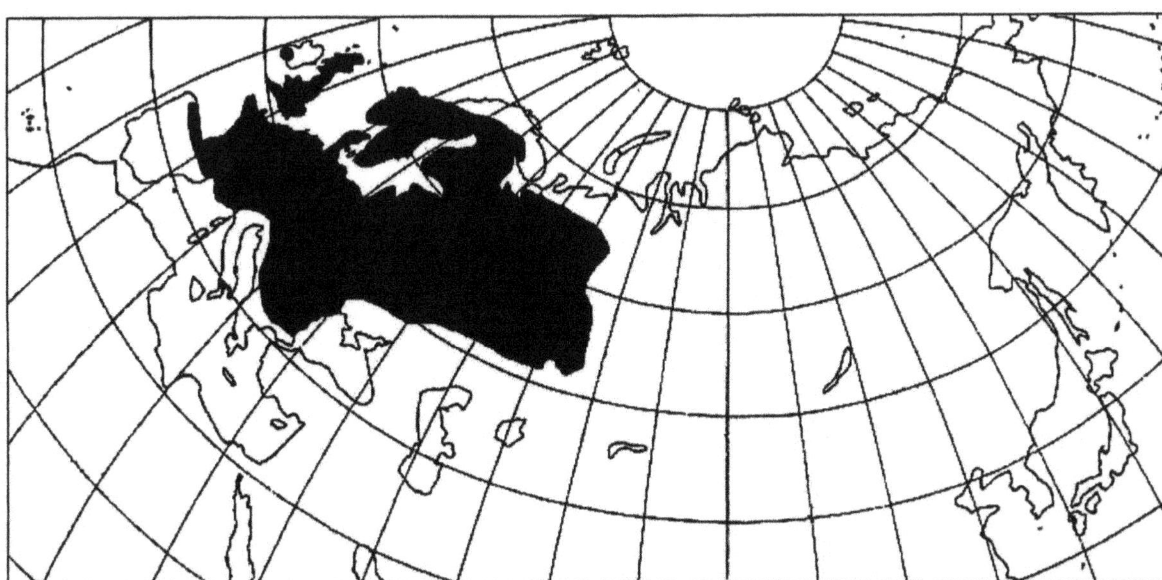

Figure 3.8: Modern Range of *Clethrionomys glareolus* (redrawn after Corbet 1978; Corbet and Harris 1991; Corbet and Ovenden 1982; Niethammer and Krapp 1982).

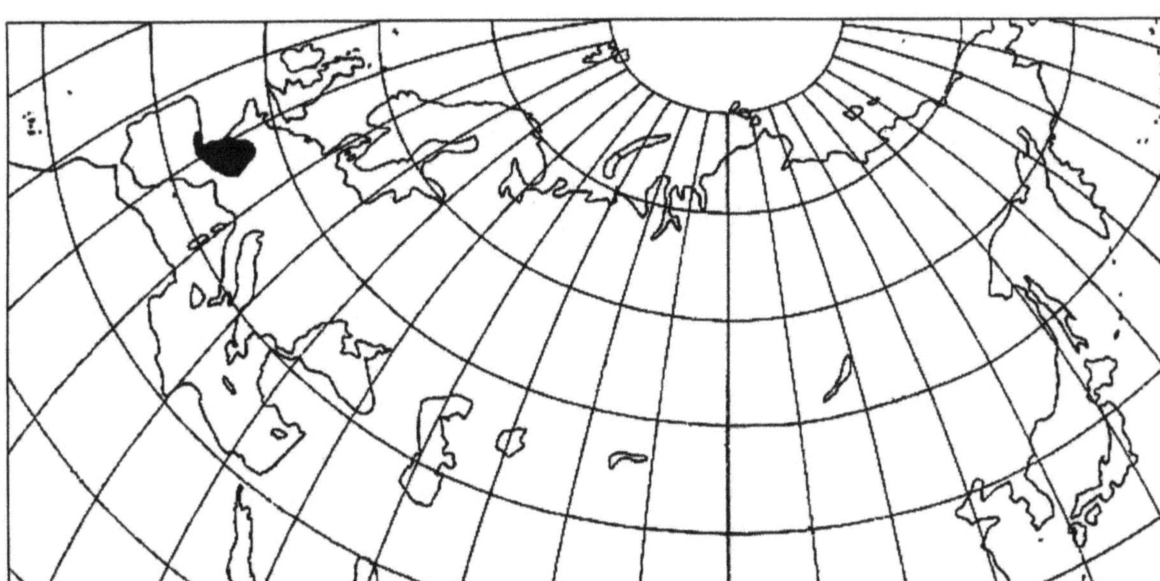

Figure 3.9: Modern Range of *Pitymys pyrenaicus* (redrawn after Corbet 1978; Corbet and Ovenden 1982; Niethammer and Krapp 1982).

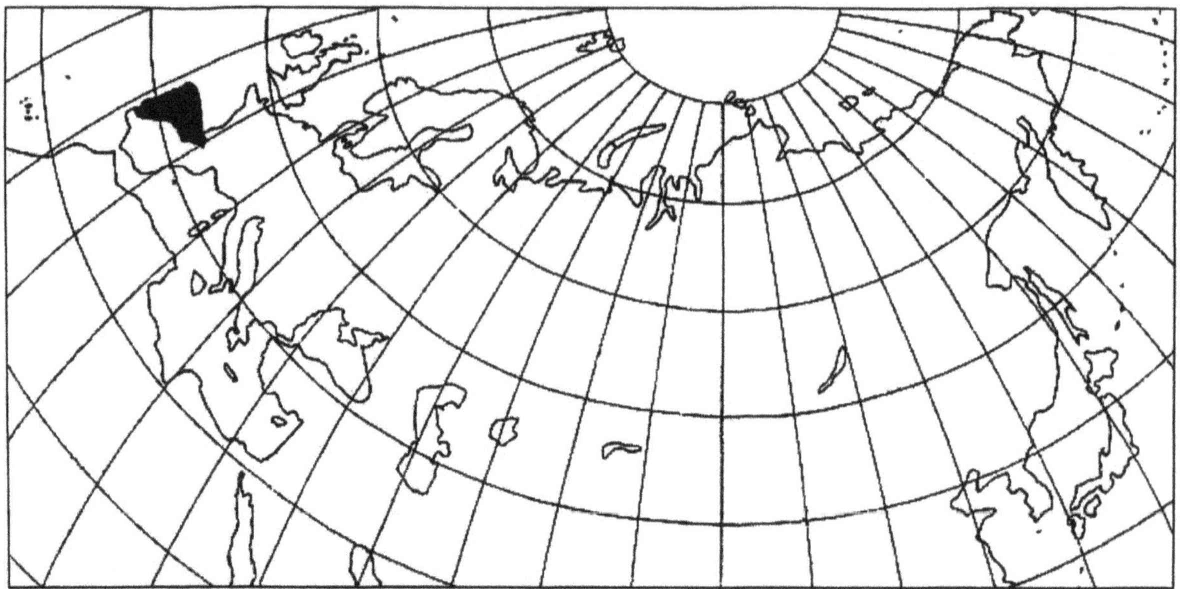

Figure 3.10: Modern Range of *Pitymys lusitanicus* (redrawn after Corbet 1978; Corbet and Ovenden 1982; Niethammer and Krapp 1982).

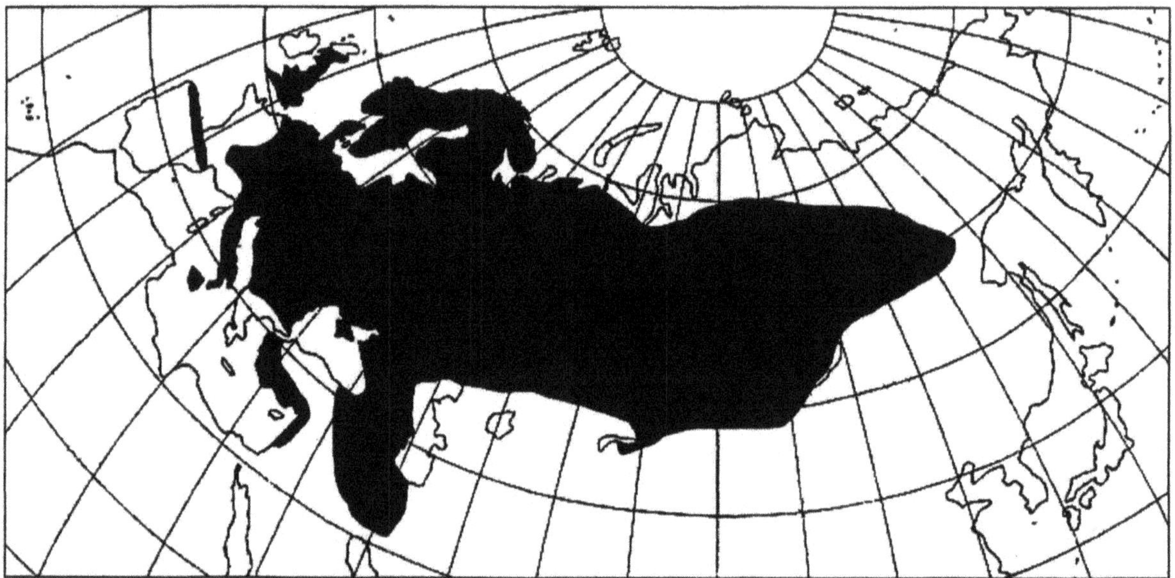

Figure 3.11: Modern Range of *Arvicola terrestris* (redrawn after Corbet 1978; Corbet and Harris 1991; Corbet and Ovenden 1982).

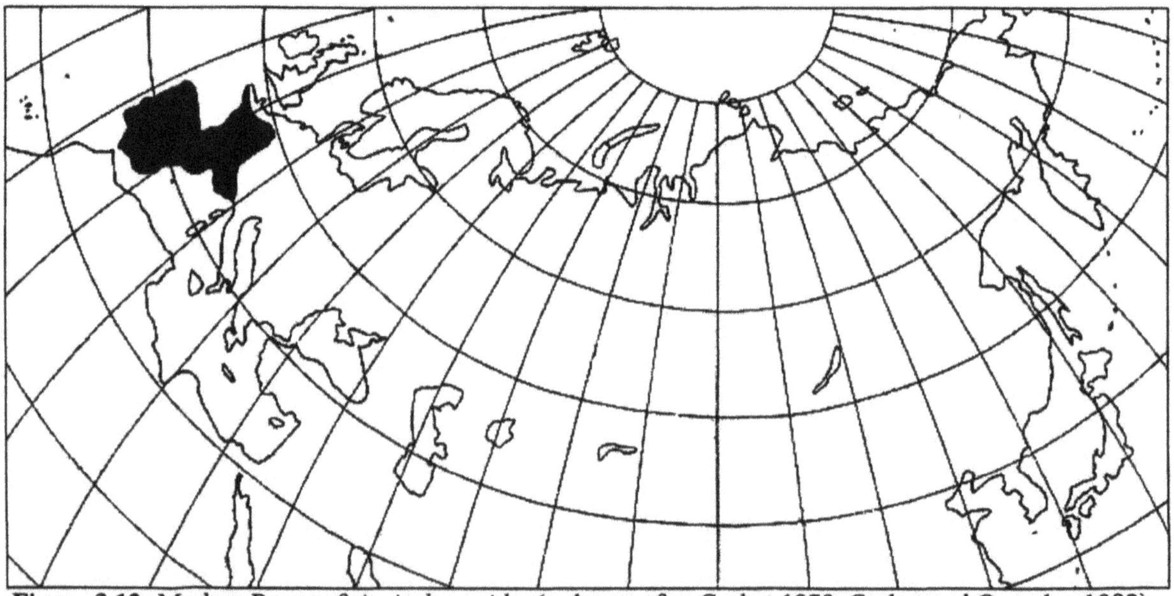

Figure 3.12: Modern Range of *Arvicola sapidus* (redrawn after Corbet 1978; Corbet and Ovenden 1982).

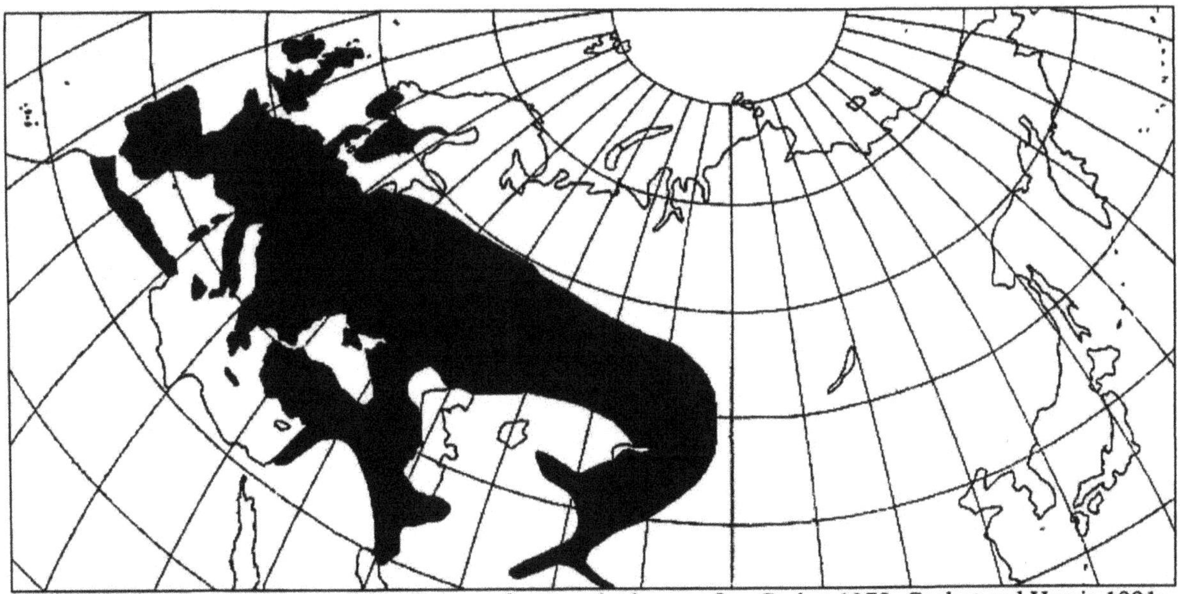

Figure 3.13: Modern Range of *Apodemus sylvaticus* (redrawn after Corbet 1978; Corbet and Harris 1991; Corbet and Ovenden 1982).

Figure 3.14: Modern Range of *Apodemus flavicollis* (redrawn after Corbet 1978; Corbet and Harris 1991; Corbet and Ovenden 1982).

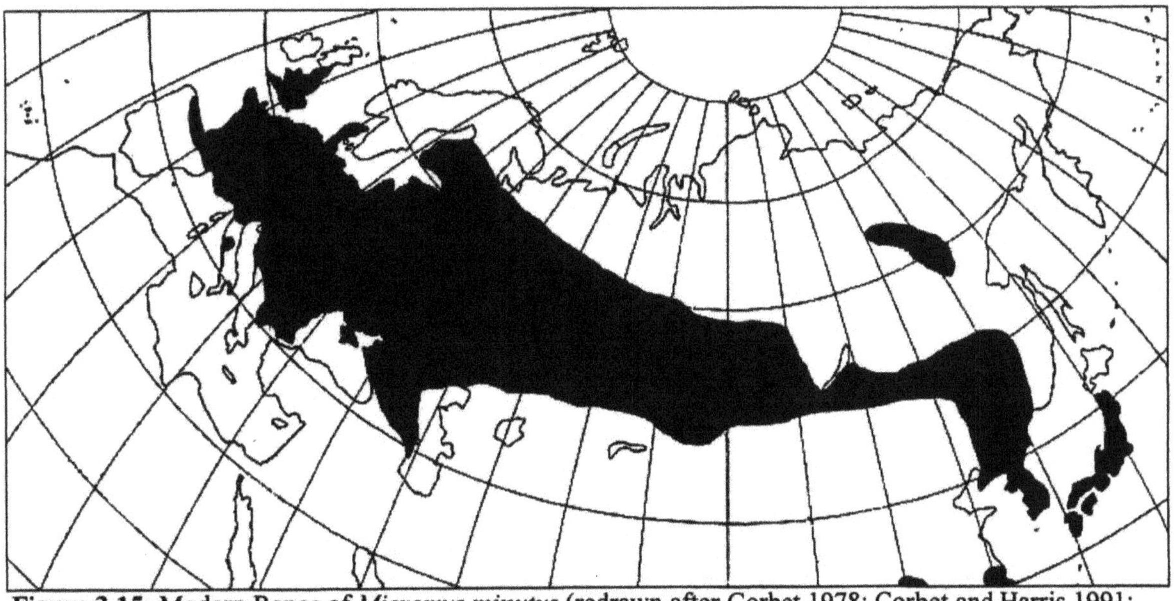

Figure 3.15: Modern Range of *Micromys minutus* (redrawn after Corbet 1978; Corbet and Harris 1991; Corbet and Ovenden 1982).

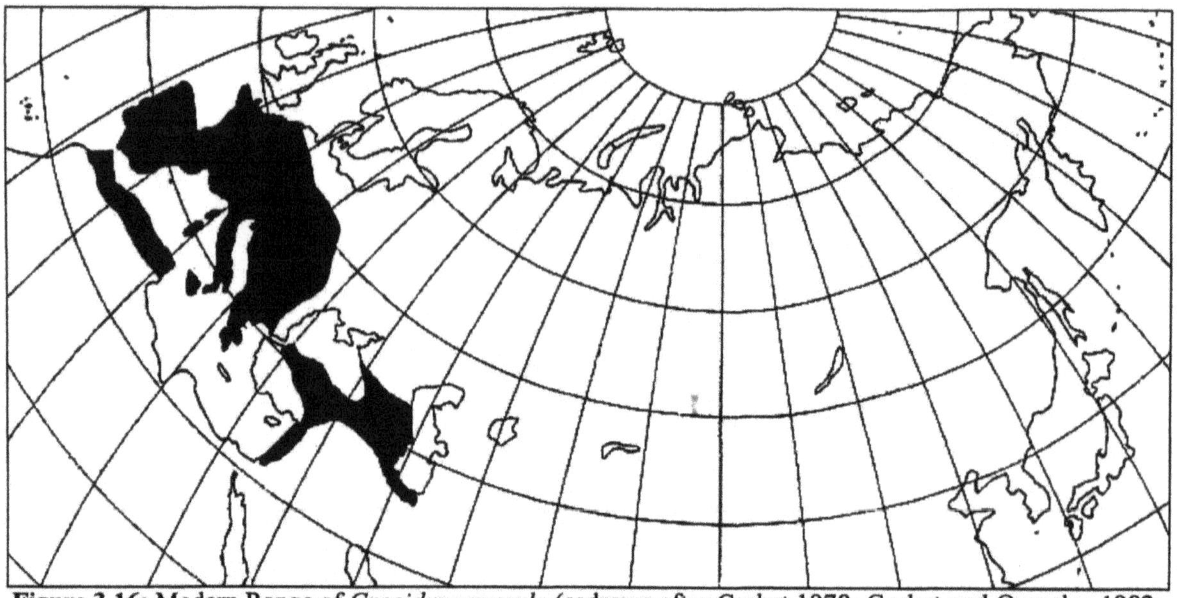

Figure 3.16: Modern Range of *Crocidura russula* (redrawn after Corbet 1978; Corbet and Ovenden 1982; Niethammer and Krapp 1990).

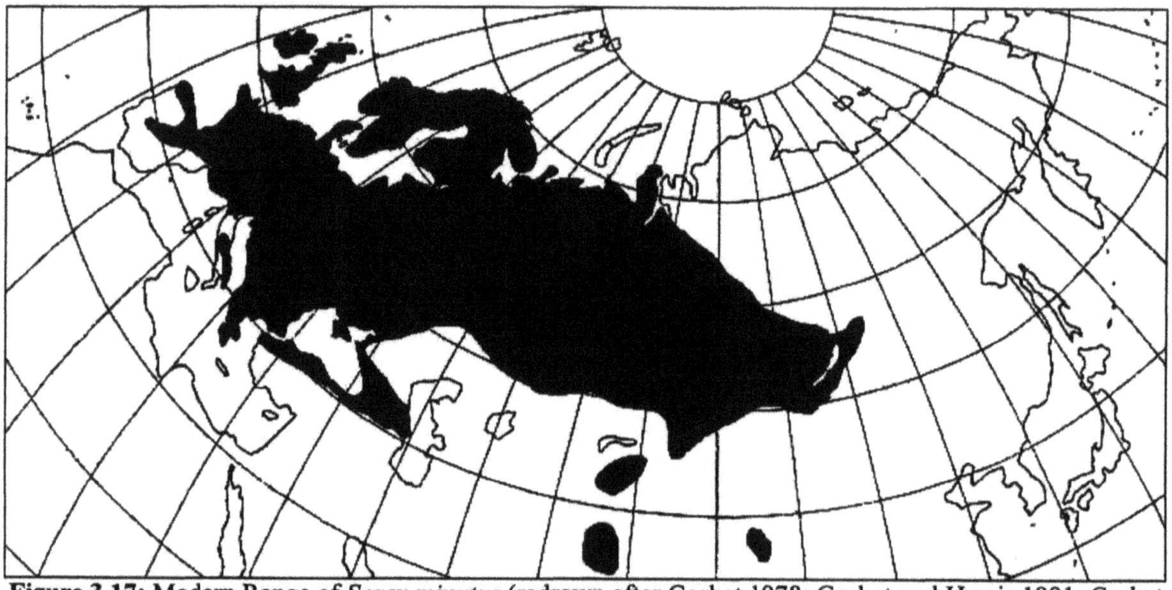

Figure 3.17: Modern Range of *Sorex minutus* (redrawn after Corbet 1978; Corbet and Harris 1991; Corbet and Ovenden 1982; Niethammer and Krapp 1990).

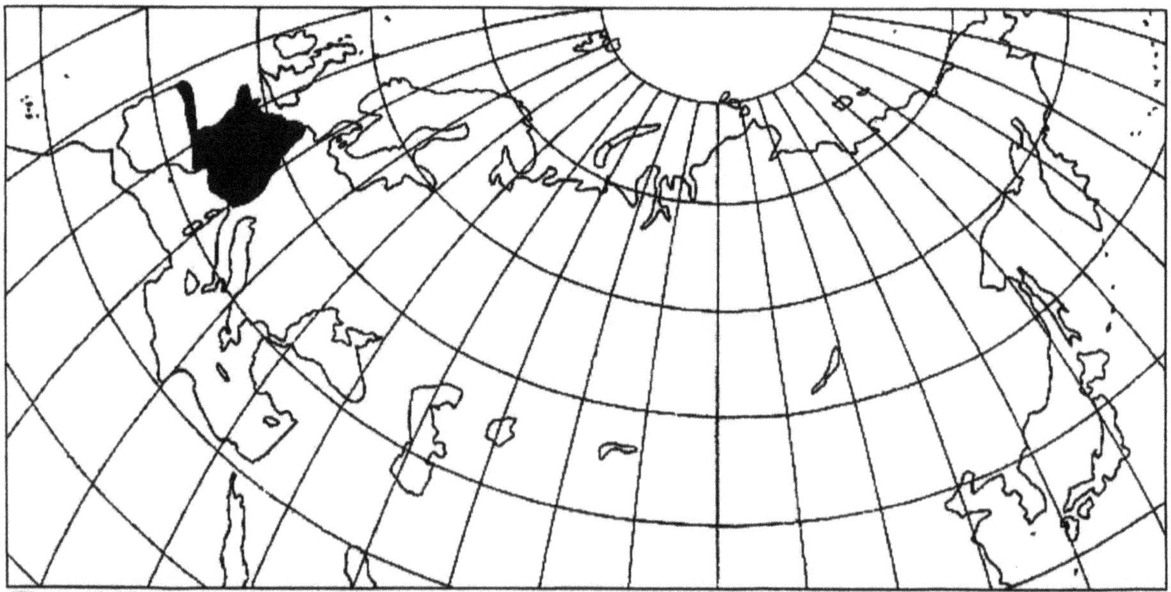

Figure 3.18: Modern Range of *Sorex coronatus* (redrawn after Corbet 1978; Corbet and Ovenden 1982; Niethammer and Krapp 1990).

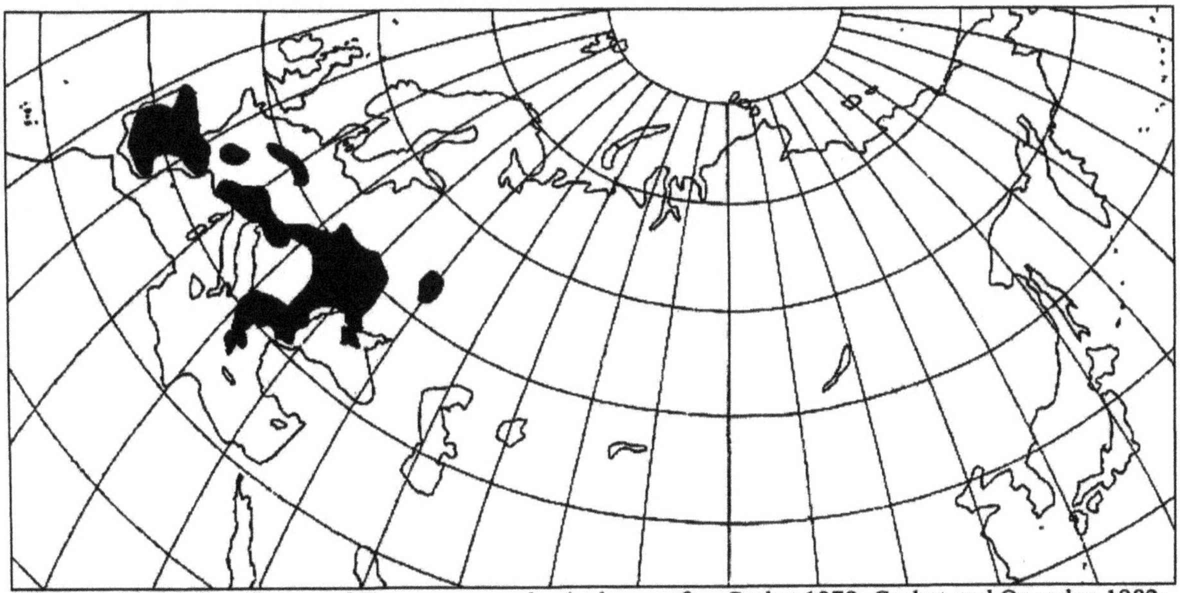

Figure 3.19: Modern Range of *Neomys anomalus* (redrawn after Corbet 1978; Corbet and Ovenden 1982; Niethammer and Krapp 1990).

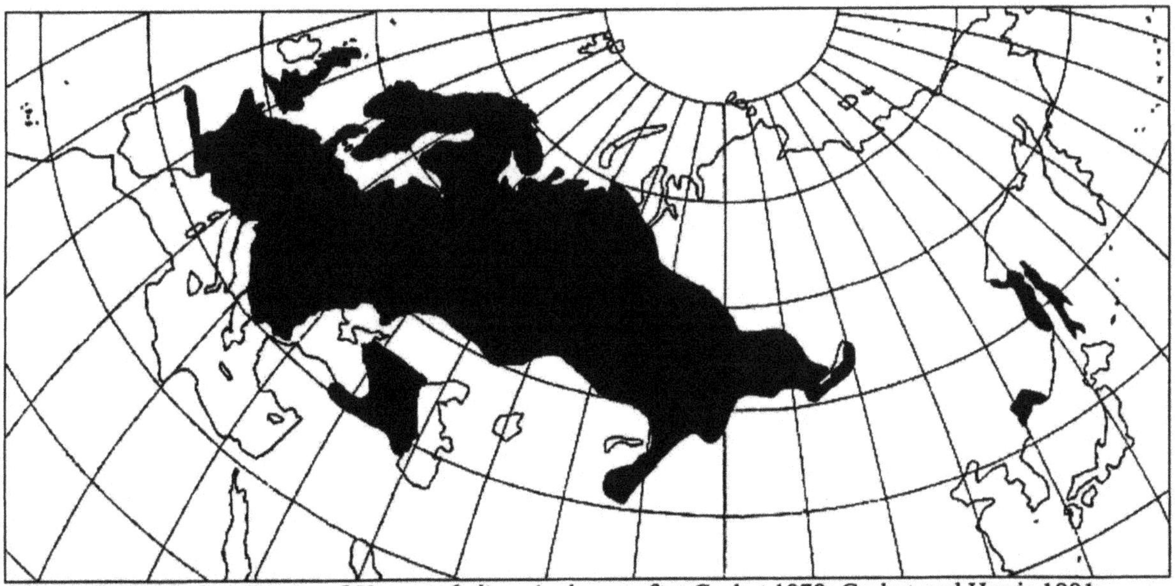

Figure 3.20: Modern Range of *Neomys fodiens* (redrawn after Corbet 1978; Corbet and Harris 1991; Corbet and Ovenden 1982).

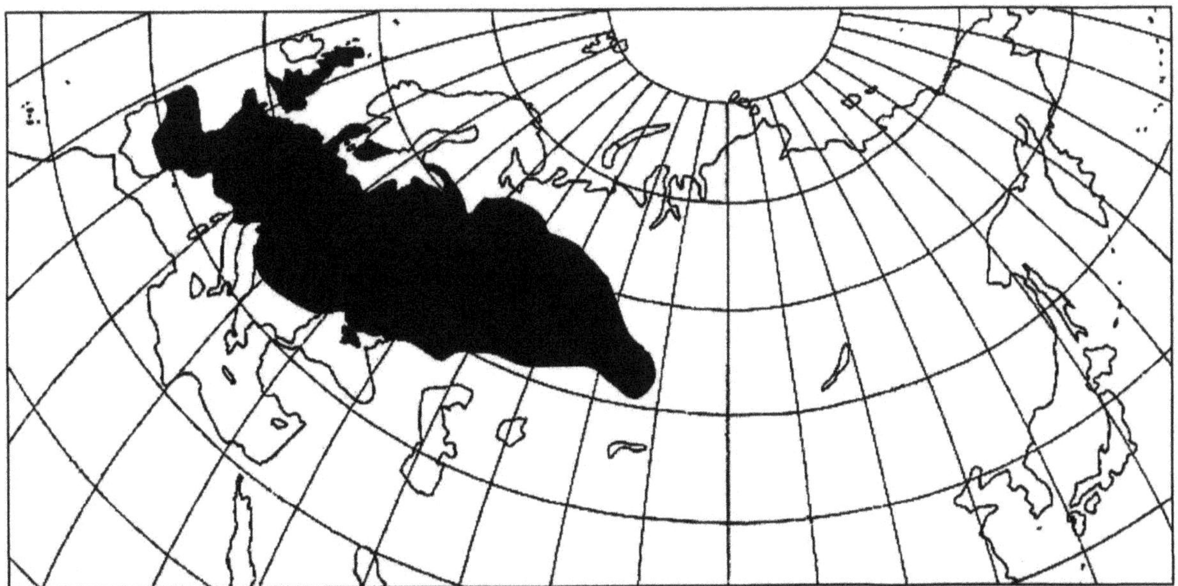

Figure 3.21: Modern Range of *Talpa europaea* (redrawn after Corbet 1978; Corbet and Harris 1991; Corbet and Ovenden 1982).

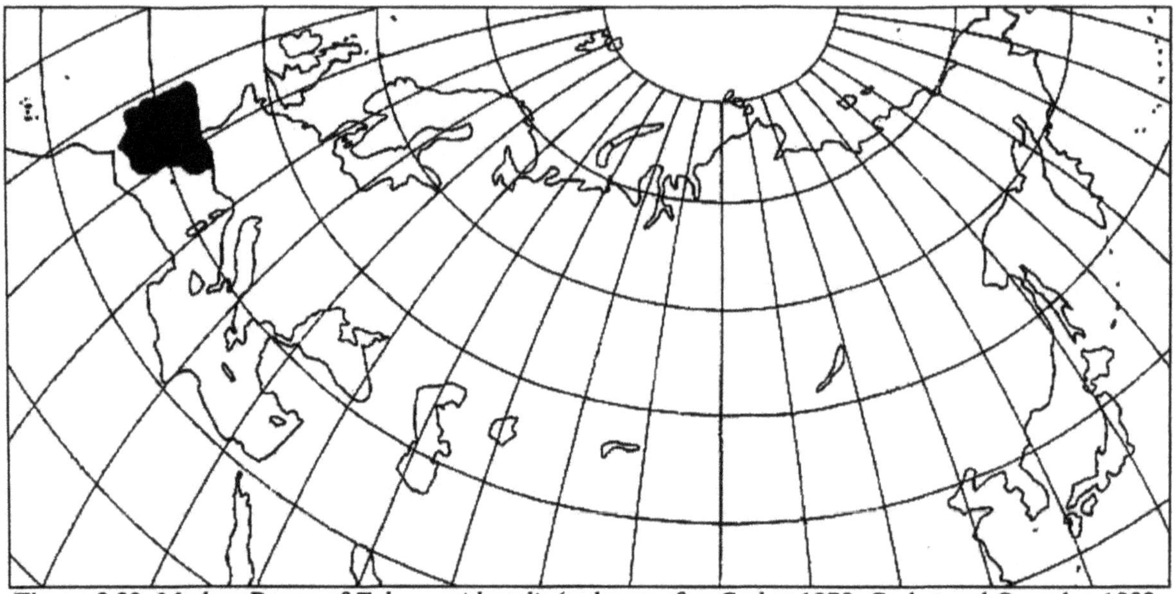

Figure 3.22: Modern Range of *Talpa occidentalis* (redrawn after Corbet 1978; Corbet and Ovenden 1982; Niethammer and Krapp 1990).

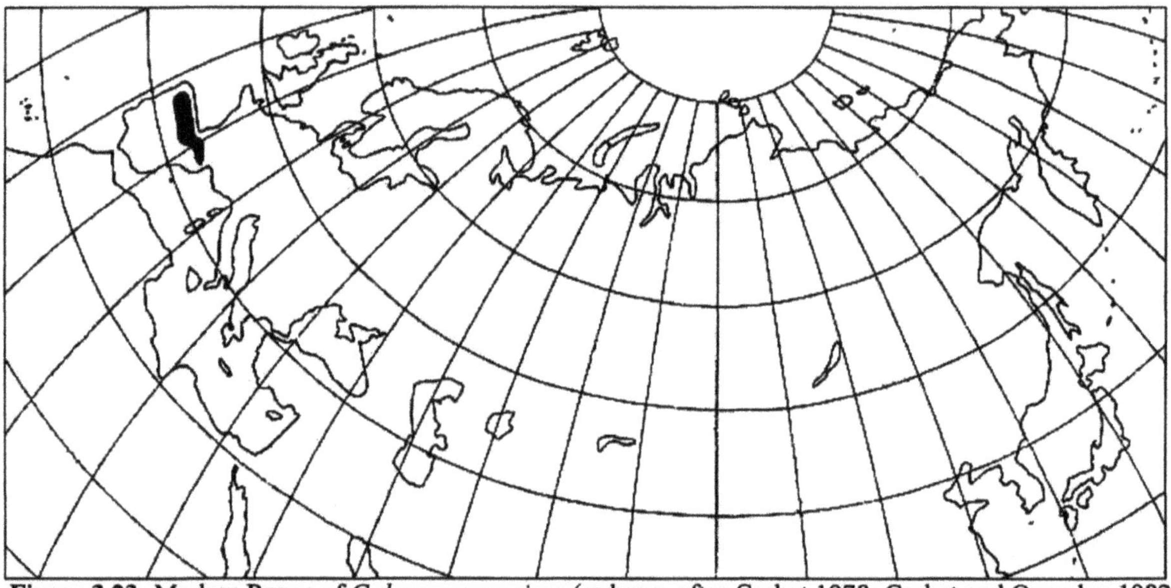

Figure 3.23: Modern Range of *Galemys pyrenaicus* (redrawn after Corbet 1978; Corbet and Ovenden 1982; Palmeirim and Hoffman 1983).

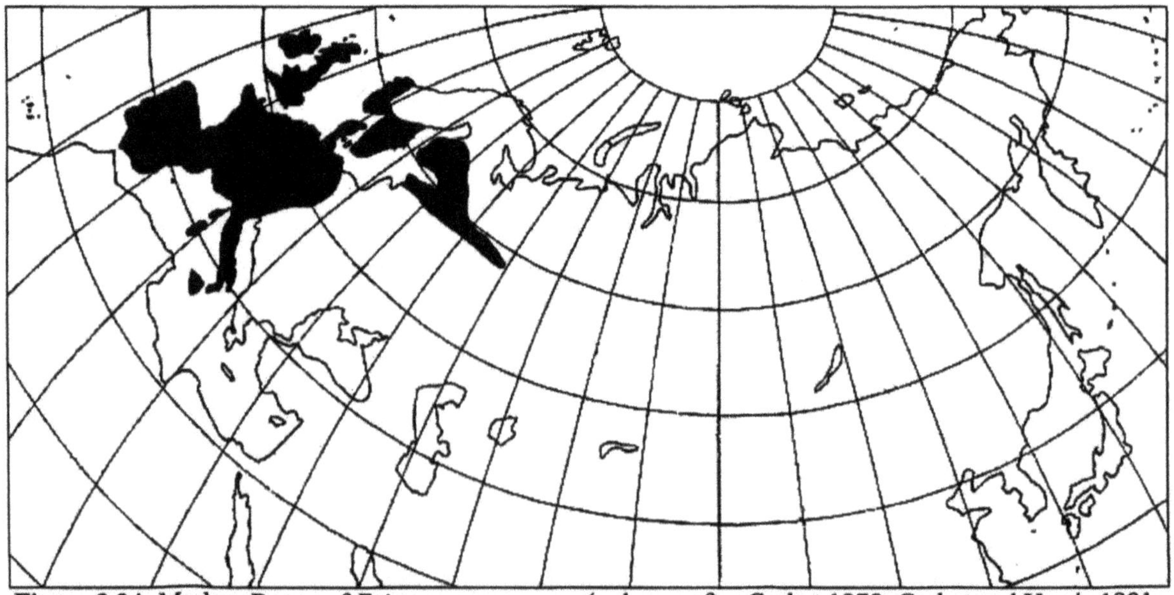

Figure 3.24: Modern Range of *Erinaceus europaeus* (redrawn after Corbet 1978; Corbet and Harris 1991; Corbet and Ovenden 1982).

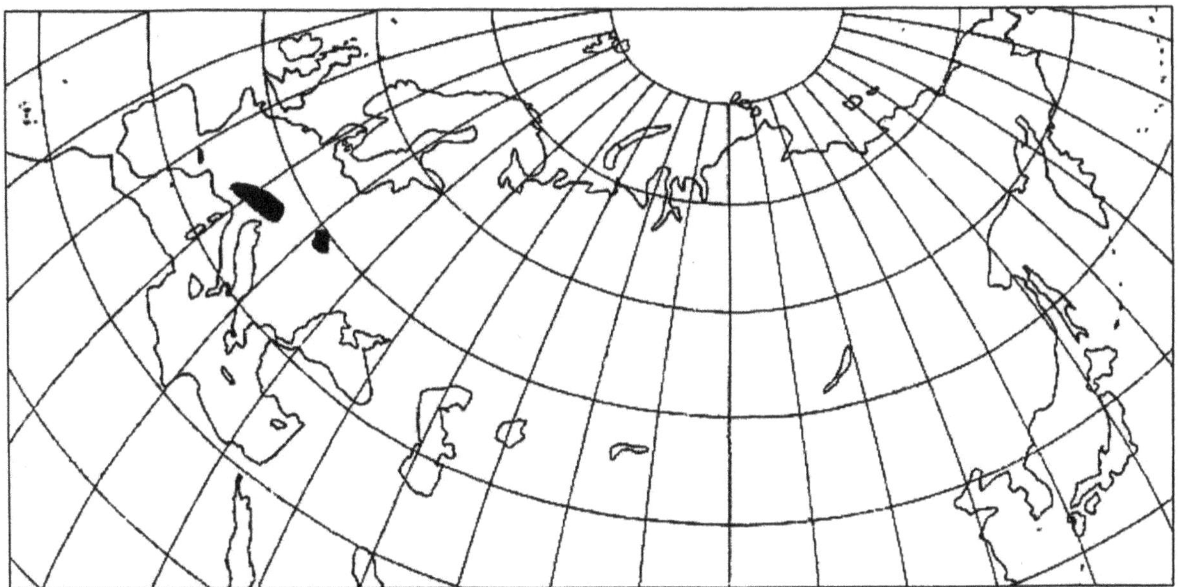

Figure 3.25: Modern Range of *Marmota marmota* (redrawn after Corbet 1978; Corbet and Ovenden 1982; Niethammer and Krapp 1978).

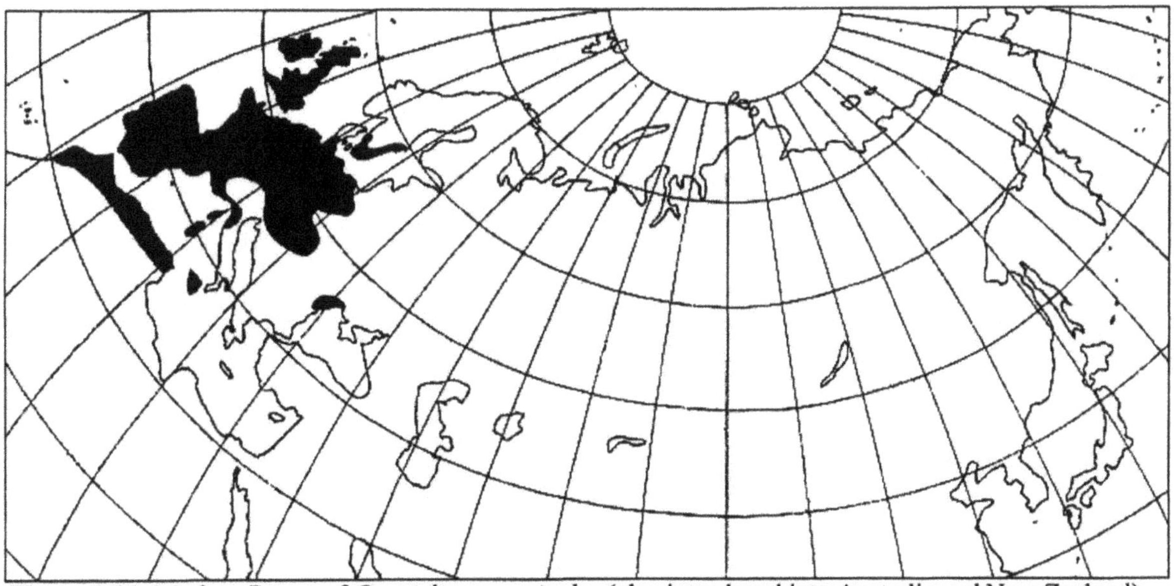

Figure 3.26: Modern Range of *Oryctolagus cuniculus* (also introduced into Australia and New Zealand) (redrawn after Corbet 1978; Corbet and Harris 1991; Corbet and Ovenden 1982).

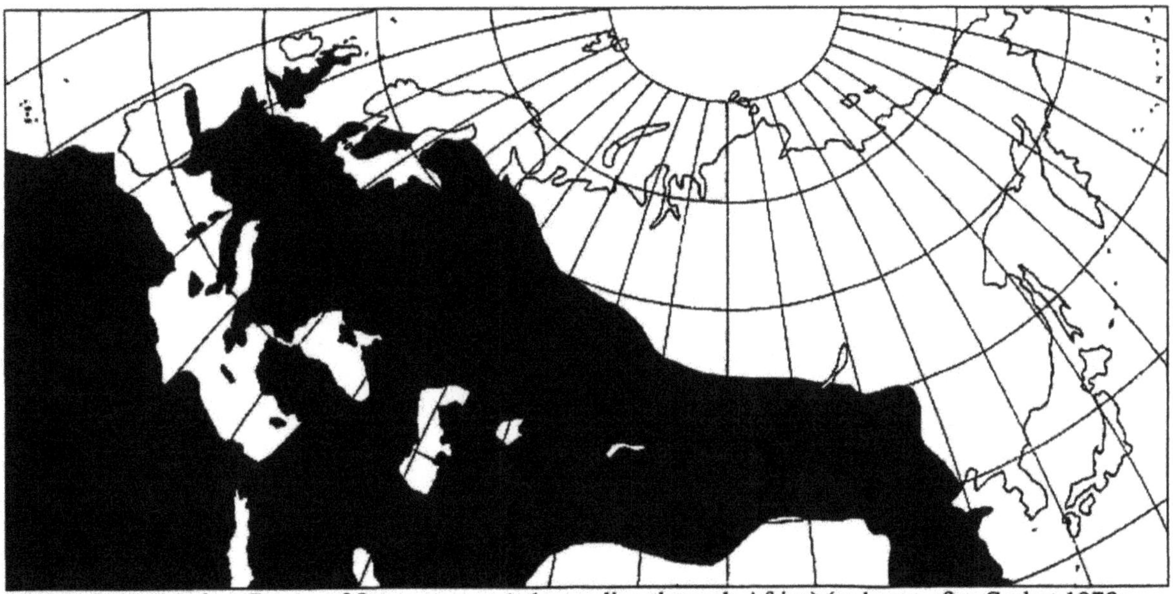

Figure 3.27: Modern Range of *Lepus capensis* (extending through Africa) (redrawn after Corbet 1978; Corbet and Harris 1991; Corbet and Ovenden 1982).

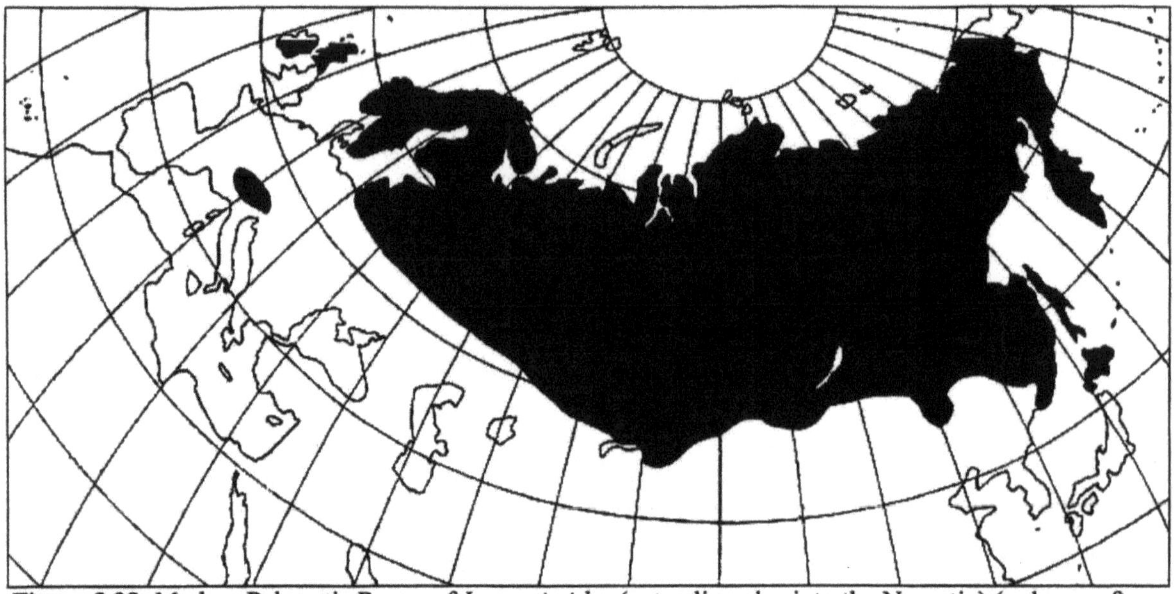

Figure 3.28: Modern Palearctic Range of *Lepus timidus* (extending also into the Nearctic) (redrawn after Corbet 1978; Corbet and Harris 1991; Corbet and Ovenden 1982).

Figure 3.29: Modern Palearctic Range of *Mustela nivalis* (extending also into the Nearctic) (redrawn after Corbet 1978; Corbet and Harris 1991; Sheffield and King 1994).

Figure 3.30: Modern Palearctic Range of *Mustela erminea* (extending also into the Nearctic) (redrawn after Corbet 1978; Corbet and Harris 1991).

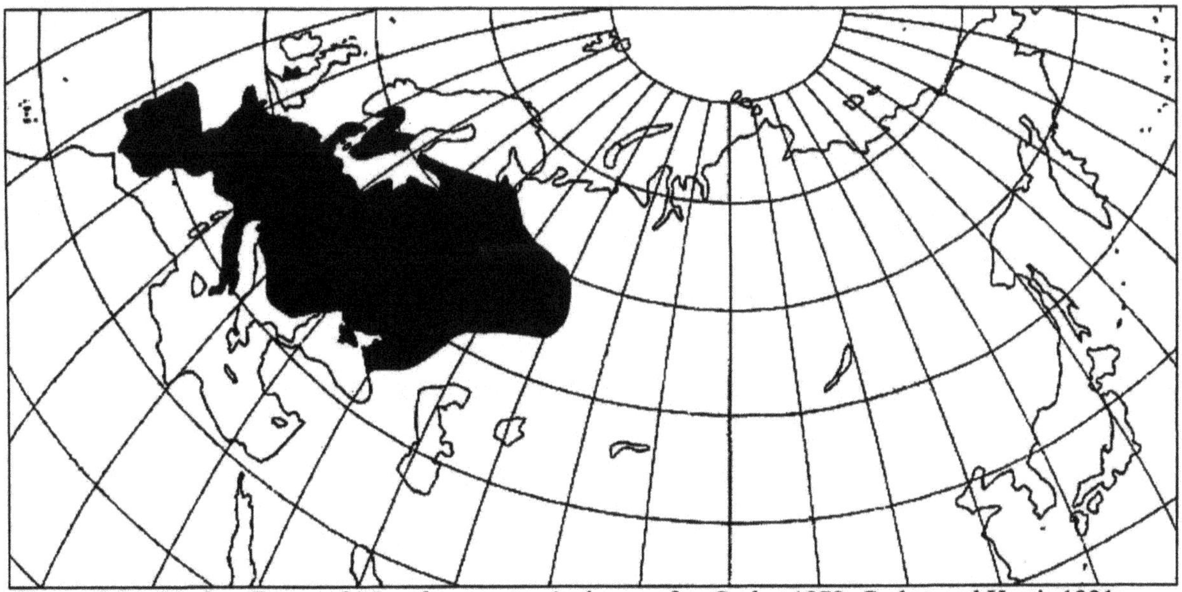

Figure 3.31: Modern Range of *Mustela putorius* (redrawn after Corbet 1978; Corbet and Harris 1991; Corbet and Ovenden 1982).

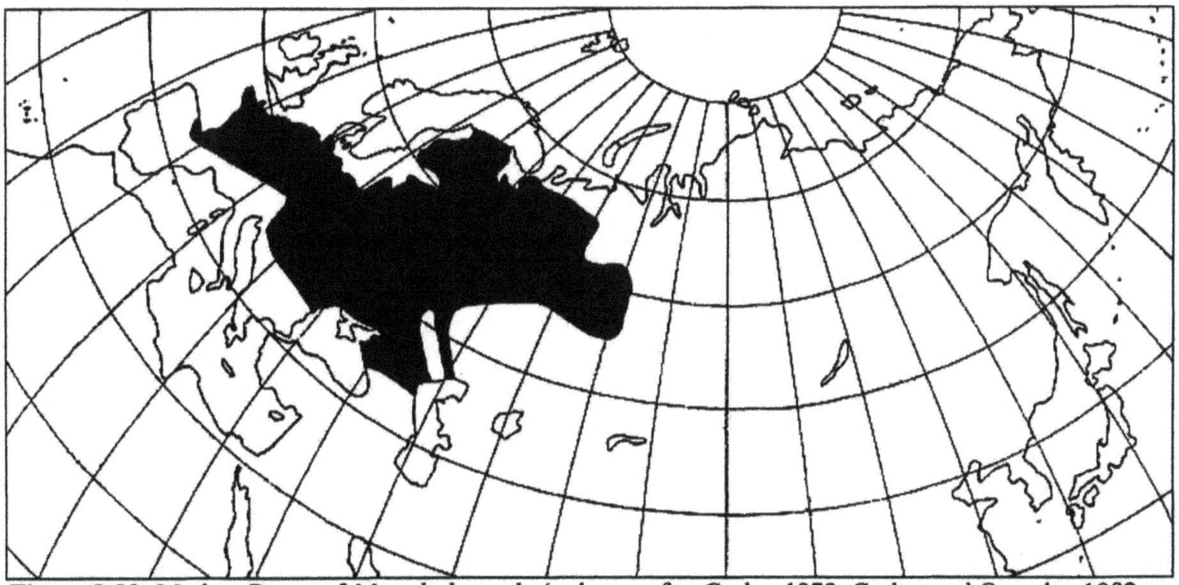

Figure 3.32: Modern Range of *Mustela lutreola* (redrawn after Corbet 1978; Corbet and Ovenden 1982; Youngman 1982, 1990).

Figure 3.33: Modern Range of *Martes martes* (redrawn after Corbet 1978; Corbet and Harris 1991; Corbet and Ovenden 1982).

Figure 3.34: Modern Range of *Martes foina* (redrawn after Corbet 1978; Corbet and Ovenden 1982).

Figure 3.35: Modern Range of *Lutra lutra* (extending also into Southeast Asia) (redrawn after Corbet 1978; Corbet and Harris 1991; Corbet and Ovenden 1982).

Figure 3.36: Modern Range of *Meles meles* (redrawn after Corbet 1978; Corbet and Harris 1991; Corbet and Ovenden 1982).

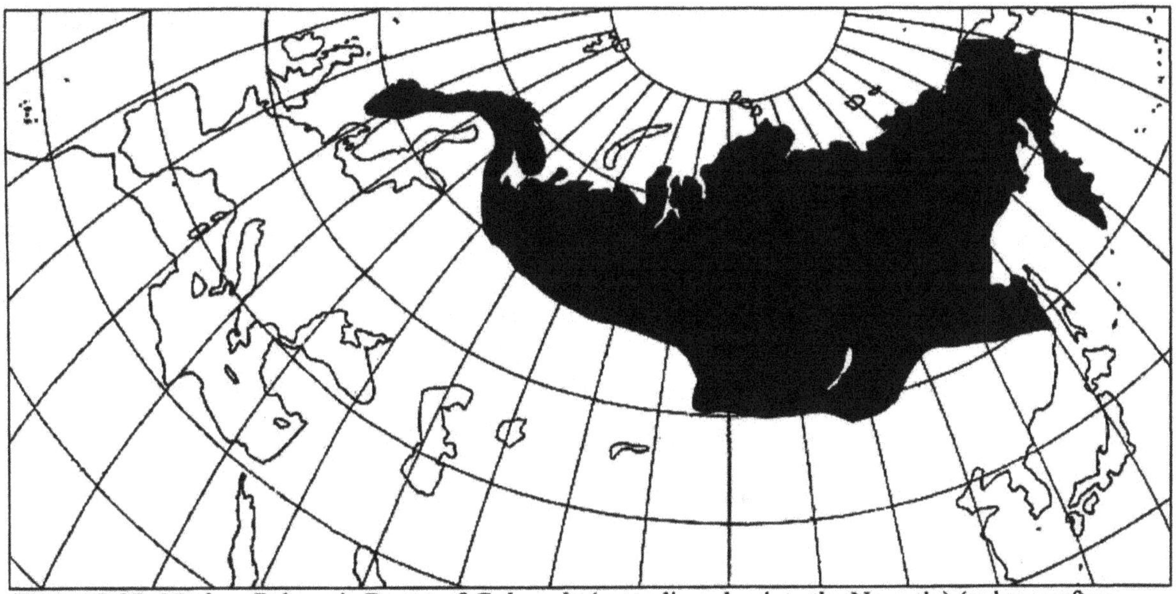

Figure 3.37: Modern Palearctic Range of *Gulo gulo* (extending also into the Nearctic) (redrawn after Corbet 1978; Pasitschniak-Arts and Larivière 1995).

Figure 3.38: Modern Palearctic Range of *Vulpes vulpes* (extending also into the Nearctic) (redrawn after Corbet 1978; Corbet and Harris 1991).

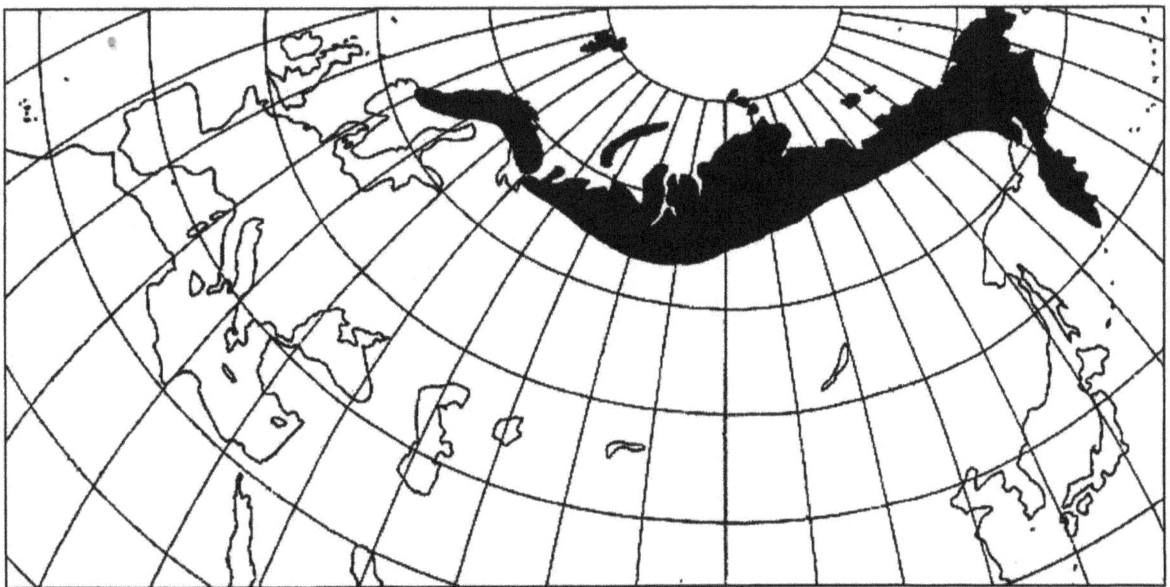

Figure 3.39: Modern Palearctic Range of *Alopex lagopus* (extending into the Nearctic) (redrawn after Corbet 1978).

Figure 3.40: Modern Range of *Felis silvestris* (extending also into Sub-Saharan Africa) (redrawn after Corbet 1978; Corbet and Harris 1991).

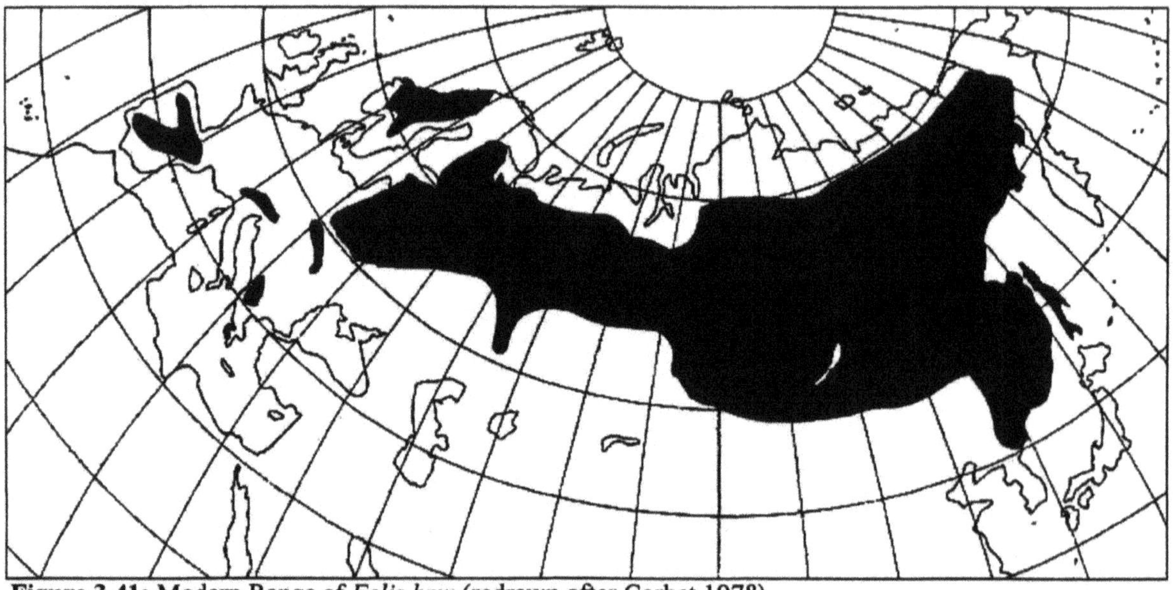

Figure 3.41: Modern Range of *Felis lynx* (redrawn after Corbet 1978).

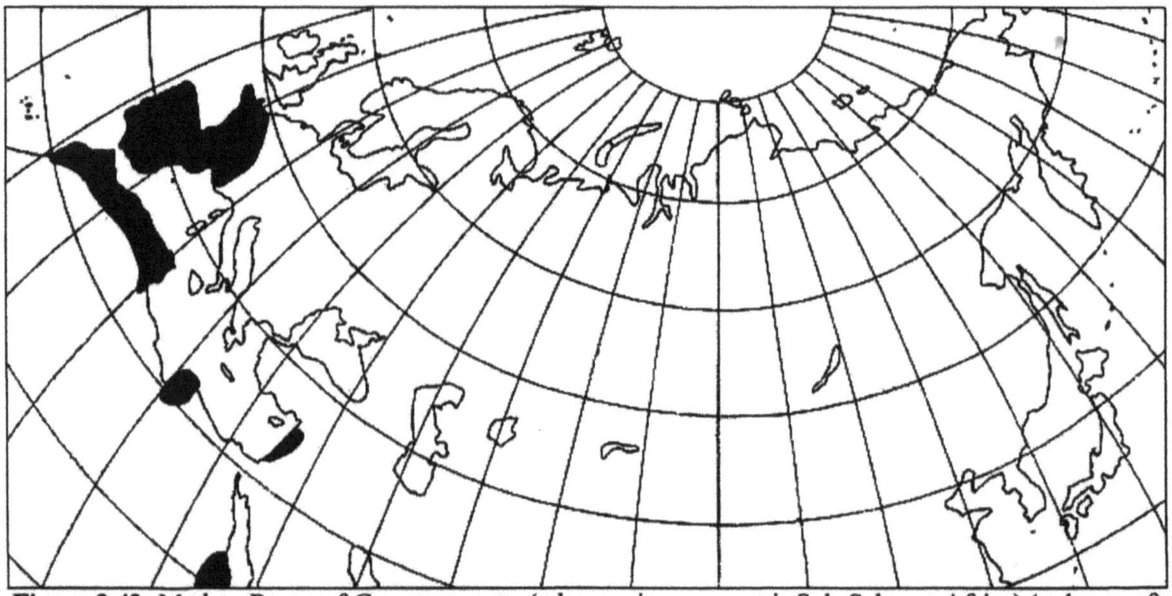

Figure 3.42: Modern Range of *Genetta genetta* (whose primary range is Sub-Saharan Africa) (redrawn after Corbet 1978; Skinner and Smithers 1990).

CHAPTER 4
MODERN MICROFAUNA OF CANTABRIAN SPAIN

The modern microfauna of the Cantabrian region and the surrounding Iberian peninsula is presented here in order to serve as a baseline for comparison to the paleomicrofauna. These data concerning the contemporary fauna comes from both trapping studies and the analysis of raptor pellet samples. New data from each of these sources collected by the author is also presented. Besides the strictly practical tasks of representing the relative microfaunal population densities, compositions, and variation across climatic, altitudinal and latitudinal clines, the relative utility and biases of these two sources of data will also be made apparent (Fulk 1976). As discussed in Chapter 2, the source of mortality for an archaeological or paleontological sample of small mammals is of extreme importance in its interpretation. This same situation is logically true for a modern sample, since the only essential difference is in the recentness of mortality. Additionally, examination of the modern small mammal fauna of a region and its relationship to the contemporary climate and environment will illustrate the relationship between Pleistocene faunas and their associated paleoenvironments (Western 1980).

Micromammalian studies in the Iberian peninsula have a substantial modern history. These had begun at least with the work earlier this century by Cabrera (1907, 1908, 1914). More recent work includes studies on the faunal distribution of the northern peninsular region by Heim de Balsac and de Beaufort (1969), Garzon-Heydt and Castroviejo (1970), Castién and Gosálbez (1992a) and Almaça (1973). Much current work concerns distributional variations related to altitude and/or latitude (Delibes de Castro 1985; Moreno and Barbosa 1992; Palomo *et al.* 1994) or habitat (Borghi *et al.* 1994; Fa *et al.* 1992). Owl dietary studies of relating to Iberian small mammal distribution are also the focus of much modern research (Brunet-Lecomte and Delibes 1984; Camacho Muñoz 1975; Campos 1978; Herrera 1973; Herrera and Jaksic 1980; Jaksic *et al.* 1982; López-Gordo *et al.* 1977; Veiga 1980).

TRAPPING STUDY OF IBERIAN FAUNA

The author sampled the terrestrial small mammals in the area surrounding the site of El Juyo and other locations in the province of Cantabria (Figure 4.1). A total of 2000 trap-nights were amassed (Table 4.1). This trapping was supplemented by the analysis of owl and hawk pellet samples from the same area. Most trapping was done with standard Museum Special snap traps (7 cm x 14 cm), supplemented where indicated below with the larger (8.3 cm x 17.6 cm) Victor Rat snap traps (both manufactured by the Woodstream Corporation, Litiz, Pennsylvania, USA). Bait was uniformly a mixture of peanut butter and rolled oats, following Nagorsen and Peterson (1980). Trap stations were at 10 m intervals, with two traps per station. Where possible, pitfall traps supplemented this array, but their use was limited by shallow soil and in some instances, constant rain. The traps were checked every morning.

While trap arrays of this configuration are effective in capturing a broad spectrum of small mammals, some exceptions do occur (Boonstra and Krebs 1978; Maddock 1992). Shrews normally feed upon a diet of mixed invertebrates, so peanut butter/rolled oats may be a less effective enticement to them. Similarly, their small size makes trapping them with standard snap traps difficult. These species seemed to be trapped more frequently during episodes of poor weather, indicating perhaps that they must expand their feeding choices when under stress (personal observation). Microtines (voles, etc.) are also more difficult to trap using this type of bait (see results below). Ligtvoet and Van Wijngaarden (1994) recommend carrot as bait for these species. Murids (Old World mice) have a more eclectic diet and are readily trapped using peanut butter/rolled oats bait. Arboreal species such as dormice (Gliridae) and fossorial species such as moles (Talpidae) tend to be overlooked by traps placed at ground level.

All captured specimens were weighed to the nearest gram and the following measurements were taken (following Nagorsen and Peterson 1980): total length, tail length, ear length and hindfoot length including nail (Table 4.2.1-4.2.8). Sex, reproductive state and/or history was recorded for each individual where possible. Some scavenging from traps did occur, particularly at Location 4 by local domestic cats and in one case by a mustelid. Scavenging also occurred at Locations 1 and 2. These individuals were not included in species tallies, since these could not be determined in most cases and few meaningful measurements could be taken. Most scavenged individuals appeared in any case to belong to *Apodemus*, which was by far the most frequently trapped genus. In addition, some *Apodemus* could not be assigned reliably to *A. sylvaticus* or *A. flavicollis*, in most cases because they were juveniles.

Table 4.1: Distribution of Trap Nights (Cantabria).

#of Trap Nights:	Loc. 1	Loc. 2	Loc. 3	Loc. 4	Loc. 5
Museum Special	357	120	353	773	292
Victor Rat	0	39	26	0	0
Pitfall	0	0	16	24	0
TOTAL	357	159	395	797	292

(Total=2000)

Table 4.2: Species Dimensions of Cantabrian Small Mammals

Table 4.2.1: *Apodemus sylvaticus* Dimensions.

#	Sex	loc.	L	Tail	Foot	Ear	Wt.
			mm	mm	mm	mm	g
94004	F	1	164	73	22	17	18
94005	F	1	167	86	23	17	13
94006	F	1	174	87	24	16	18
94007	M	1	170	86	21	16	17
94008	M	1	181	91	23	17	17
94009	M	1	189	98	23	15	26
94010	F	1	*164*	*65*	24	18	23
94011	M	1	162	75	21	15	14
94012	F	1	162	69	23	16	16
94013	F	1	156	79	23	14	12
94014	M	1	173	87	22	15	19
94015	M	1	182	85	25	18	24
94016	F	1	174	88	21	17	15
94017	F	1	165	82	23	16	15
94018	M	1	174	85	24	16	20
94019	F	1	180	89	24	17	25
94020	M	1	183	88	23	17	20
94022	F	1	177	82	21	17	23
94023	M	1	208	105	24	18	29
94024	F	1	*135*	*47*	24	17	21
94025	F	1	146	72	23	16	12
94027	M	1	181	97	24	18	17
94030	M	1	*160*	*68*	22	18	20
94034	M	2	192	87	22	18	32
94038	F	3	176	87	22	17	14
94050	F	4	179	90	21	17	25
94051	F	4	185	95	24	16	24
94052	M	4	180	94	24	16	18
94053	M	4	172	85	23	16	18
94054	M	4	164	81	24	16	16
94055	F	4	*151*	*54*	22	16	22
94056	F	4	180	90	23	16	22
94059	F	4	*165*	*78*	22	17	22
94060	M	4	174	89	24	16	18
94066	M	4	165	83	23	17	18
94067	M	4	185	94	23	17	20
94071	M	4	195	97	24	17	32
94072	M	4	183	96	24	17	21
94073	F	4	170	88	24	16	18
94081	F	4	180	85	23	16	20
94090	M	4	183	97	24	19	18
94093	F	4	200	103	23	17	24
94094	M	4	183	96	23	16	24
95001	M	5	189	100	24	17	22
95004	M	5	191	98	24	16	24
95005	M	5	190	104	24	17	26
95008	M	5	188	92	23	19	24

Table 4.2.2: *Apodemus flavicollis* Dimensions.

#	Sex	loc.	L	Tail	Foot	Ear	Wt.
			mm	mm	mm	mm	g
94001	M	1	213	104	24	17	32
94002	F	1	163	81	22	17	17
94003	F	1	*159*	*60*	23	17	29
94021	F	1	208	102	24	17	29
94031	M	1	194	102	24	18	26
94032	M	1	161	86	24	17	24
94035	F	2	*153*	*62*	23	17	22
94036	F	3	202	103	22	17	28
94037	F	3	194	92	24	15	27
94039	M	3	200	102	23	17	26
94040	M	3	217	109	25	18	34
94041	M	3	207	107	24	18	29
94044	M	3	210	104	25	17	29
94045	F	3	211	108	24	18	32
94046	M	3	191	101	25	17	22
94061	M	4	191	100	24	17	23
94062	M	4	186	92	24	16	23
94063	F	4	176	91	23	16	16
94064	F	4	178	89	23	17	19
94065	F	4	171	87	23	17	18
94068	M	4	*140*	*54*	24	17	19
94069	M	4	174	91	24	16	17
94074	M	4	187	97	25	16	20
94075	M	4	183	92	25	18	24
94076	F	4	183	89	24	17	21
94084	F	4	197	101	25	17	27
94085	F	4	165	87	22	15	16
94091	F	4	*171*	*78*	24	18	32
94105	M	4	*160*	*67*	24	17	25
95002	M	5	190	95	25	17	29
95007	F	5	*155*	*73*	22	17	18

Table 4.2.3: *Rattus rattus* Dimensions.

#	Sex	loc.	L	Tail	Foot	Ear	Wt.
			mm	mm	mm	mm	g
94086	M	4	264	147	31	22	56

Table 4.2.4: *Microtus agrestis* Dimensions.

#	Sex	loc.	L	Tail	Foot	Ear	Wt.
			mm	mm	mm	mm	g
95003	M	5	140	41	20	14	34

Table 4.2.5: *Sorex coronatus* Dimensions.

#	Sex	loc.	L	Tail	Foot	Ear	Wt.
			mm	mm	mm	mm	g
94042	F	3	117	46	13	7	7
94043	F	3	114	46	13	7	7
94092	M	4	116	40	13	7	9
94098	F	4	111	48	13	6	7
94103	M	4	110	41	12	5	7

Table 4.2.6: *Sorex minutus* Dimensions.

#	Sex	loc.	L mm	Tail mm	Foot mm	Ear mm	Wt. g
94047	F?	3	105	41	11	7	3.5
94048	F?	3	100	40	11	7	3

Table 4.2.7: *Neomys fodiens* Dimensions.

#	Sex	loc.	L mm	Tail mm	Foot mm	Ear mm	Wt. g
94070	F	4	137	48	17	7	17
94099	F	4	128	52	18	8	12
94100	F	4	138	50	18	7	11
94104	M	4	132	52	17	6	10

Table 4.2.8: *Crocidura russula* Dimensions.

#	Sex	loc.	L Mm	Tail mm	Foot mm	Ear mm	Wt. g
94079	F	4	118	37	12	9	9
94080	M	4	94	37	8	8	6.5
94102	M	4	101	39	13	8	6.5
94106	M	4	112	39	13	9	10
94107	M	4	102	38	12	7	7
94108	M	4	105	36	13	8	9
94109	M	4	99	36	13	9	7

Loc. = Trapping location
L = Body length from nose to end of last tail vertebra (mm)
Tail = Length of tail from base to end of last vertebra (mm)
Foot = Length of hind foot, including nail (mm)
Ear = Length of pinnae, excluding fur (mm)
Wt. = Mass in grams

Italics denote that a measurement is incomplete due to pre-mortem injury or postmortem scavenging.

Figure 4.1: Map of Cantabria Showing Trapping Locations 1-5 and Roost Locations A-C.

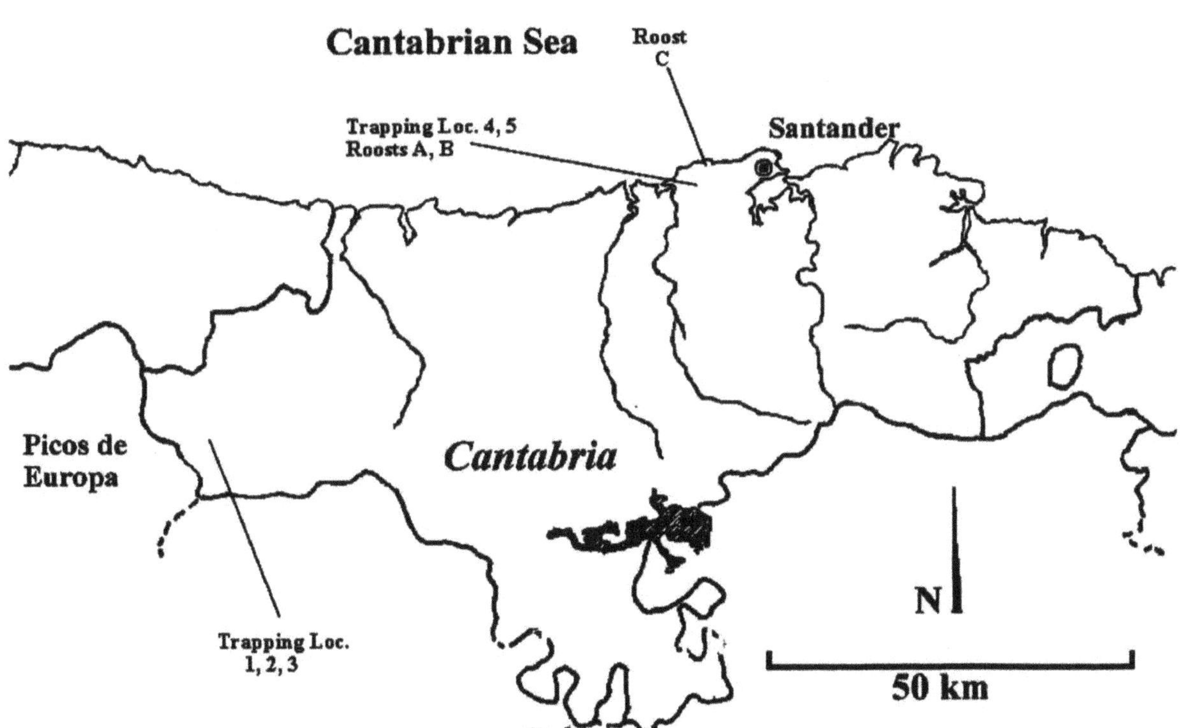

Table 4.3: Species Distribution among Trapping Locations (Cantabria).

Taxon	Loc.1 Scrub Oak Forest	Loc.2 Stream Edge	Loc.3 Mature Oak Forest	Loc.4 Forest/ Meadow/ Pasture	Loc.5 Meadow	Total
Apodemus (indeterminate)	3	0	0	13	0	16
Apodemus flavicollis	6	1	8	14	2	31
Apodemus sylvaticus	23	1	1	18	4	47
Rattus rattus	0	0	0	1	0	1
Sorex coronatus	0	0	2	3	0	5
Sorex minutus	0	0	2	0	0	2
Neomys fodiens	0	0	0	4	0	4
Crocidura russula	0	0	0	6	1	7
Total	32	2	13	59	8	114

Trapping location 1: Liebána, Cantabria, Spain; about 3 km W-NW of Potes, altitude 550 m, August 27-30, 1994. The sampled habitat was scrub oak forest on a very steep slope, with shallow topsoil and many patches of exposed bedrock. Ground cover was sparse. Short pasture was nearby. No open water was nearby, and conditions previous to and during this cycle of trapping were very dry.

Trapping location 2: Liebána, Cantabria, Spain; about 0.3 km W of Turieno, altitude 340 m, September 2-10, 1994. The habitat was along the Rio Deva, which at this point is a shallow stream lined with gallery deciduous forest. Some brushy ground cover was available. Nearby habitats include human habitations and short pastures. No mole tunnels were noted, though the topsoil allows burrowing. One Pyrenean desman (*Galemys pyrenaicus*; Insectivora, Talpidae, Desmaninae) was observed in the stream habitat but none were trapped.

Trapping location 3: Liebána, Cantabria, Spain; about 1 km N of Agüebanes, altitude 820 m, September 6-10, 1994. The traps were placed in a mature oak forest, with moderate ground cover and slope. No open water is present at this location, and conditions had been dry in general. The topsoil allows burrowing, but entrances to mole tunnels were only infrequently encountered.

Trapping location 4: Igollo, Cantabria, Spain; adjacent to cave of El Juyo, altitude 60 m, September 23-30, 1994. This habitat is isolated pockets of forest (primarily eucalyptus and oak) surrounded by tall pasture and meadows. Areas of low brush are also present. A small source of open water is present. The topsoil is very suitable for burrowing, and the entrances to mole tunnels have been frequently encountered (though none of these animals were trapped). Weather conditions during the cycle of trapping were extremely wet, with moderate to heavy rain each day.

Trapping location 5: Igollo, Cantabria, Spain; about 0.2 km N of the cave of El Juyo, altitude 50 m, April 6-14, 1995. The area trapped lay within the confines of a large, long-abandoned limestone quarry now converted to meadow. Some patches of low brush, scree slope, and open short grass are also present. The shallow topsoil is generally unsuited to burrowing. The weather was dry to moist previous to and during this time of trapping.

The results of these trapping sessions are illustrative of the contemporary Cantabrian microfauna as well as processes of microfaunal mortality. Foremost is the high proportion of *Apodemus sylvaticus* and *A. flavicollis*, two species normally associated with significant forest cover (Table 4.3). Even in these mixed forest/cropland settings dominated by the latter cover type, these two species dominated all trapped taxa. Basing environmental reconstruction upon trapping results alone for all the locations sampled would lead to the false picture that all areas had primarily forest cover, even though this varied greatly among locations. Microfaunal samples from Pleistocene Spanish sites (Chapters 5 and 8) have only tiny proportions of *Apodemus*. This proportion generally increases through the Azilian period when the ameliorated climate of the Holocene was reached. Microfaunal diversity as disclosed by trapping is at a high variance from that indicated from raptor pellet studies (discussed below).

RAPTOR PELLET STUDIES OF IBERIAN FAUNA

Roosting locations were scouted for principally in the immediate area of the site of El Juyo itself. This area was chosen for its proximity to the site, thereby removing any differences in terrain or altitude (except for relative sea level) from affecting a direct comparison with the Pleistocene fauna recovered. The immediate environs of the site also offer relatively undeveloped territory. The author has also observed barn owls (*Tyto alba*), Peregrine falcon (*Falco peregrinus*) and ever-present buzzards (*Buteo buteo*) actively hunting in this area. Roost locations A, B and C therefore immediately surround Trapping Location 5 and are nearby to Trapping Location 4.

Roost location A: El Juyo Quarry Site-2

Sampling occurred on July 21, 1994. This narrow fissure (Figure 4.2) was very difficult to access. Roosting birds perch high above the base of the fissure, which is sloped and allows pellets to spill eventually onto the scree slope below.

Sampling of the accumulated pellets indicated that two species of raptor utilized this roost successively. A large mass (over 1 kg) of disaggregating barn owl pellets was found to underlay a few freshly deposited, distinctive pellets of the buzzards (*Buteo buteo*) that had apparently just occupied this site. About 150 barn owl pellets are estimated to make up this sample. Intact pellets had the typical black, polished exterior and size (again 4.5 cm long and 2.5 cm wide) typical to this species. Care was taken to separate these two sources of pellets.

Subsequent return to this site (sampled again on July 4 and August 16, 1995) yielded 43 fresh pellets of *Buteo buteo*. These pellets were therefore accumulated during this season, while the season of occupation for the barn owls at this same roost is unknown. No evidence of additional owl habitation at this site was detected. This species of hawk is ubiquitous in the daytime sky above the El Juyo cave site, and is the most commonly observed diurnal raptor in the Cantabrian region.

All the pellet samples from the El Juyo quarry come from roosts in natural fissures exposed by the limestone quarrying that had gone on at this location (see Trapping Location 5). The habitats available within a 1 km radius for these predators are primarily agricultural, with a mix of high pastures and some cropland. Isolated patches of deciduous forest, hedgerows, and brush make up less than 20% of the total area. Bare rock, even in the quarry, makes up only a tiny percentage of the overall area. Human habitations consist primarily of farm buildings and well-spaced homes. No significant bodies of water exist within this radius, excepting some small springs. Altitude within this area does not climb above 121 m.

Roost location B: El Juyo Quarry Site-3 & 3A
This tiny cave/large fissure (Figure 4.3) is located high up the quarry wall and contains two separate roosting sites. Sampling occurred on July 22, 1994. All intact pellets had the exterior black sheen typical of those egested by a barn owl. Site 3 was the more interior, which allowed most of its pellets to tumble from a high ledge into a deep, barely accessible fissure. Only one intact pellet was recovered, which measured 3.8 cm x 2.1 cm, plus disaggregating bones and fur. Less than 20 pellets were originally present.

Site 3A consists of a secure ledge near the entrance of the cave, from which pellets drop to a small, sheltered area of scree. Many whole and partially intact pellets were recovered, a typical one of which measured 3.7 cm x 2.5 cm. Return visits to this location have yielded no additional signs of owl occupation. All indications point to these two adjacent roosts being used at the same time, perhaps alternately.

Roost location C: Cemetery Site-1
This isolated site is located in the upper interior of a large mausoleum, about 4 m above the floor on a stone ledge. The cemetery is located about 0.5 km from the Cantabrian coast, on the top of a 40 m hill, 4 km N-NE of the site of El Juyo. A 1 km hunting radius would encompass a significant portion of open sea, so this zone may have been shifted southwards to increase the usable area. This entire zone consists of open terrain with a variety of cover types: high pasture, meadows, cropland, and coastal scrub and exposed rocks. Scattered buildings and isolated patches of trees are found throughout this area. No significant open freshwater lies within this zone. The highest point is 49 m a.s.l.

Sampling occurred on September 15, 1994. This site was used for nesting the previous summer, since some of the pellets were still relatively fresh. This sample therefore is the only one derived from barn owls for which the season of accumulation can be stated. Evidence that barn owls occupied this roost comes from the size and appearance of the intact pellets and the recovery of fresh feathers identified to this species. The pellet total is about 10. No additional use of this roost has yet been made by the owls.

The failure of trapping alone to represent fully the small mammal fauna of Cantabria is apparent (Table 4.4). The taxa *Micromys minutus*, *Mus musculus*, *Arvicola terrestris*, *Pitymys lusitanicus*, *Talpa europaea* and *Crocidura suaveolens* were not acquired in 2000 trap nights, and the taxa *Rattus rattus* and *Microtus agrestis* are represented by one individual each. These taxa account for 341 (36.5%) of the total prey items taken by the local raptors. The only species of small terrestrial mammal known to be present in Cantabrian lowlands that were not taken by these raptors sampled are *Talpa occidentalis*, *Neomys anomalus*, *Clethrionomys glareolus*, *Microtus arvalis*, *Arvicola sapidus* and *Rattus norvegicus*. Barn owls prove far more effective at sampling the local Cantabrian fauna than do standard trapping techniques.

Tyto alba also proved to be a more effective sampler of the local small fauna than the diurnal raptor *Buteo buteo*. Direct comparison of the two predators at Roost Location A is possible in a setting where the exact hunting range was utilized in succession. Among small terrestrial mammals, *T. alba* left behind 12 prey species as opposed to 5 for *B. buteo*. While total sample MNI at this location was much higher for the former predator than the latter (554 vs. 69), these prey items come from more closely allied numbers of pellets (ca. 150 vs. 43). While both species hunt on the wing within comparable hunting radii, *B. buteo* has fewer small mammal species from which to choose: only those species active during the daytime can be taken. Far more small mammal species are primarily nocturnal, thus having activity patterns coinciding with *T. alba*. *B. buteo* is also a greater destructor of small bones (Chapter 2), so some prey items may have been rendered unidentifiable and therefore reduced apparent dietary biodiversity.

OTHER MICROFAUNAL STUDIES OF NORTHERN SPAIN

Altitudinal gradation of the small fauna of Cantabrian Spain has been studied further by Delibes de Castro (1985). The study area in this case was an 18 by 30 km region covering the north and south slopes of the eastern Cantabrian mountains. These altitudinal gradients were divided into 250 m sections, from below 500 m to 1700 m. A total of 1115 small mammals were trapped from July through October. The northern slope has an Atlantic climate and yielded a total of 24 species, while the southern slope has a more continen-

Figure 4.2: El Juyo Quarry Site 2 (Roost A).

Figure 4.3: El Juyo Quarry Site 3 and 3A (Roost B).

Figure 4.4: Environs of El Juyo Overlooking Nearby Quarry.

Table 4.4: Modern Small Fauna of the El Juyo Area from *Tyto alba* and *Buteo buteo* Pellets.

	Location					
	A (*Tyto alba*)	A (*Buteo buteo*)	B (3) (*Tyto alba*)	B (3A) (*Tyto alba*)	C (*Tyto alba*)	**Total**
Muridae						
Apodemus spp.	67[1]	1	2	38	3	111
M. minutus	8	0	1	4	1	14
R. rattus	4	0	0	5	0	9
M. musculus	7	0	0	1	0	8
M. agrestis	27	6	4	17	5	59
Microtidae						
A. terrestris	52	0	15	8	8	83
P. lusitanicus	90	30	2	42	0	164
Insectivora						
T. europaea	3	0	0	0	0	3
S. coronatus	77	4	5	17	1	104
N. fodiens	2	0	1	0	0	3
C. russula	196	27	4	110	10	347
C. suaveolens	1	0	0	0	0	1
Other						
Chiroptera	2	0	0	0	0	2
Amphibia	2	0	0	0	0	2
Aves	16	1	0	7	0	24
Invertebrates	X[2]	X		X		
TOTAL	554	69	34	249	28	934

[1]MNI counts for mammals and amphibians based upon sided mandibles or maxillae; Aves based upon synsacra.
[2]"X" denotes presence. Not included in totals.

Figure 4.5: Areal Extent (%) of Habitat Types in the Modern Environs of El Juyo Derived from Barn Owl Diet

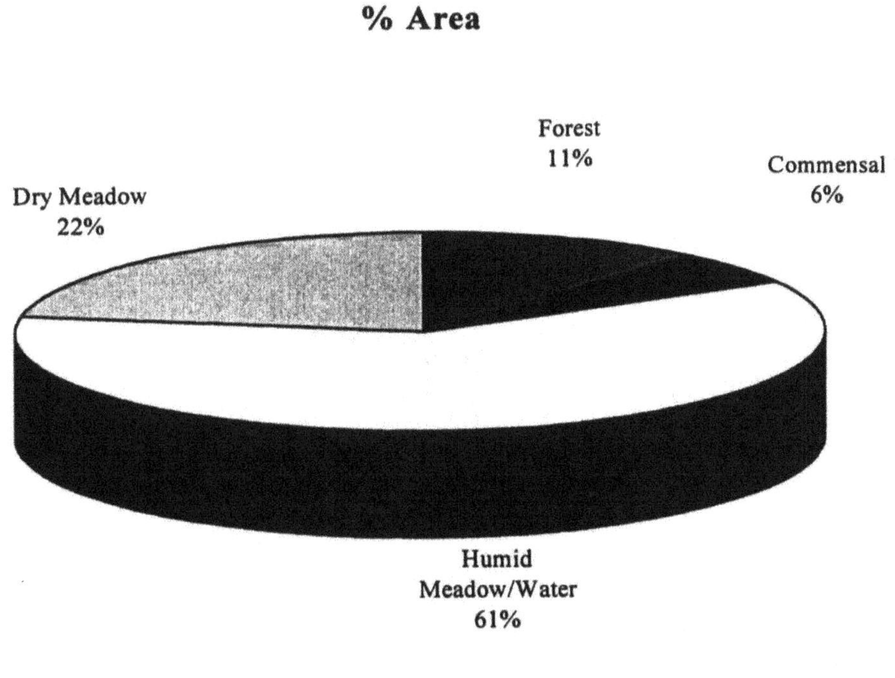

tal climate with some Mediterranean forest areas and yielded 21 species. Four altitudinal groups were indicated:

1.} Species decreasing in abundance as altitude increases:
Apodemus spp., *Mus musculus*, *Rattus rattus*, *R. norvegicus*, *Arvicola sapidus*, *Clethrionomys glareolus*, *Pitymys duodecimcostatus*, *Crocidura russula*.

2.} Species preferring median altitudes:
Glis glis, *Eliomys quercinus*, *Pitymys lusitanicus*, *Microtus agrestis*.

3.} Species increasing in abundance as altitude increases:
Pitymys savii (=*P. pyrenaicus*), *Microtus nivalis*, and possibly *Sorex minutus* and *Neomys fodiens* (but sample sizes for these two taxa were small).

4.} Remaining species:
Talpa europaea, *T. caeca* (= *T. occidentalis*), *Neomys anomalus*, *Galemys pyrenaicus* and *Arvicola terrestris*.

Delibes de Castro (1985) also noted that non-altitudinal factors may affect these species' distributions. *Microtus nivalis* was always associated with its typical biotope (rocky substrates) even at lower altitudes. Changes in altitude may affect species distribution only indirectly, in that changing plant communities over an altitudinal cline may be the proximate cause for a species' presence or absence. One would therefore not expect *T. europaea* and *T. caeca* (= *T. occidentalis*) to be affected as greatly by changes in altitude, given their fossorial lifestyle. Similarly, species associated with open water (*Arvicola terrestris* and *Galemys pyrenaicus*) may be less associated with changes in altitude. Delibes de Castro (1985:55) also notes that these results are only seasonally applicable, since cooler conditions could push the altitudinal range of some species downwards for part of the year.

Latitude also affects the distribution pattern of small mammals, even over relatively small changes in range. Moreno and Barbosa (1992) examined a total of 3000 barn owl pellets from 20 localities in the province of Huesca (Spain). Altitudes ranged from 420 to 981 m covering sites in the Ebro Valley, the Pyrenean foothills and the Pyrenees themselves. A total of 9744 small mammals were recovered from the seasonally-pooled pellets, so yearly cyclic changes in distribution were averaged out. A total of 17 (12 rodent, 5 insectivore) species were found. These fell into three groupings:

1.} Species distribution related to latitude:
Apodemus sylvaticus, *Mus spretus*, *Pitymys pyrenaicus*, *Clethrionomys glareolus*, *Sorex coronatus* and *Neomys fodiens*.

2.} Species distribution related to altitude:
Microtus agrestis, *Suncus etruscus* and *Crocidura russula*.

3.} Species distribution related to neither:
Eliomys quercinus, *Rattus rattus*, *R. norvegicus*, *Microtus nivalis*, *M. cabrerae*, *Pitymys duodecimcostatus*, *Sorex minutus*.

While some of these species are found only in the Mediterranean zone of Spain, many of these species are distributed in Cantabrian Spain. *A. sylvaticus* distribution directly correlated with latitude, with a clear positive gradient to the north of this province (Moreno and Barbosa 1992:172-173). This distribution also yields a positive correlation with precipitation and was unrelated to temperature. This result is in keeping with the overall range of this species across most of temperate Europe and into Asia (see Chapter 3). *A. sylvaticus* distribution was also related to structural features of the vegetation, *i.e.* cover. *P. pyrenaicus* has a range centering upon the Pyrenees, so a positive correlation with latitude reflects penetration northward back into its main range. *S. coronatus* and *N. fodiens* fall into a similar category. Each has the Pyrenees/Cantabrian Cordillera as the southerly limit to its modern range, so increasing density correlating with more northerly latitude in this region simply means a movement into the primary range. *M. agrestis* increased in density with increasing altitude, which is the general trend for this species in Spain (Moreno and Barbosa 1992:173). *C. russula* also increased in density with increasing altitude, which goes against the general trend for the Crocidurinae. In this case however, a positive correlation with human habitations my be reflected instead of a preference for higher altitudinal zones. *S. minutus* prefers higher altitudes (>1000 m) in the Pyrenean portion of its range. Since all study sites were lower than this limit, its distribution was less likely to be affected by latitude and altitude and more likely to be positively correlated with humidity. The author, though dealing with much smaller trapping samples, did not encounter this species until trapping above 800 m in the Cantabrian Cordillera. Similarly, *Microtus nivalis* would be expected to show a positive correlation with altitude had this study been able to encompass higher montane zones and their much greater proportion of suitable rocky substrates. Other species such as *R. rattus* and *R. norvegicus* are also found in modern Cantabrian Spain, but with their relatively recent introduction as commensals, their distributional habits are of little use in the interpretation of Pleistocene faunas.

Brunet-Lecomte and Delibes (1984) present barn owl pellet data (Table 4.5) for the Duero Basin (Castilla-Léon, Spain) which was divided into multiple biomes (montane, transitional and flatlands). The most northerly (temperate, montane) zone contains multiple roost sites from the Cantabrian Cordillera, although most of these are from the southern slopes of this range. The Cantabrian subsample came from 33 localities and yielded 5684 individual prey items. The peaks in this zone reach over 2000 m, with regular peaks of 1000 m. Precipitation averages 600 mm/year, with some years reaching 1000 mm. Annual temperature averages <10^0 C. Less than 30% of this studied zone is under cultivation, with most of the remainder consisting of pasture and deciduous forest (mostly Pyrenean oak *Querqus pyrenaica*, beech *Fagus sylvatica* and Sweet chestnut *Castanea sativa*), and some Scots pine (*Pinus sylvestris*). Most roost sites were churches, abandoned buildings and natural shelters (Brunet-Lecomte and Delibes 1984:214-215).

Table 4.5: Microfauna from Barn Owl Pellets, Cantabrian Cordillera.

Taxon	# of Loc.	% Total Biomass	% Total
Eliomys quercinus	4	0.1	0.4
Apodemus sylvaticus	30	14.3	21.3
Mus spp.	23	6.7	6.0
Rattus spp.	1	<0.1	0.2
Microtus arvalis	21	5.5	8.3
M. agrestis	18	5.1	7.7
Arvicola terrestris	4	0.5	1.7
A. sapidus	5	0.4	1.3
Pitymys spp.	29	17.9	21.3
Talpa spp.	7	0.4	1.6
Sorex spp.	19	14.2	7.6
Sorex minutus	4	0.1	<0.1
Neomys fodiens	2	0.1	<0.1
N. anomalus	6	0.4	0.3
Crocidura suaveolens	1	<0.1	<0.1
C. russula	30	33.1	19.7
Chiroptera	2	0.1	0.1
Aves	14	0.8	2.2
Reptiles	1	<0.1	<0.1
Amphibia	5	0.3	0.2
Arthropods	5	0.1	<0.1

# of Loc. = # of localities of total (33) at which a given taxa was found.
% Total = % of numerical total of prey items.
% Total = % of total biomass of prey items.

(Following Brunet-Lecomte and Delibes 1984:216)

PALEOENVIRONMENTAL INTERPRETATION FROM BARN OWL DIET

Paleoenvironmental interpretation from microfaunal remains follows from the taphonomic agency responsible for their introduction into El Juyo: the egested pellets of *Tyto alba*. Research by previous authors also indicates that the diet of this predator is a relatively unbiased sample of the small mammalian prey within an area and hence a clear picture of the environment at the time of deposition. Some exceptions due to prey behavior do occur but are correctable given the present state of information about barn owl behavior. Forest species tend to be underrepresented, especially those dwelling arboreally such as members of Gliridae (*Muscardinus avellanarius*, *Eliomys quercinus*, etc.). An upper limit of about 100 g for individuals taken also limits taxa in the diet (Marti 1973b). The lower size limit includes arthropods. Within this size range, this predator takes prey in close proportion to what is available. Specifically, the prey biomass taken from the habitats available within the hunting range of the owl's daytime roost is reflective of their relative abundance.

Other authors disagree with this assessment of barn owl diet and its representation of the surrounding habitat proportions (Fast and Harrison 1975; Grayson 1981). They, however, fail to consider the relative sizes of the prey species taken. Evans and Emlen (1947) found that barn owl diet did not reflect the relative proportions of woodland and open fields for the bird studied at a site in California, U.S.A. The hunting range of this bird was directly observed to cover 165 acres, 140 (84.8%) of which were open grain/alfalfa fields and 25 (15.2%) were cottonwood/oak/willow forest. The micromammalian prey species recovered from the pellets are shown in Table 4.6. While the number of prey items recovered from the two major habitat types do not reflect their relative abundances, their total biomasses do. Prey species associated with open habitats contributed 307 prey items (42.9%) and forest-associated species contributed 408 prey items (57.1%) to the total diet. This relationship is inverted when prey biomass is taken into account. Forest species contributed 6,920 g (24.7%) to the diet, while open species contributed 21,105 g (84.8%). This relationship closely matches the proportions of these two major biotopes in the hunting range.

This discrepancy between numbers of prey items and associated habitat area is explained by the differences in average prey mass. The largest numeric contributor to diet (283 individuals) is *Mus* (house mouse), which averages only 15 g each. The largest biomass contributor to diet (16,405 g) is *Thomomys* (pocket gopher), which averages 85 g each. A single individual of *Thomomys* is an average night's hunting take. Over five *Mus* need be taken to match this total, so continued hunting is necessary on average to feed the owl adequately (Marti 1973b). Provisioning of nestlings will only increase this total. The high number of *Mus* taken cannot be explained through greater ease of capture, since they live in the more difficult (closed) hunting habitat and one would normally expect their underrepresentation in the owl diet.

Table 4.6: *Tyto alba* Mammal Diet from California, U.S.A.

Taxa	Habitat Pref.	Total Prey Items	Ave. Mass (g)	Biomass in Diet (g)
Mus	Woods	283	15	4,245
Thomomys	Open	193	85	16,405
Microtus	Open	110	40	4,400
Peromyscus	Woods	92	20	1,840
Reithrodontomys	Woods	19	15	285
Rattus	Woods	8	65	520
Sorex	Woods	6	5	30
Lepus	Open	4	75	300

	Woods		Open		Total
Prey Items	408	(57.1%)	307	(42.9%)	715
Prey Biomass	6,920 g	(24.7%)	21,105 g	(75.3%)	28,025 g
Habitat Area	25 a.	(15.2%)	140 a.	(84.8%)	165 a.

(Source: Evans and Emlen 1947:9)

Other researchers have detected the correlation between barn owl diet and the relative amounts of prey habitat types within their hunting ranges. Lovari *et al.* (1976) studied barn owl diet from seven roost sites in a central Italian Mediterranean environment. Proportions of tree cover ranged from 60% to 2%, with varying proportions of scrubland, cropland and grassland making up the remainders. Two of the most abundantly taken prey taxa, *Apodemus* spp. and *Pitymys savii* show, respectively, significant positive and negative correlations with forest cover. The biomass of *Apodemus* ranged from 50.9% to 27.6% across the range from most to least forest cover, while *P. savii* ranged in an inverse pattern from 8.0% to 56.8% of the biomass taken. The results of this research again show that the biomasses of prey species taken can be highly reflective of the habitat types in the barn owl's hunting range. One must consider additionally that some species have primary and secondary habitat choices. These include most importantly *Apodemus* spp. (*A. sylvaticus* and *A. flavicollis* in this area) which can also be found in open areas, so interpretation must includet this variability.

Recent analysis of owl pellets from the vicinity of El Juyo (below) also indicates a strong correlation between the biomass of barn owl prey and the amounts of vegetation types present. Table 4.7 indicates the total terrestrial small mammal prey of barn owls from pellet Locations A and B. These barn owl roosting sites are within 150 m of the El Juyo cave entrance. The surrounding area is primarily farmland and pasture, with few houses and little open water (Figure 4.4).

This environment (within an arbitrary 1 km hunting radius) is reflected by the total biomass (as approximated by Prey Units) derived from each habitat type. The slight deciduous forest cover (primarily introduced eucalyptus stands and hedgerows) makes up less than 20% of total cover. Forest species (*Apodemus*) make up 11.3% of prey biomass (Figure 4.5). The low biomass of commensal species (5.6%) is in keeping with the sparse structures in the area. Both humid and dry meadow favoring species make up the bulk of diet (83.2%). This reflects closely the areal extent of cleared agricultural/pastoral land in the area. The mix of humid and dry meadow favoring species is explicable in terms of modern land use practices. This coastal region is quite humid and receives over one meter of rain annually. Areas of drier meadow (or with vegetation mimicking that of dry meadow) is created by frequent mowing for hay and grazing. Open water within 1 km is restricted to small springs and intermittent small inundated areas. These are reflected by the presence of *Arvicola terrestris* and *Neomys fodiens*. The former species can also be found away from open water in humid meadows and is a frequent component of the small fauna along this humid coastal strip.

Also significant is the lack of some species. Arboreal dormice species such as *Glis glis* are absent. These can be expected to show up in low numbers in barn owl diet where tree cover is more extensive. *Clethrionomys glareolus*, a forest-dwelling species, is also absent. The lowland setting is reflected by the absence of *Pitymys pyrenaicus*, *Galemys pyrenaicus* and *Microtus nivalis* which are found at generally higher altitudes in the Cantabrian Mountains (Chapter 3).

Table 4.7: Modern Terrestrial Micromammal Prey of Barn Owls from the Environs of El Juyo.

	MNI	%	Prey Units[1]	Biomass	%
Forest					
Apodemus spp.	107	13.2	1.0	107	11.3%
Total					**11.3%**
Commensal					
R. rattus	9	1.1	5 0	45	4.8%
M musculus	8	1.0	1.0	8	0.8%
Total					**5.6%**
Humid Meadow/Water Edge					
A. terrestris	75	9.3	5.0	375	39.7%
P. lusitanicus	134	16.5	1.0	134	14.2%
S. coronatus	99	12.2	0.5	49.5	5.2%
T. europaea	3	0.4	5.0	15	1.6%
N. fodiens	3	0.4	0.75	2.25	0.2%
Total					**60.9%**
Dry Meadow					
C. russula	310	38.3	0.5	155	16.4%
M. agrestis	48	5.9	1.0	48	5.1%
M. minutus	13	1.6	0.5	6.5	0.7%
C. suaveolens	1	0.1	0.5	0 5	0.1%
Total					**22.3%**
Total	810		100.0%	945.75	**100.0%**

[1]Prey Units are based upon 1 Unit = one 20 g individual and considers juvenile prey contribution to diet (Southern 1954).

In comparison, modern barn owl prey from superficial deposits at Amalda, an upland site in the Basque County (Chapter 8) are quite different from those near El Juyo. This fauna (Pemán 1990:229) is dominated by *Apodemus* and has present the arboreal *Glis glis* and forest-dwelling *Clethrionomys glareolus*. The region surrounding Amalda has forest cover approaching 40%. This upland site also has few *Arvicola*. *Pitymys pyrenaicus* is abundant, reflecting the higher altitude of this setting. The open meadow component is reflected most abundantly by *Crocidura* spp., *Sorex* spp., and *Pitymys lusitanicus*. True alpine species (*G. pyrenaicus* and *Microtus nivalis*) are absent.

In summary, the modern microfauna of northern Spain as sampled by barn owls proves to be a reliable indicator of the small terrestrial mammal species present and their relative biomass. These relative biomasses in turn are a reliable indicator of the local biotopes within an owl's hunting range, falling broadly into forest cover, dry grassland, humid grassland, wetlands, water edges, scrub land, heathland, alpine meadow, open rock/scree, and commensal settings. When examined in this framework (Chapter 5), the Upper Paleolithic small mammal fauna of El Juyo will be used to interpret the Lower Magdalenian paleoenvironment. This analysis will be extended to other sites where possible in Chapter 8. In addition, the contemporary small mammal fauna of Spain have been illustrated to show the large changes that have occurred after the Holocene transition.

CHAPTER 5
EL JUYO MICROFAUNA AND PALEOENVIRONMENTAL INTERPRETATION

This chapter details the smaller microfaunal remains (Rodentia, Insectivora and *Mustela nivalis* of Carnivora) from El Juyo and the interpretation of paleoenvironment from their relative proportions. The remains of larger microfauna (larger Carnivora and Lagomorpha) are presented elsewhere (Chapter 9). This analytical separation is not only due to the size and taxonomic differences of these species, but to the probable manner of introduction into sites of each size category. The smaller taxa usually entered sites as the bones in egested nocturnal raptor pellets. The larger taxa usually entered as a result of human hunting for food or furs or as large terrestrial carnivore prey remains. An exception to these patterns are the remains of hedgehog (*Erinaceus europaeus*), which fall into neither site introduction category. These individuals most likely entered sites on their own, owing to their commensal habits (Chapter 3). The inclusion of their remains therefore most likely results from natural mortality. This interpretation is reinforced by the pattern of their skeletal recovery at El Juyo (Chapter 2), which makes their introduction to the cave as human prey less likely. The remains analyzed below are only those within the prey size range of barn owls.

EL JUYO

The microfaunal remains from El Juyo are listed in Table 5.1. The elements tallied are the identified teeth for each taxa, with n = 7862. This total includes teeth (n = 376) identified in an earlier study of El Juyo by Castro Bernardez (1986). These teeth have been identified to species level in most cases. Where this was not possible (as for the teeth of *Apodemus* not being separable into *A. sylvaticus* or *A. flavicollis*), metrical analysis was attempted in order to estimate the proportions of the possible species present within the generic totals. This biometrical analysis is presented in Chapter 6. In most cases postcranial microfaunal remains could not be identified to species or even genus level, so they could not be used in paleoenvironmental interpretation. The postcranial remains of *Talpa* have also been analyzed metrically to supplement species-level identification to *T. europaea* and *T. occidentalis* (Chapter 6).

The number of teeth identifiable to each taxa also varies and is a factor in paleoenvironmental interpretation. Each molar of *Arvicola* is identifiable to that genus. Confusion with other microtine species is impossible due to the much greater size of this genus: adult body mass approaches 300 g, while the other microtines in question approach only 40-60 g. In raw numbers of identifiable molars, *Arvicola* is overrepresented in comparison to other microtine species. This problem is circumvented by using the MNI (Minimum Number of Individuals) for each taxa, where counts are based upon the most abundant (sided) tooth for each species.

This identifier tooth varied among species. Both *Pitymys pyrenaicus* and *P. lusitanicus* are identifiable by $^3M^3$. Both $_1M_1$ and $_2M_2$ are identifiable to the genus *Pitymys*, but are not separable to species level even with metrical analysis (Chapter 6). MNI counts for each level were therefore derived for *Pitymys*, with separate MNI counts for *P. lusitanicus* and *P. pyrenaicus*. These latter counts were not included in level totals and calculations derived from them, since this procedure would have counted some individuals more than once. *Pitymys* generic counts were used in biomass calculations, with the relative proportions of these two species estimated from their individual counts. In one case only (Level 4R) did the MNI for one of these species (*P. pyrenaicus*) exceed that derived for *Pitymys* at large. In this one case the MNI for *P. pyrenaicus* was used in place of undesignated *Pitymys*, since no *P. lusitanicus* was detected for that level. Since these two species have similar adult mass, estimates of their relative proportions do not affect biomass calculations.

Among the genus *Microtus*, two of the four species present are identifiable by their $_1M_1$: *Microtus nivalis* and *M. oeconomus*. No special manipulation was necessary to calculated the relative proportions of these two taxa. The $_1M_1$ of *M. arvalis* and *M. agrestis* is not consistently separable morphologically and must be grouped together. The $^2M^2$ of *M. agrestis* is separable from all the *Microtus* species present and allows estimation of the relative proportions of *M. arvalis* and *M. agrestis* within the counts of their grouped sample. The relative biomass of *M. agrestis* was estimated by subtracting its MNI from that of the combined MNI for *M. agrestis* and *M. arvalis*. This MNI for *M. agrestis* was not included in overall totals, since this would have resulted in some individuals being counted twice (as per calculations for *Pitymys*). *M. agrestis* and *M. arvalis* are of similar adult mass, so estimation of their relative proportions does not affect overall biomass calculations.

The other molars of *Microtus* are not separable among this genus and in some cases not from *Pitymys*. These molars are grouped together under *Microtus*/*Pitymys* for the purpose of estimating relative abundance of these smaller taxa to *Arvicola* and to the other species present. These identified teeth are not included in MNI totals, except for a singular molar from Level 1 whose presence indicates the existence without double-counting of one individual *Microtus*/*Pitymys*. These other taxa (*Neomys fodiens*, *N. anomalus*, *Sorex coronatus*, *S. minutus*, *Apodemus*, *Pliomys lenki*, *Talpa* and *Mustela nivalis*) generally have all their molars (and other teeth if present) identifiable to the taxonomic level indicated. Larger taxa (*Erinaceus europaeus*, *Oryctolagus cuniculus*, *Mustela putorius* and *Vulpes vulpes*) are excluded from this analysis since they are too large in general to be barn owl prey.

Prey units (following Southern 1954) are based upon one unit equaling one 20 g individual. This somewhat abstracted figure considers the normal proportion of juvenile small mammals in barn owl diet (which normally makes up the bulk). Hence, most rodents (*Microtus* spp. and *Pitymys* spp.) equal one prey unit. Larger taxa (*Arvicola* cf. *terrestris*,

Table 5.1: Identified Microfaunal Teeth from El Juyo.

	11		10		9		8		7		6/9		6		5		4R		4S		4		3		2/4		2		1		Total	Total
	NIT	MNI	NIT	MNI	NIT	MNI	NIT	MNI	NIT	MNI	NIT	MNI	NIT	MNI	NIT	MNI	NIT	MNI	NIT	MNI	NIT	MNI	NIT	MNI	NIT	MNI	NIT	MNI	NIT	MNI	NIT	MNI
Talpa spp.	3	1	x		11	2	11	3	22	2	1	1	107	9	x		11	1	101	8	32	4	x		1	1	x				300	32
Erinaceus europaeus					10	2	33	4	20	2			11	2			8	2	23	3	17	2									122	17
Neomys fodiens													1	1					9	2											10	3
Neomys anomalus													7	1					2	1											9	2
Sorex coronatus					2	2	17	3	1	1			81	13					35	6	33	5									169	30
Sorex minutus	2	1											5	2					19	5	12	3									38	11
Oryctolagus cuniculus																			1	1											1	1
Vulpes vulpes							4	1					1	1					4	1	1	1									10	4
Mustela nivalis	6	1							1	1			3	1					8	2	1	1									19	6
Mustela putorius					1	1							2	2																	3	3
Apodemus							3	1					3	2					54	11	9	3	2	1			1	1			72	19
Arvicola cf. *terrestris*	3	2	3	2	39	7	112	15	46	9	1	1	459	45			89	12	1583	153	301	27	1	1	6	2	2	1	2	1	2647	278
Pliomys lenki																			1	1											1	1
Pitymys	2	1			1	1	22	12	5	2			27	13			4	[2]	177	69	40	13	1	1							279	112
P.pyrenaicus	1	[1]			1	[1]	14	[8]	2	[1]			21	[11]			4	3	86	[46]	20	[12]									149	3[83]
P.lusitanicus							1	[1]											12	[7]											13	[8]
Microtus nivalis																			7	5	4	2									11	7
M.oeconomus	1	1			1	1	13	7	7	5	1	1	94	49			16	10	210	113	46	28	1	1	3	2	1	1			394	219
M.agrestis/arvalis			1	1	7	6	20	10	7	4			59	30	1	1	10	7	196	107	85	44			4	3	3	2			393	215
M.agrestis					1	[1]	7	[4]	2	[1]			14	[10]			1	[1]	51	[28]	22	[14]			1	[1]					99	[60]
Microtus/Pitymys	8	*	2	*	30	*	150	*	41	*	6	*	510	*			103	*	1746	*	515	*	3	*	6	*	2	*	1	1	3123	1
Total	26	7	6	3	104	22	407	56	154	26	9	3	1405	171	1	1	246	35	4325	488	1138	133	8	4	21	8	9	5	3	2	7862	964

NIT = Number of Identified Teeth
MNI = Minimum Number of Individuals
"x" denotes presence of a species, not included in totals.
Numbers in brackets [ ] are not included in totals (see text).

Talpa and *Mustela nivalis*) are therefore only 5 prey units each, since smaller individuals among these species are preferentially preyed upon by barn owls. Shrew species are each less than one prey unit. As described in Chapter 4, the relative proportions of species within barn owl dietary biomass reflects the relative proportions of habitats associated with these species.

LEVELS 10 AND 11

Due to small sample size (NIT = 32), statistical interpretation of these levels is limited at this time (Table 5.2). The small sample size is due to the limited areal exposure of these two levels obtained by the modern excavations (<1 m^2 for each) and the location of this excavated area far away from the owl roost in the rear of the cave vestibule. Owl-introduced microfauna were trampled or rolled across the entire excavated extent of the vestibule to reach this point. Additional excavations of these levels in areas closer to the introduction point beneath the owl roost may yield much larger samples of microfauna.

The species present in Levels 11 and 10 are consistent with what has been recovered from the upper, more completely excavated levels of El Juyo (Figure 5.1). Species common to Cantabrian Spain today include the microtines *Arvicola* cf. *terrestris*, *Microtus agrestis*/*arvalis* and *Pitymys pyrenaicus*. An important indicator of cold climate is the presence in Level 11 of *Microtus oeconomus*. No species strongly indicative of forested conditions have yet been recovered from these levels, a pattern shown (below) to be typical of El Juyo during the Lower Magdalenian.

LEVELS 8 AND 9

The present extent of excavation of Level 9 includes only areas distant from the primary source of introduction beneath the owl roost in the northeast corner of the cave. Though large portions of this level have been excavated and processed, few microfaunal remains have been recovered (NIT = 104). Level 8 is characterized by the dense accumulation of marine molluscan shells, primarily *Patella vulgata* and *Littorina littorea* and dense amounts of highly fragmentary fish remains. Human occupation may have been very intense during this time, and Level 8 may have been accumulated quite rapidly. Owl occupation may therefore have been more sporadic, contributing to the relatively low amounts of microfauna identified from this level as well (NIT = 407). This low microfaunal recovery rate is not due to generally poorer preservation, since fish remains are at their most abundant in these levels and are inherently more fragile than those of terrestrial mammalian microfauna.

Levels 8+9 are dominated by humid meadow/water edge species, which represent 83.6% of the total biomass units (Table 5.3). Typical species of this selection of related habitat types are *Talpa*, *S. coronatus*, *P. lusitanicus*, *M. oeconomus* and *Arvicola* cf. *terrestris*. Dry meadow habitats (Figure 5.2) had a much smaller areal coverage (9.1%) as represented by the presence of *Microtus agrestis*/*arvalis*.

Since the combined MNI is 16 (based upon ${}_1M_1$) for these two taxa with similar dental morphology, the presence of both species is likely: the MNI for *M. agrestis* alone (based upon ${}^2M^2$) is only 5. Since both of these species have similar associated primary habitats (with *M. arvalis* being the more generalist and found in most habitats), paleoenvironmental interpretation is unaffected by the necessity of estimating their relative biomass units.

The presence of very slight forest cover (0.6%) within the hunting radius of a barn owl is indicated by the presence of *Apodemus* in very low amounts. The presence of some forest cover is reinforced by the presence of *Erinaceus europaeus* in Levels 8 and 9. This species prefers forest edge habitats and so can be found preferentially in areas lacking dense forest cover. Due to the general harshness of the climate, this cover would have been in isolated patches in more sheltered areas (including the sinkhole surrounding the cave entrance).

The overall climate and habitat represented by this microfauna is very open, humid and cold. This last element is indicated by the presence in significant amounts of *Microtus oeconomus*, which today is found much farther north in Europe (Chapter 3). This species also reinforces the interpretation of open, humid plant cover, since in drier habitats *M. oeconomus* is outcompeted by the more dominant *M. arvalis* (Ligtvoet and Van Wijngaarden 1994). Combining information from pollen and macrobotanical remains (Chapter 7), the plant community was made up primarily of heathlands with some areas of grasses.

LEVEL 7

Level 7 has been nearly completely excavated and analyzed for its microfauna, but this relatively small accumulation has yielded a small sample (NIT = 154). Carbon from this level has been dated to 14,440+/-180 B.P. (Freeman *et al.* 1988:12). Level 7 is characterized by the presence of two basin-shaped hearths and may represent an early phase of the Level 6 occupation.

The paleoenvironment of Level 7 shows the same pattern as the previous accumulations (Table 5.4). The microfaunal biomass profile (Figure 5.3) is dominated by humid meadow/water edge species (84.6%), with generalist (9.8%) and dry meadow species (5.6%) contributing far smaller amounts. No forest species were detected in the microfauna of this level, but small sample size may also account for this lack. Since *Erinaceus europaeus* is present in Level 7, some forest must have been available in the close vicinity of El Juyo. The character of this level like those accumulated beneath it is that of cold, humid coastal heath with some areas of grassland and patches of forest, perhaps in more sheltered areas.

LEVEL 6/9

The few remains in this grouping (Table 5.5) are from an area of unclear contact between these two larger levels and may therefore have come from either. The taxa identified (*Talpa*, *Arvicola* cf. *terrestris*, *M. oeconomus* and indeterminate *Microtus*) are consistent with the findings of either level. All of these taxa indicate open and humid conditions, with *M. oeconomus* additionally indicating cold conditions.

Table 5.2: Owl-Introduced Microfauna from Levels 10 and 11.

10+11	Identified Teeth	%Identified Teeth	MNI	%MNI	Prey Units	Biomass Units	%Biomass Units	Habitat Association	%Habitat Area
Talpa	3	9.4	1	10.0	5	5	14.9	Hum Meadow	
S.minutus	2	6.3	1	10.0	0.5	0.5	1.5	Hum Meadow	
M.oeconomus	1	3.1	1	10.0	1	1	3.0	Humid Meadow	
Arvicola cf. *terrestris*	6	18.8	4	40.0	5	20	59.7	Hum Mead/Water	79.1
M.agrestis/arvalis	1	3.1	1	10.0	1	1	3.0	Dry Meadow	3.0
Mustela nivalis	6	18.8	1	10.0	5	5	14.9	Generalist	
P.pyrenaicus	1	3.1	[1]	*	[1]	*	3.0	Generalist	17.9
Pitymys	2	6.3	1	10.0	1	1	[3.0]	*	
Microtus/Pitymys	10	31.3	*	*	*	*	*	*	
Total	32	100.2	10	100.0		33.5	100.0		100.0

Figure 5.1: Paleoenvironment of El Juyo Levels 10 and 11.

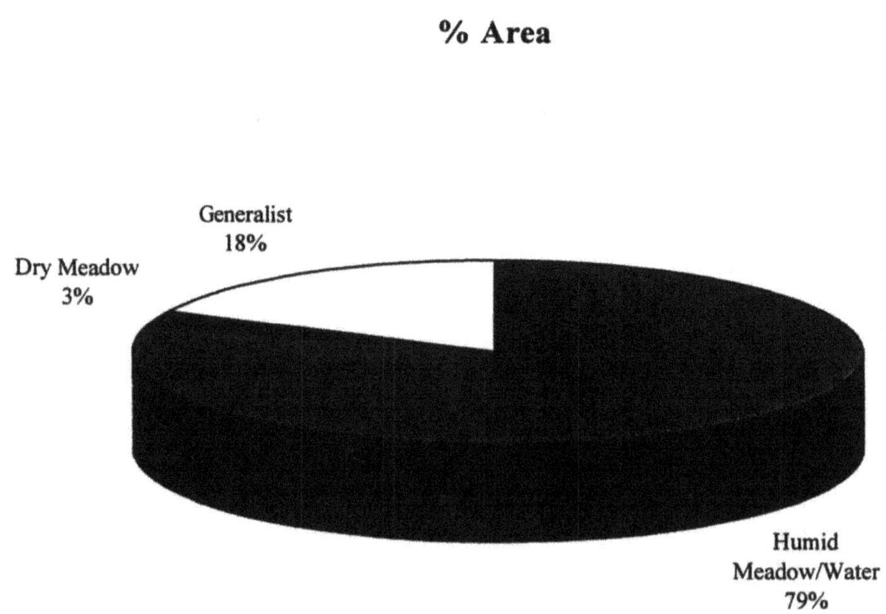

Table 5.3: Owl-Introduced Microfauna from Levels 8 and 9.

8+9	Identified Teeth	%Identified Teeth	MNI	%MNI	Prey Units	Biomass Units	%Biomass Units	Habitat Association	%Habitat Area
Apodemus	3	0.6	1	1.4	1	1	0.6	Forest	0.6
Talpa	22	4.8	5	7.1	5	25	14.2	Hum Meadow	
S.coronatus	19	4.1	5	7.1	0.5	2.5	1.4	Hum Meadow	
P.lusitanicus	1	0.2	[1]	*	[1]	*	0.7	Hum Meadow	
M.oeconomus	14	3.0	8	11.4	1	8	4.6	Hum Meadow	
Arvicola cf. *terrestris*	151	32.6	22	31.4	5	110	62.7	Hum Mead/Water	83.6
M.agrestis/arvalis	27	5.8	16	22.9	1	16	9.1	Dry Meadow	
M.agrestis	8	1.7	[5]	*	[1]	*	[2.8]	Dry Meadow	9.1
P.pyrenaicus	15	3.2	[9]	*	[1]	*	6.7	Generalist	6.7
Pitymys	23	5.0	13	18.6	1	13	[7.4]	*	
Microtus/Pitymys	180	38.9	*	*	*	*	*	*	
Total	463	99.9	70	99.9		175.5	100.0		100.0

Figure 5.2: Paleoenvironment of El Juyo Levels 8 and 9.

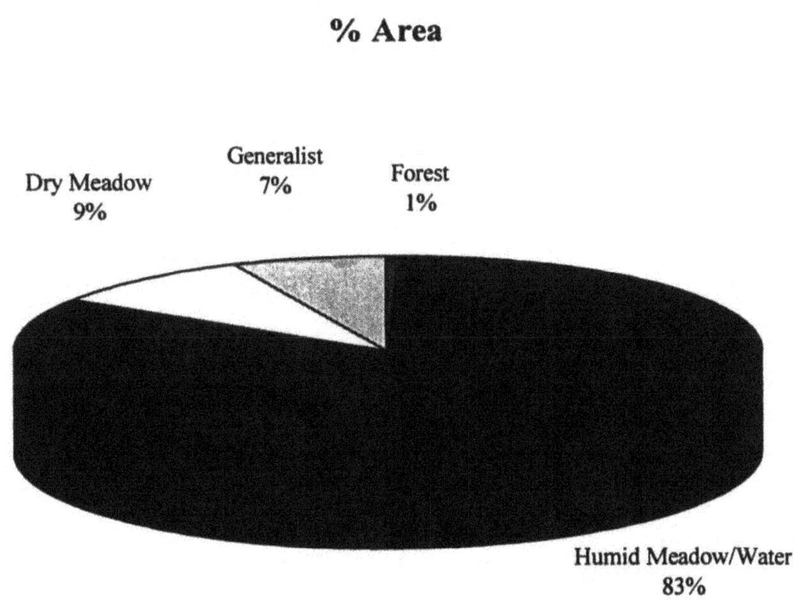

Table 5.4: Owl-Introduced Microfauna from Level 7.

7	Identified Teeth	%Identified Teeth	MNI	%MNI	Prey Units	Biomass Units	%Biomass Units	Habitat Association	%Habitat Area
Talpa	22	16.4	2	8.3	5	10	14.0	Hum Meadow	
S coronatus	1	0.7	1	4.2	0.5	0.5	0.7	Hum Meadow	
M oeconomus	7	5.2	5	20 8	1	5	7.0	Hum Meadow	
Arvicola cf. *terrestris*	46	34.3	9	37.5	5	45	62.9	Hum Mead/Water	84.6
M.agrestis/arvalis	7	5.2	4	16.7	1	4	5.6	Dry Meadow	
M agrestis	2	1.5	[1]	*	[1]	*	[1.4]	Dry Meadow	5.6
Mustela nivalis	1	0.7	1	4.2	5	5	7.0	Generalist	
P.pyrenaicus	2	1.5	[1]	*	[1]	*	2.8	Generalist	9.8
Pitymys	5	3.7	2	8.3	1	2	[2.8]	*	
Microtus/Pitymys	41	30.6	*	*	*	*	*	*	
Total	111	99 8	24	100.0		71.5	100.0		100 0

Figure 5.3: Paleoenvironment of El Juyo Level 7.

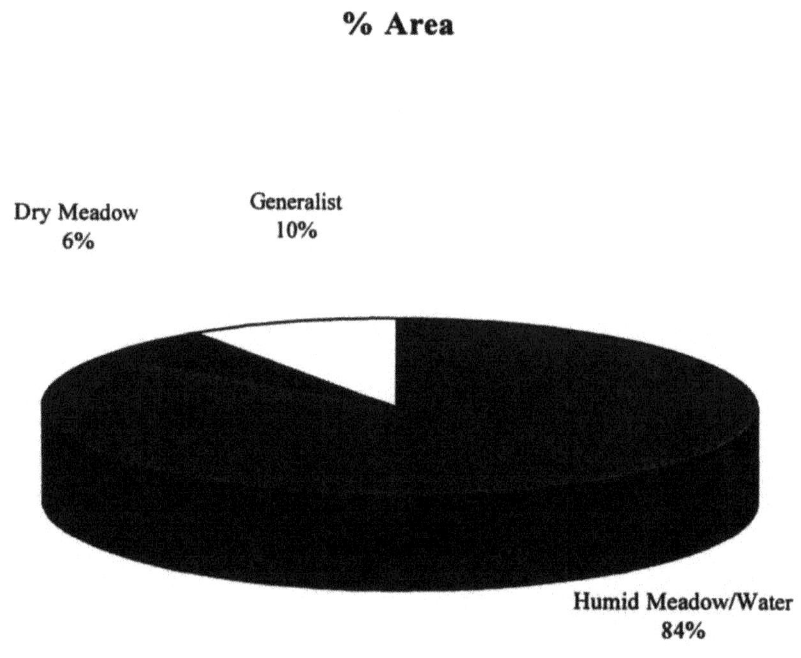

Table 5.5: Owl-Introduced Microfauna from Level 6/9.

	NIT	MNI
Talpa	1	1
Arvicola cf. *terrestris*	1	1
Microtus oeconomus	1	1
Microtus/Pitymys	6	*
Total	9	3

LEVEL 6

The cultural deposits of Level 6 are characterized by the intense exploitation of *Cervus elaphus* (Klein *et al.* 1981) and the construction of multiple building foundations of limestone rubble (Freeman *et al.* 1988). Marine mollusc exploitation was significant but not as intensive as during the formation of Level 8 (Krupa 1994). Level 6 has also yielded a dense accumulation of identified microfaunal teeth (n=1405). This large accumulation results partially from the large areal extent excavated on the northern side of the vestibule, downslope from the point beneath the owl roost. Barn owl occupation was also intense during accumulation and could have alternated seasonally with human occupation or over greater periods of time. With the evidence of substantial building within the vestibule during this occupation, it is possible that abandoned structures served as temporary owl roosts.

The climate of Level 6 times continued to be cool and wet, with humid coastal heath interspersed with small forest patches and some grasslands (Table 5.6). Forest-associated taxa (*Apodemus*) reappear in small amounts (0.5%). *Erinaceus europaeus* also reappears in small amounts. Humid meadow/ water edge species continue to dominate the faunal profile (Figure 5.4). These include for the first time *Neomys fodiens* and *N. anomalus*. Each of these shrew species is normally associated with water edge habitats, but they can also be found away from water in humid vegetation settings. (*N. fodiens* is found both by trapping and in owl pellets in the vicinity of El Juyo today.) *Arvicola* cf. *terrestris* dominates the total biomass units of humid meadow species. Cold, humid climate is reinforced by the large amounts of *Microtus oeconomus*.

Both *M. arvalis* and *M. agrestis* were probably present, since their combined MNI is 30, while the MNI for *M. agrestis* alone is only 10. One third of the *M. agrestis/arvalis* biomass units are therefore estimated to belong to the former species. Biometrical analysis (Chapter 6) also supports the interpretation of both species' presence. These species indicate the presence of some areas of dry grassland species: no more than 7.9% of the total land area, since *M. arvalis* is also a generalist species and is associated with other habitat types.

LEVEL 5

Level 5 is a culturally-sterile travertine layer. This deposit covered most of the excavated vestibule area and had to be cut through by the Magdalenian inhabitants before their construction of the Sanctuary (4S). The single identified microfaunal molar comes from *Microtus agrestis/arvalis* (Table 5.7). The presence of *Talpa* is also indicated by the recovery of a single postcranial remain. More microfauna may have been contained in this layer but was not recovered. All of these taxa were common during the Lower Magdalenian period and in modern Cantabrian Spain. All are typical of open environments.

Table 5.7: Owl-Introduced Microfauna from Level 5.

	NIT	MNI
Talpa	X	
Microtus agrestis/arvalis	1	1
Total	1	1

LEVEL 4R

Level 4R is treated separately from the remainder of Level 4 and 4S and is essentially a large feature accumulated at the same time. These deposits represent most probably the mixed fill dug out during the Lower Magdalenian construction of the Sanctuary area (Level 4S) and dumped to its southern edge. This large pit was then filled in with imported sediments laid out in multiple tumuli and surmounted with a semicircular wall and large capping slab (Freeman and González Echegaray 1981; Freeman *et al.* 1988). Recent analysis has also indicated the presence of some human (hand and foot) bones from the lower portions of the mound complex. The previously deposited materials through which the Magdalenian inhabitants dug include at least Levels 5, 6, 8 and 9 and possibly deeper. These mixed remains have been analyzed separately from those of the rest of this level. The low density of identified teeth (NIT = 246) is consistent with Levels 8 and 9 being the origin of the majority of these sediments, as is the composition of the sediments themselves. Analysis of these remains therefore serves more to reaffirm the origin of these deposits than to illuminate paleoenvironment.

Due to their uncertain origin, little can be inferred from them. The fauna present are in general agreement with those of the other levels of the site (Table 5.8). Humid meadow species dominate total biomass units (88.3%), with some dry meadow (8.2%) and generalist (3.5%) biomass present (Figure 5.5). Also typical is the lack of any forest species among the owl-introduced microfauna, although *Erinaceus europaeus* does occur in small amounts (NIT=8).

LEVEL 4/4S

The sediments of Level 4 accumulated on the surface of Level 5 (travertine) in conjunction with the construction of the Sanctuary (4S). Carbon from Level 4 has been dated to 13,920+/-240 B.P. (Freeman *et al.* 1988:12). This broad deposit on the western half of the vestibule has yielded a dense microfauna sample (NIT = 1138). The Sanctuary area (Level 4S) has the densest amount of identified teeth recovered from El Juyo (NIT = 4325). This latter accumulation is a likely result of the location of this very large feature on the eastern, interior side of the vestibule near the owl roosting area. During large stretches of time in the construction of this tumulus complex, the unfinished dugout structure served as a lowpoint in the site in which tumbling owl pellets could accumulate and be incorporated into tumuli sediments. Hum-

Table 5.6: Owl-Introduced Microfauna from Level 6.

6	Idetntified Teeth	%Identified Teeth	MNI	%MNI	Prey Units	Biomass Units	%Biomass Units	Habitat Association	%Habitat Area
Apodemus	3	0.2	2	1.2	1	2	0.5	Forest	0.5
Talpa	107	7.7	9	5.4	5	45	11.9	Hum Meadow	
S coronatus	81	5.8	13	7.8	0.5	6.5	1.7	Hum Meadow	
S minutus	5	0.4	2	1.2	0.2	0.4	0.1	Hum Meadow	
M oeconomus	94	6.8	49	29.5	1	49	13.0	Hum Meadow	
Arvicola cf. *terrestris*	459	33.0	45	27.1	5	225	59.6	Hum Mead/Water	
N.fodiens	1	0.1	1	0.6	0.75	0.75	0.2	Wetlands/Water	
N anomalus	7	0.5	1	0.6	0.75	0.75	0.2	Wetlands/Water	86.7
M agrestis/arvalis	59	4.2	30	18.1	1	30	7.9	Dry Meadow	
M agrestis	14	1	[10]	*	[1]	*	[2.6]	Dry Meadow	7.9
Mustela nivalis	3	0.2	1	0.6	5	5	1.3	Generalist	
P.pyrenaicus	21	1.5	[11]	*	[1]	*	3.4	Generalist	4.7
Pitymys	27	1.9	13	7.8	1	13	[3.4]	*	
Microtus/Pitymys	510	36.7	*	*	*	*	*	*	
Total	1388	100.0	166	99.9		377.4	99.8		99.8

Figure 5.4: Paleoenvironment of El Juyo Level 6.

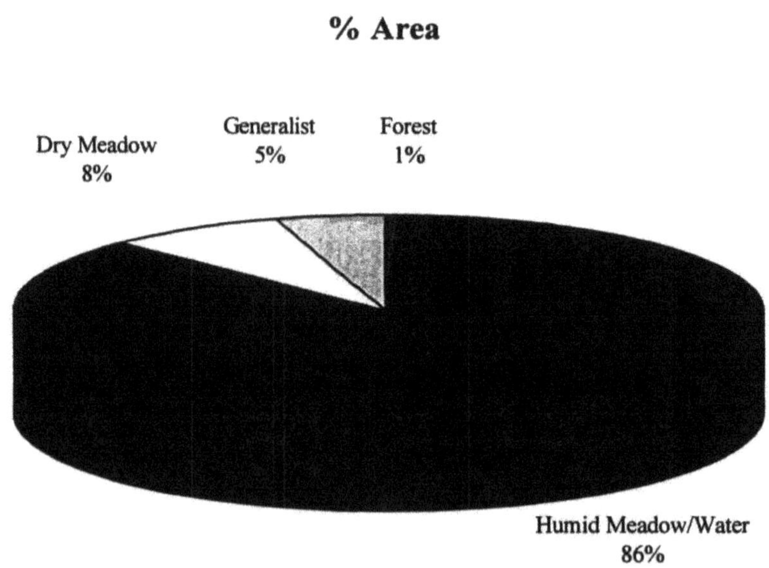

Table 5.8: Owl-Introduced Microfauna from Level 4R.

4R	Identified Teeth	%Identified Teeth	MNI	%MNI	Prey Units	Biomass Units	%Biomass Units	Habitat Association	%Habitat Area
Talpa	11	4.6	1	2.9	5	5	5.9	Humid Meadow	
M.oeconomus	16	6.7	10	29.4	1	10	11.8	Humid Meadow	
Arvicola cf. *terrestris*	89	37.4	12	35.3	5	60	70.6	Humid Mead/Water	88.3
M.agrestis/arvalis	10	4.2	7	20.6	1	7	8.2	Dry Meadow	
M agrestis	1	0.4	1	2.9	[1]	*	[1.2]	Dry Meadow	8.2
P pyrenaicus	4	1.7	3	8.8	1	3	3.5	Generalist	3.5
Pitymys	4	1.7	[2]	*	[1]	*	*	*	
Microtus/Pitymys	103	43.3	*	*	*	*	*	*	
Total	238	100.0	34	99.9		85	100.0		100.0

Figure 5.5: Paleoenvironment of El Juyo Level 4R.

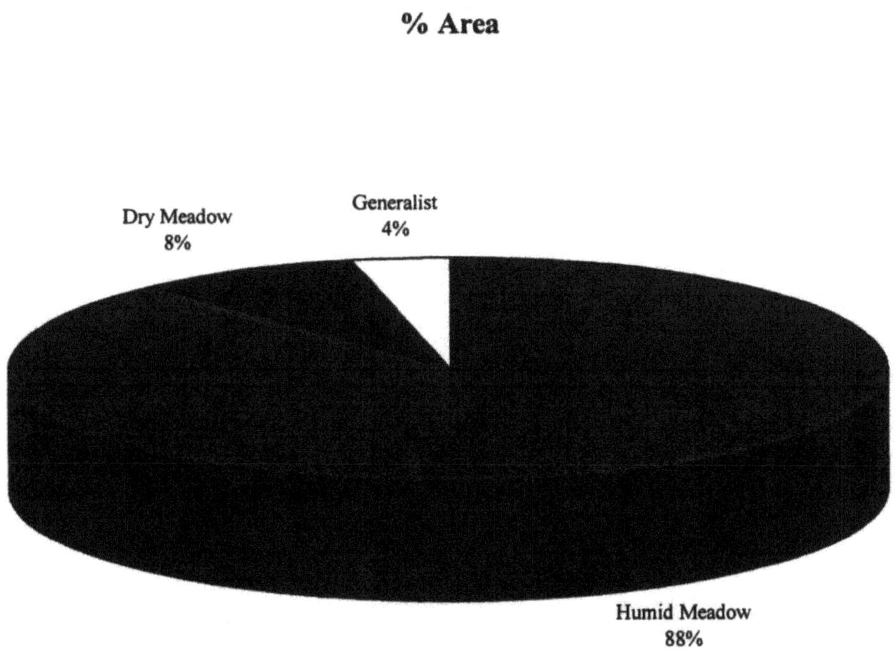

Table 5.9: Owl-Introduced Microfauna from Level 4/4S.

4+4S	Identified Teeth	%Identified Teeth	MNI	%MNI	Prey Units	Biomass Units	%Biomass Units	Habitat Association	%Habitat Area
Microtus nivalis	11	0.2	7	1.1	1	7	0.5	Alpine	0.5
Apodemus	63	1.2	14	2.3	1	14	1.0	Forest	
Pliomys lenki	1	0.02	1	0.2	1	1	0.1	(Interstadial)	1.1
Talpa	133	2.5	12	2.0	5	60	4.3	Humid Meadow	
S coronatus	68	1.3	11	1.8	0 5	5.5	0.4	Humid Meadow	
S minutus	31	0 6	8	1.3	0.2	1.6	0.1	Humid Meadow	
M oeconomus	256	4.7	141	23.0	1	141	10.2	Humid Meadow	
P.lusitanicus	12	0.2	[7]	*	[1]	*	0.6	Humid Meadow	
Arvicola cf. *terrestris*	1884	34.8	180	29.4	5	900	65.2	Humid Mead/Water	
N.fodiens	9	0.2	2	0.3	0.75	1.5	0 1	Wetlands/Water	
N anomalus	2	0.04	1	0.2	0.75	0.75	0 05	Wetlands/Water	81.0
M agrestis/arvalis	281	5.2	151	24 6	1	151	10.9	Dry Meadow	
M.agrestis	73	1.3	[42]	*	[1]	*	[3 0]	Dry Meadow	10.9
Mustela nivalis	9	0.2	3	0.5	5	15	1.1	Generalist	
P pyrenaicus	106	2.0	[58]	*	[1]	*	5.3	Generalist	6.4
Pitymys	217	4.0	82	13.4	1	82	[5 9]	*	
Microtus/Pitymys	2261	41.7	*	*	*	*	*	*	
Total	5417	100.2	613	100.1		1380.35	99.9		99.9

Figure 5.6: Paleoenvironment of El Juyo Level 4/4S.

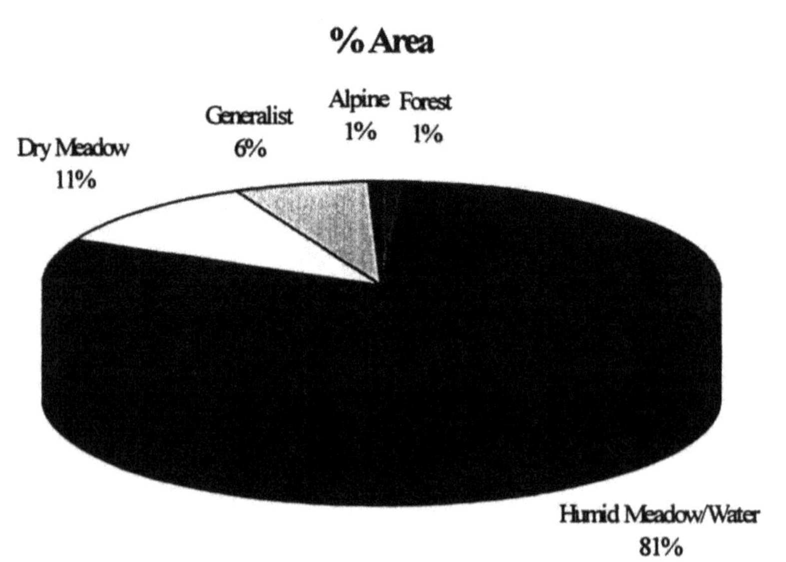

an habitation remains during the time of construction are relatively sparse (represented by Level 4). Levels 4 and 4S also contain the largest concentration of identified microfaunal teeth (NIT = 5463 or 69.5% of the site total). Occupation of El Juyo during the time of their accumulation may have been more intensive by barn owls than by humans, with human usage of the vestibule switched to sporadic occupation during episodes of tumulus construction with long gaps in between.

Open, humid heath continued to dominate the landscape with small patches of forest in more sheltered areas (Table 5.9 and Figure 5.6). In addition to *Apodemus*, forest species are represented by singular appearance of *Pliomys lenki*. This occurrence represents the latest survival in Cantabrian Spain of this now extinct species. The abundance of *P. lenki* in Pleistocene deposits correlates with interstadial conditions and the presence of extant forest species (Bartolomei *et al.* 1975) in its European distribution and hence was likely a forest species. The presence of some forest cover is also supported by the continued presence of *Erinaceus europaeus* (NIT = 40).

Humid, open vegetation species still dominate the total biomass units (81.0%): *Talpa, Sorex coronatus, S. minutus, Microtus oeconomus, Pitymys lusitanicus* and *Arvicola* cf. *terrestris*. Also present are two water-associated shrews, *Neomys fodiens* and *N. anomalus*. Dry meadow biomass increased slightly from Level 6 (up to 10.9%). *Microtus arvalis* is likely present in addition to *M. agrestis*, since the combined MNI is 151 compared with 42 for the latter species. *M. agrestis* biomass units (3.0%) are therefore estimated at 27.8% of the total (10.9%) for dry meadow species (42/151 x 10.9%).

Climate during the time of Level 4/4S shows some deterioration. The alpine species *Microtus nivalis* first appears during the time of their deposition. The presence of *M. nivalis* indicates the downward altitudinal shift of the exposed alpine rock zone occupied by this species with the concomitant downward spread of alpine glaciation. Since El Juyo is a coastal lowland site 60 m above the current sea level, the presence of an alpine species in its fauna is highly significant. This species did not need to dwell at the site itself, but within the normal hunting radius of a barn owl. Elevation does not exceed 200 m a.s.l. within a 1 km radius of the site. Even within 5 km of El Juyo (a maximal estimate of hunting radius), elevation does not climb above 600 m. Present range of *Microtus nivalis* is currently no closer than 20 km from El Juyo among the peaks of the Cantabrian Cordillera (Chapter 3). The alpine zone must have intruded to much lower altitudes during the time of Level 4/4S deposition. Cold conditions are further reinforced by the continued presence of *M. oeconomus*, which comprises 10.2% of the total biomass units.

LEVEL 3

Level 3 is the first to have been accumulated after the collapse of the main entrance and is composed primarily of travertine deposits. It is nearly sterile of cultural remains, but those found thus far indicate an ephemeral Bronze Age penetration into the nearly-sealed cave. With the entrance closed to a small crevice, the former vestibule ceased to be a viable owl roosting site. The site could have been penetrated (as it continues to be today) by microfauna and small terrestrial predators. The scant microfaunal remains (NIT = 8 for Level 3) recovered from this and succeeding levels are thus a natural consequence of the changed spatial layout.

The taxa present are consistent with those found in the paleolithic levels (Table 5.10). *Apodemus*, *Pitymys* and *M. oeconomus* have been identified from molars, and the presence of *Talpa* has been determined from postcranial remains. While *Apodemus*, *Pitymys* and *Talpa* are common to contemporary Cantabrian Spain, *M. oeconomus* underwent a cliserial range shift northward as the Pleistocene ended and the warmer conditions of the Holocene prevailed. Its presence in Level 3 therefore indicates a vestigial population soon to become locally extinct, or the possibility of mixing of materials from Level 4 which was accumulated during the Pleistocene.

Table 5.10: Owl-Introduced Microfauna from Level 3.

	NIT	MNI
Talpa	X	
Apodemus	2	1
Arvicola cf. *terrestris*	1	1
Pitymys	1	1
Microtus oeconomus	1	1
Microtus/Pitymys	3	*
Total	8	4

LEVELS 2/4 and 2

The remains from Level 2/4 come from a contact area and hence indicate the possible mixing of microfauna from both accumulations. These remains (NIT = 21) are not included in the analysis of paleolithic climate, since they may have been introduced during the much later deposition of Level 2 (NIT = 9). This mostly sterile level has yielded scant remains from a Late Roman use of the cave. Taxa typical to modern Cantabrian Spain (*Talpa*, *Arvicola* and *Microtus agrestis/arvalis*) are present (Table 5.11). *M. oeconomus* is still present in Level 2, possibly indicating an even later local survival for this species. The local environment must have included significant moist areas, with mostly open conditions still present.

Table 5.11: Owl-Introduced Microfauna from Levels 2/4+2.

	2/4		2	
	NIT	MNI	NIT	MNI
Talpa	1	1	x	
Apodemus			1	1
Arvicola cf. *terrestris*	6	2	2	1
Microtus oeconomus	3	2	1	1
Microtus agrestis/arvalis	4	3	3	2
Microtus agrestis	1	[1]		
Microtus/Pitymys	6	*	2	*
Total	21	8	9	5

LEVEL 1

Level 1 was accumulated into recent times. No cultural remains have been recovered from this sterile, silty/sandy deposit. Only the remains of *Arvicola* (Table 5.12) and one smaller microtine (*Pitymys* or *Microtus*) have been recovered from this level (NIT = 3). These taxa are consistent with the modern fauna of Cantabrian Spain and open environments.

Table 5.12: Owl-Introduced Microfauna from Level 1.

	NIT	MNI
Arvicola cf. *terrestris*	2	1
Microtus/Pitymys	1	1
Total	3	2

COMPARISON OF LEVELS

The Lower Magdalenian levels of El Juyo are not characterized by a complete microfaunal turnover compared with modern Cantabrian Spain: most changes reflect shifted emphasis among the species present. While microtines (*Microtus*, *Arvicola*, *Pitymys*) are a large part of the modern fauna, they dominate the Lower Magdalenian levels at the expense of murids (*Apodemus* and *Micromys*, plus introduced commensal taxa *Mus* and *Rattus*). Shifts have also occurred among the microtines present. While *Microtus nivalis* is present in the Cantabrian Cordillera today, it does not intrude into the coastal lowlands. *Arvicola terrestris* dominates the Lower Magdalenian microfauna and is present in lower proportions in contemporary Cantabrian Spain. *M. oeconomus* is found today in Spain. Its presence there indicates a large cliserial shift of fauna zones southward through Europe during the Pleistocene. Lacking from the Lower Magdalenian fauna is the microtine *Clethrionomys glareolus*, which is common in Cantabrian Spain today and is associated with forested environments. Lowland Cantabrian Spain is also more densely populated by *Pitymys lusitanicus* among this genus. The El Juyo microfauna is dominated by *P. pyrenaicus* at the expense of the former species. This indicates another altitudinal shift in species representation, since *P. pyrenaicus* shows some correlation with increasing altitude (Delibes de Castro 1985).

All the insectivore species present in the Lower Magdalenian levels of El Juyo are also common to modern Cantabrian Spain: *Sorex minutus*, *S. coronatus*, *Neomys fodiens*, *N. anomalus*, *Talpa* spp., and *Erinaceus europaeus*. *Crocidura russula* and *C. suaveolens* are lacking from the Lower Magdalenian microfauna. Both species are associated with dry meadow conditions, and their complete absence is further indication of the humid Lower Magdalenian conditions. *Galemys pyrenaicus* (Pyrenean desman) is also absent from the El Juyo microfauna. This alpine species is not found today in the coastal lowlands, but is found in some Pleistocene levels of other lowland sites (Chapter 8).

Microfauna among the Lower Magdalenian levels of El Juyo shows very little change over the span of their accumulation (Table 5.13 and Figure 5.7). Open, cold, and humid preferring species dominate through all levels, with small amounts of forest species and dry grassland species present. Forest cover is much higher even in the heavily deforested landscape of modern Cantabrian Spain. The most significant shift is in the appearance of the alpine species *Microtus nivalis* in Level 4/4S, indicating marked deterioration of climate during its accumulation.

Table 5.13: Comparison of El Juyo Environments by Biomass Units.

	4+4S	4R	6	7	8+9	10+11	Modern
Commensal	0.0%	0.0%	0.0%	0.0%	0.0%	0.0%	5.6%
Alpine	0.5%	0.0%	0.0%	0.0%	0.0%	0.0%	0.0%
Forest	1.1%	0.0%	0.5%	0.0%	0.6%	0.0%	11.3%
Humid Meadow/Water	81.0%	88.3%	86.7%	84.6%	83.6%	79.1%	60.9%
Dry Meadow	10.9%	8.2%	7.9%	5.6%	9.1%	3.0%	22.3%
Generalist	6.4%	3.5%	4.7%	9.8%	6.7%	17.9%	0.0%
MNI	613	33	166	24	70	10	810
NIT	5417	238	1388	111	463	32	*

Figure 5.7: Comparison of El Juyo Environments by Biomass Units.

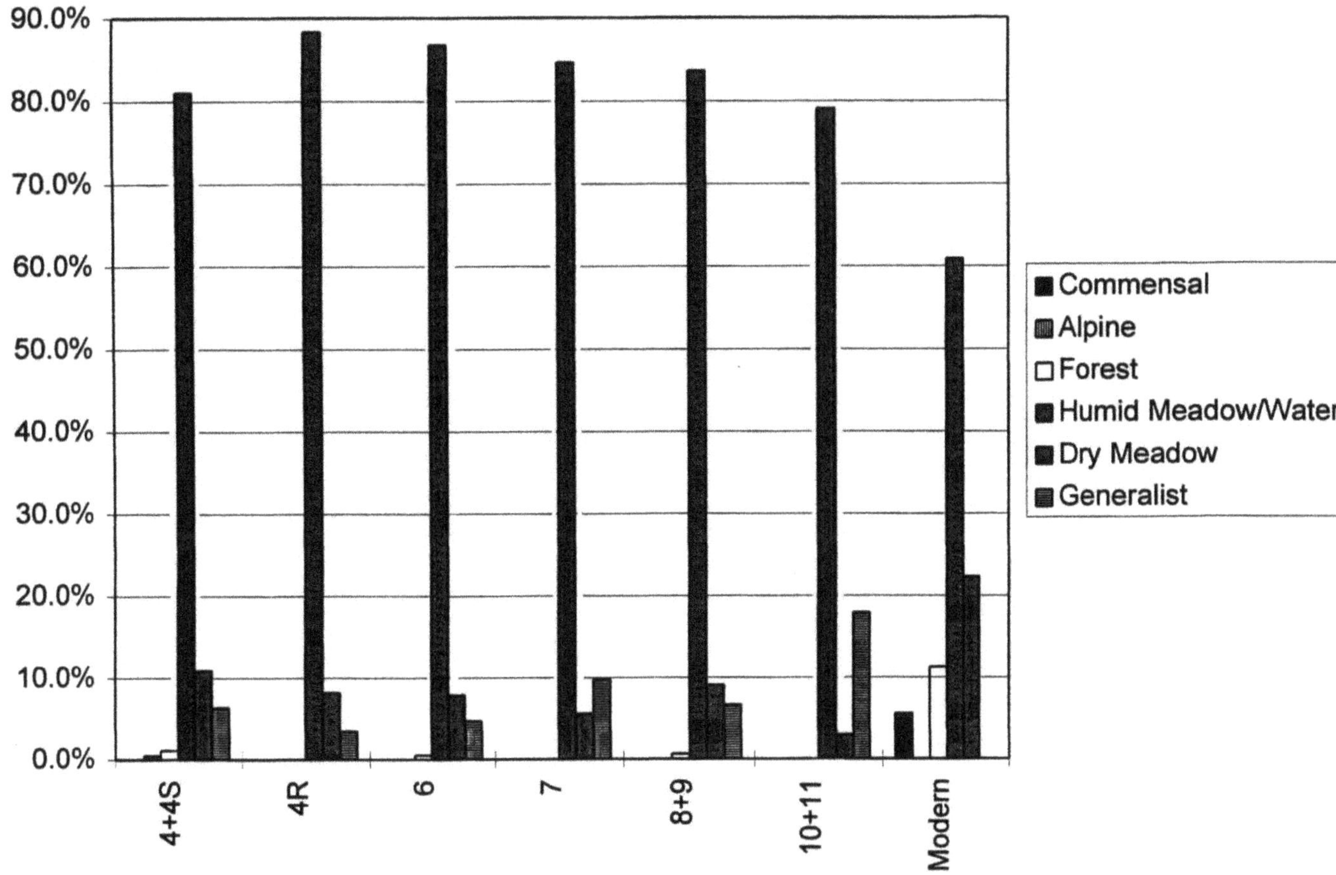

CHAPTER 6
BIOMETRICAL ANALYSIS

Elements of the El Juyo microfaunal sample were examined metrically. Measurement of selected elements aids taxonomic identification (Brothwell and Jones 1978; Ruprecht 1971), separation among the species of a genus, and quantifies microevolutionary changes in populations (Berry 1978). Small mammalian evolutionary change is critical to the structure of Pleistocene chronology as built upon the succession of evolutionary forms (Chaline 1972, 1974, 1985; Heinrich 1982, 1987; Hinton 1911, 1926; Janossy 1975; Stuart 1980, 1981, 1982; Tchernov 1968, 1975, 1979, 1982; Von Koenigswald 1973). Size changes may also reflect changes in environmental parameters such as rainfall or temperature (Klein 1991) and can be compared with modern data from small fauna populations existing across varying climatic regimes.

Not every tooth was chosen for analysis. For example, some microtine teeth are so similar that they cannot be distinguished reliably, especially the $^1M^1$ of the microtines present. Multiple species would therefore be represented in the metrical data. Changes in this data between levels or sites could indicate actual changes in the size of a single taxon or shifts in the relative abundances of multiple taxa. Where an element could be identified to only two species, metrical analysis was performed to attempt separation of these two species using this criterion. Separation to species level was less difficult for members of other taxa, especially the shrew species identified. Metrical analysis of these was limited by the generally low sample sizes, but these are presented below.

Small mammal postcranial remains present an even greater difficulty in identification to lower taxonomic levels (Graham and Saunders 1978). Factors include a paucity of previous research, their greater destructibility than teeth and the close similarity among related taxa. These elements are typically ignored in archaeological analyses (for exceptions see Andrews 1990 and Eastman 1995). A taphonomic analysis of these is presented in Chapter 2, where these elements have been grouped according to size class where species identification was not possible. Further research into modern small faunal postcranial remains may allow these archaeological remains to be put to greater use (Brown and Twigg 1969; Dunmire 1955; Graham and Saunders 1978; Ventura 1992; Ventura and Gosálbez 1986). Since teeth tend to be the most durable and identifiable skeletal elements, no loss of identified species is likely: any MNI derived from postcranial remains will be generally lower than that derived from teeth.

The postcranial remains of *Talpa* are an exception to the above situation. This taxa has highly specialized front limbs evolved for a fully fossorial lifestyle. *Talpa* humeri are one of the most easily identified elements encountered (in owl pellets and archaeological assemblages) and are highly robust. Metrical analysis of this element is presented below and is of value in distinguishing between *T. europaea* and *T. occidentalis*. The postcranial remains of Chiroptera are an additional exception. The teeth of these taxa in their European range can be so small and delicate as to elude recovery even with fine screening methods. In their case, identification from the relatively large and robust distal humerus is more common (Felten *et al.* 1973) than from other elements (de Paz and Benzal 1989).

***Arvicola* spp.**

Two taxa may be represented by the teeth identified to this genus: *A. terrestris* and *A. sapidus*. The former species of water vole currently inhabits a range extending across most of temperate and boreal Europe into Asia. The latter species is found only in the Iberian peninsula and southern France (Chapter 3). Their ranges therefore overlap in the Cantabrian region (Ventura and Gosálbez 1988, 1989; Zabala 1983). Separation of materials between these two taxa is important when reconstructing paleoenvironment due to this great difference in modern range. The presence of *A. terrestris* indicates a more extreme climate than *A. sapidus*, which is found in a relatively restricted and warmer region. Across their ranges, both species are normally associated with water edges/humid meadows and thus humid conditions.

Table 6.1: Maximum Length (mm) of $_1M_1$ of *Arvicola* spp. from El Juyo.

Level	N	Mean	St. Dev.	Range
11	1	4.02	*	*
9	4	3.60	0.28	3.44-4.01
8	10	3.86	0.37	3.47-4.65
7	4	3.74	0.23	3.60-4.09
6	44	3.76	0.22	3.36-4.42
4R	13	3.70	0.33	3.23-4.24
4S	93	3.72	0.23	3.11-4.43
4	13	3.71	0.19	3.51-4.16
Total	182	3.73	0.24	3.11-4.65

Table 6.2: Maximum Length (mm) of $^3M^3$ of *Arvicola* spp. from El Juyo.

Level	N	Mean	S.D.	Range
11	1	2.80	*	*
9	7	2.37	0.20	2.19-2.72
8	12	2.26	0.17	1.97-2.57
7	2	2.48	0.06	2.43-2.52
6	42	2.39	0.23	1.92-2.79
4R	12	2.36	0.16	2.06-2.62
4S	155	2.38	0.19	1.89-2.99
4	30	2.34	0.24	1.98-2.86
2/4	1	2.14	*	*
Total	262	2.37	0.20	1.89-2.99

Figure 6.1.1: Maximum Length Measurements Taken on Microtine Molars.

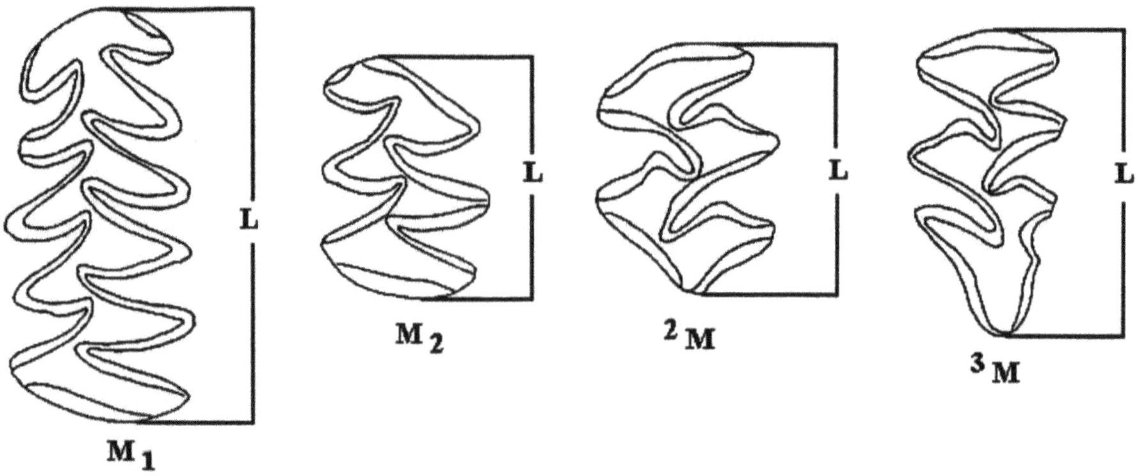

Figure 6.1.2: Maximum Length Measurements Taken on *Apodemus* Molars.

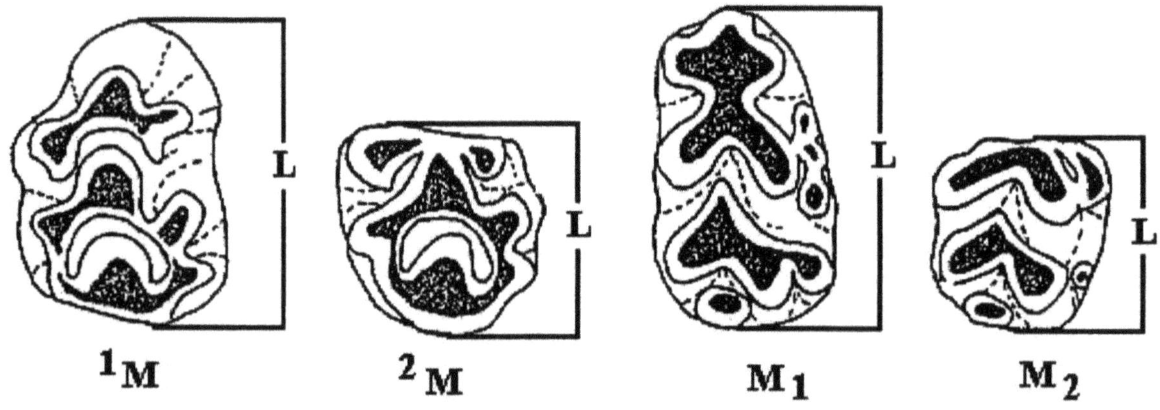

Figure 6.1.3: Maximum Length Measurements Taken on Shrew Species (Molar Length and Ramus Height).

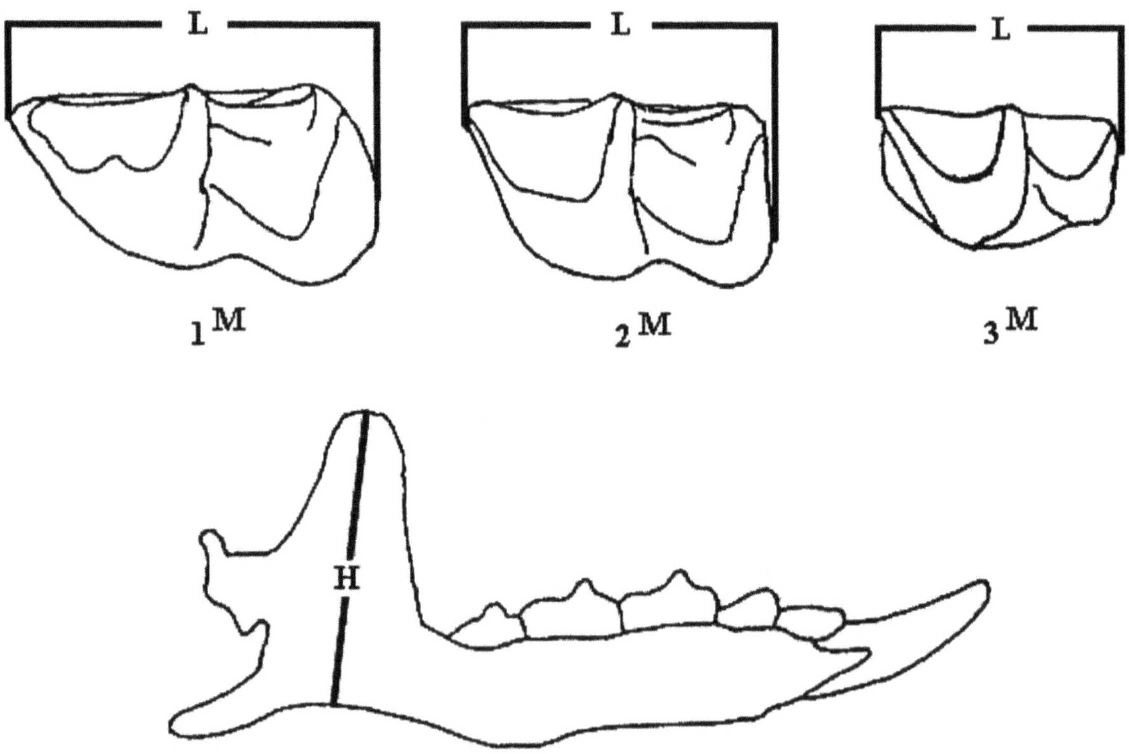

Figure 6.1.4: Maximum Length Measurement Taken on *Talpa* Humerus.

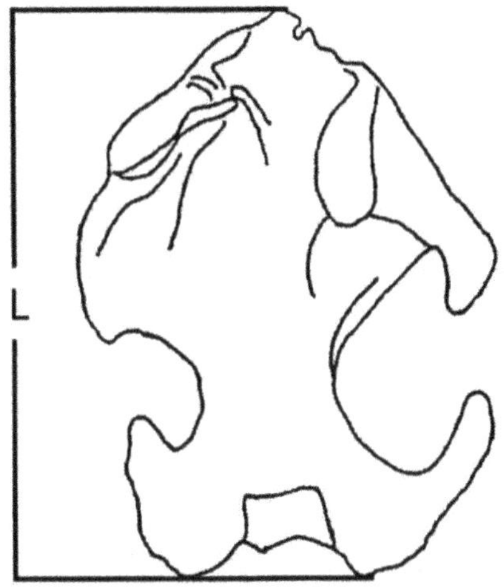

Figure 6.1.5: Maximum Length Measurements Taken on the Molars and Mandible of *Erinaceus europaeus*.

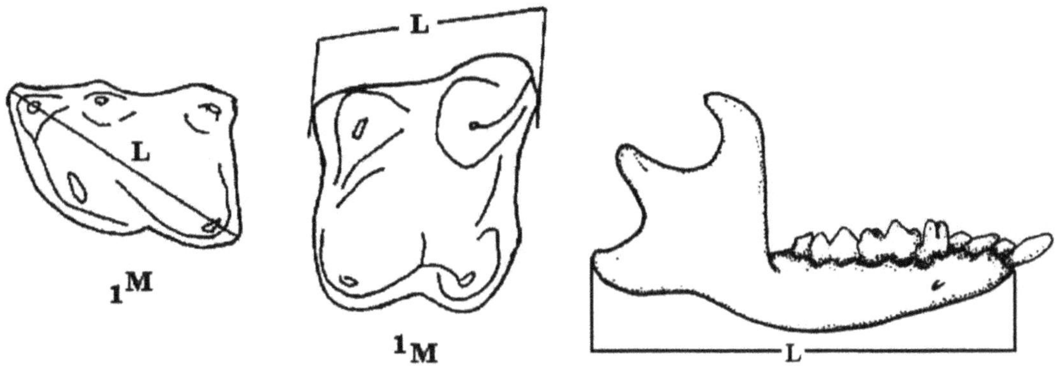

Figure 6.1.6: Maximum Length Measurements Taken on the Molars of *Mustela nivalis*.

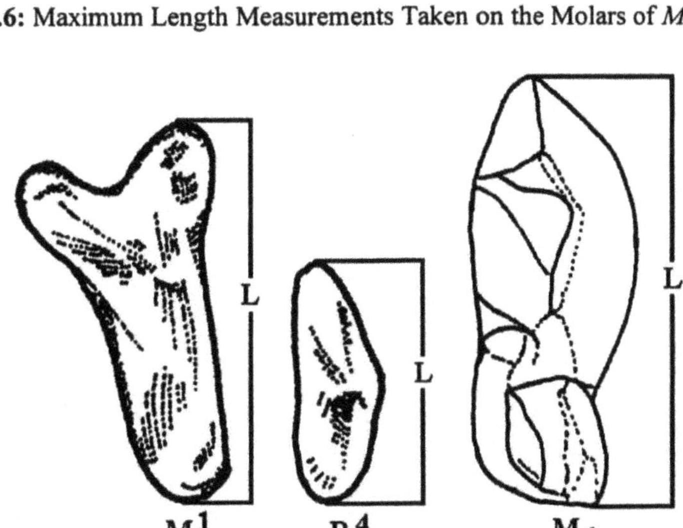

Table 6.3: Comparison of ${}_1M_1$ Length (mm) of *Arvicola.*

Site	n=	Mean	S.D.	Range	*t*	p=
El Juyo (Total)	182	3.73	0.24	3.11-4.65	*	*
A. terrestris (Spain-Modern)[1]	37	3.53	0.17	3.24-4.03	6.16	**<0.0001**
A. sapidus (Spain-Modern)[2]	14	4.87	0.19	4.50-5.20		
A. terrestris (Europe-Modern)[2]	26	4.55	0.21	4.10-4.90		
A. t. terrestris (Poland-Modern)[3]	20	4.18	0.17	3.90-4.44		
A . t. exitus (Poland-Modern)[3]	25	3.91	0.12	3.68-4.10		
A. terrestris (Poland-Holocene)[3]	25	4.05	0.13	3.80-4.41		
A. terrestris (Poland-Late Glacial)[3]	25	4.00	0.19	3.65-4.34		
A. terrestris (Poland-Upper Pleniglacial)[3]	14	4.04	0.17	3.75-4.40		
A. terrestris (Poland-Lower Pleniglacial)[3]	10	4.00	0.17	2.77-4.24		
Arvicola (La Fage)[4]	39	3.58	0.18			
Arvicola (Nestier)[4]	30	3.78	0.23			
Arvicola (Combe Grenal I)[4]	33	3.44	0.12			
Arvicola (Combe Grenal II)[4]	27	3.87	0.13			
Arvicola (Columbière 3)[4]	31	4.30	0.21			
Arvicola (Aitzbitarte-Aurignacian)[5]	12	4.0		3.4-4.3		
Arvicola (Aitzbitarte-Solutrean)[5]	16	4.0		3.8-4.2		
Arvicola (Aitzbitarte-Magdalenian)[5]	22	4.0		3.8-4.3		
Arvicola (Aitzbitarte-Azilian)[5]	18	4.0		3.8-4.3		
Arvicola (Urtiaga-Magdalenian)[5]	18	4.1		3.6-4.4		
Arvicola (Urtiaga-Azilian)[5]	24	4.1		3.8-4.5		
Arvicola (Ekain-Lower Magdalenian)[6]	37	3.97	0.24	3.5-4.5	-5.37	**<0.0001**
Arvicola (Ekain-Upper Magdalenian)[6]	19	3.89	0.17	3.6-4.2	-3.74	**0.0009**
Arvicola (Ekain-Azilian)[6]	20	4.02	0.21	3.6-4.6	-5.74	**<0.0001**

[1]El Juyo Quarry Site-2 (Chapter 4)
[2]Niethammer and Krapp (1982)
[3]Nadachowski (1982:52)
[4]Chaline (1972:344)
[5]Altuna (1972:222-226)
[6]Zabala (1984)

Percentage values (p) in bold indicate statistical significance at or equal to 0.05.

Table 6.4: Comparison of ${}^3M^3$ Length (mm)of *Arvicola.*

Site	n=	Mean	S.D.	Range	*t*	p=
El Juyo (Total)	262	2.37	0.20	1.89-2.99	*	*
A. terrestris (Spain-Modern)[1]	32	2.21	0.18	1.85-2.66	4.70	**<0.0001**
A .t. terrestris (Poland-Modern)[2]	20	2.64	0.11	2.38-2.80		
A. t. exitus (Poland-Modern)[2]	25	2.44	0.14	2.19-2.71		
A. terrestris (Poland-Holocene)[2]	25	2.51	0.16	2.18-2.78		
A. terrestris (Poland-Late Glacial)[2]	25	2.54	0.20	2.18-3.01		
A. terrestris (Poland-Upper Pleniglacial)[2]	10	2.63	0.16	2.35-2.86		
A. terrestris (Poland-Lower Pleniglacial)[2]	10	2.63	0.13	2.45-2.83		

[1]El Juyo Quarry Site-2 (Chapter 4)
[2]Nadachowski (1982:53)

Skeletal separation between these two taxa is difficult (Ventura 1991), due to relatively recent evolutionary divergence (Chaline and Graf 1988). Most useful measurements for separating these taxa (Corbet *et al.* 1970; Reichstein 1963) are not feasible for subfossil material lacking intact crania, mandibles and socketed tooth rows. *Arvicola sapidus* averages the larger of the two in body mass and overall dimensions. Both species are significantly larger than the other microtine species present in Europe (except for the recently introduced muskrat, *Ondatra zibethicus*), so each molar can be identified to this genus.

The $_1M_1$ and $^3M^3$ were chosen for measurement (Figure 6.1.1) on the practical grounds of there being the most comparative information available for these two molars (Table 6.1 and 6.2). In addition, minor morphological variations in them when correlated with length may in the future allow separation of these two species using isolated teeth. No significant difference was found for either molar between the levels of El Juyo. Any fluctuations in the relative proportions of the two possible species were therefore minor.

Histograms of tooth length for $_1M_1$ and $^3M^3$ (Figures 6.2 and 6.3, respectively—see end of chapter) do not show a bimodal distribution for either tooth, indicating that probably one taxon dominates the overall profile. Such distributions might also be produced where two population means were very close, but the actual values (discussed below) indicate strongly that most individuals present at El Juyo belong to *A. terrestris*. Outlying values at the upper end might also belong to the larger species (*A. sapidus*). Comparison of numerical values (Table 6.3) indicates that the dominant taxon is *Arvicola terrestris*. The $_1M_1$ of *A. sapidus* averages consistently larger than that of *A. terrestris*. A modern sample mean for the former species is 4.87 mm. While covering a broad range, sample means for the latter species range from 3.53 mm to 4.55 mm, with most figures clustering at the middle of this range around 4.00 mm. The total mean for El Juyo is 3.73 mm, which falls well within the range for *A. terrestris*. The majority of individuals present therefore belong to this species. The range of $_1M_1$ measurements from El Juyo is broad enough (3.11-4.65 mm) to allow the presence of *A. sapidus* in very small amounts. Since *A. sapidus* has a more southerly modern range, the dominance of *A. terrestris* is expectable in a Pleistocene environment.

Comparison with *Arvicola* populations from other archaeological sites runs into the same problem of division between these two species: other sample means for $_1M_1$ length may also represent the presence of two species. Since these sample means cluster at or below 4.00 mm, these site faunas are probably also dominated by *A. terrestris*. The presence of this species is also more likely in these colder climatic regimes, due to the more southerly modern range of *A. sapidus*.

Comparison of $^3M^3$ length among populations of *A. terrestris* shows great variation for this species. This tooth tends to get smaller from the Pleistocene to the Holocene, which would correlate with a general increase in temperature. Body size may therefore follow Bergmann's Rule in this case and average larger under colder conditions. The largest populations come from Pleistocene Poland, with intermediate values from Holocene and modern populations there. The smallest population (mean = 2.21 mm) comes from modern Spain, which is also the warmest climate among those sampled. The modern Spanish sample has a significantly smaller mean ($t = 4.70$, $P < 0.0001$) than the El Juyo Pleistocene population from the same location. At least locally in Cantabrian Spain, *A. terrestris* has decreased significantly in $^3M^3$ size since the Lower Magdalenian period.

Microtus oeconomus

This species is clearly identifiable by its $_1M_1$, the anterior portion of which has a loop pattern distinct from that of the other members of *Microtus* (and *Pitymys*) present or potentially present in El Juyo. All measurements (Figure 6.1.1) presented in Table 6.5 therefore are assignable to this species only. No significant differences in $_1M_1$ length were found among the levels of El Juyo, nor was the overall distribution of these bimodal (Figure 6.4). Other molars from *M. oeconomus* ($^1M^1$, $^2M^2$, $^3M^3$, $_2M_2$, and $_3M_3$) are not clearly distinguishable from *M. agrestis/arvalis*, and in some cases not from those of *Pitymys*.

Comparison of *M. oeconomus* $_1M_1$ length with other locations (Table 6.6) indicates the El Juyo sample mean is much smaller in this regard, probably reflecting at least in part smaller body size for this population. This measurement also shows large variation among all the sampled populations and no clear correlation with climate. The El Juyo sample is significantly different from that of the Lower Magdalenian levels of Ekain ($t = -3.65$, $p = 0.002$), a Cantabrian site located in the nearby Basque province of Guipúzcoa. The largest sample mean (2.98 mm) comes from modern Poland, while samples from populations living under more extreme (Pleistocene) climatic conditions are smaller. No correlation with Bergmann's Rule is therefore indicated for *M. oeconomus*. No comparison with modern Spain could be made, since this species' range has retreated northward after the warming of the Holocene.

Table 6.5: Maximum Length (mm) of $_1M_1$ of *Microtus oeconomus* from El Juyo.

Level	N	Mean	S.D.	Range
11	1	2.40	*	*
9	1	2.27	*	*
8	7	2.52	0.15	2.36-2.82
7	1	2.57	*	*
6/8	1	2.57	*	*
6	44	2.51	0.11	2.20-2.84
4R	13	2.54	0.17	2.22-2.81
4S	100	2.57	0.15	2.08-2.92
4	27	2.51	0.14	2.23-2.74
2/4	2	2.51	0.06	2.47-2.55
Total	197	2.54	0.14	2.08-2.92

Table 6.6: Comparison of $_1M_1$ Length (mm) of *Microtus oeconomus.*

Site	n=	Mean	S.D.	Range	*t*	p=
El Juyo (Total)	197	2.54	0.14	2.08-2.92	*	*
Poland (Modern)[1]	30	2.98	0.13	2.74-3.37		
Poland (Late Glacial)[1]	16	2.84	0.09	2.64-2.96		
Poland (Upper Pleniglacial)[1]	30	2.77	0.14	2.50-3.19		
Poland (Upper Pleniglacial)[1]	30	2.65	0.09	2.47-2.89		
Germany (Modern)[2]	15	2.71		2.05-2.90		
Aitzbitarte (Aurignacian)[3]	45	2.77	0.21	2.5-3.1		
Aitzbitarte (Solutrean)[3]	103	2.72	0.14	2.3-3.1		
Aitzbitarte (Magdalenian)[3]	140	2.70	0.09	2.4-3.0		
Aitzbitarte (Azilian)[3]	29	2.73	0.26	2.5-3.0		
Ekain (Lower Magdalenian)[4]	15	2.73	0.20	2.2-3.0	-3.65	**0.0024**

[1]Nadachowski (1982:76)
[2]Niethammer and Krapp (1982:377)
[3]Altuna (1972:234)
[4]Zabala (1984:327)

Table 6.7: Maximum Length (mm) of $_1M_1$ of *Pitymys* spp. from El Juyo.

Level	N	Mean	S.D.	Range
11	1	2.63	*	*
9	1	2.36	*	*
8	9	2.47	0.12	2.22-2.66
7	1	2.41	*	*
6	8	2.55	0.14	2.25-2.70
4R	2	2.52	0.30	2.30-2.73
4S	47	2.53	0.12	2.27-2.76
4	11	2.56	0.13	2.35-2.72
Total	80	2.53	0.13	2.22-2.76

Table 6.8: Maximum Length (mm) of $_2M_2$ of *Pitymys* spp. from El Juyo.

Level	N	Mean	S.D.	Range
11	1	1.42	*	*
8	6	1.39	0.08	1.24-1.47
7	1	1.38	*	*
6	4	1.45	0.06	1.39-1.51
4S	41	1.37	0.08	1.20-1.50
4	18	1.36	0.07	1.22-1.46
Total	71	1.38	0.07	1.20-1.51

Table 6.9: Comparison of $_1M_1$ Length (mm) of *Pitymys* spp.

Site	n=	Mean	S.D.	Range	*t*	p=
El Juyo (Total)	80	2.53	0.13	2.22-2.76	*	*
P. lusitanicus (France-Modern)[1]	60	2.65				
P. lusitanicus (France-Modern)[1]	60	2.63				
P. lusitanicus (France-Modern)[1]	8	2.63				
P. pyrenaicus (France-Modern)[1]	20	2.77				
P. pyrenaicus (France-Modern)[1]	16	2.85				
P. pyrenaicus (Spain/France-Modern)[2]	8	2.68	0.09	2.58-2.83	-4.14	**0.0026**

[1]Chaline (1972:342)
[2]Niethammer and Krapp (1982:444)

***Pitymys* spp.**

Two taxa may be represented by the measurements (Figure 6.1.1) for the molars ($_1M_1$ and $_2M_2$) presented in Table 6.7 and 6.8: *Pitymys lusitanicus* and *P. pyrenaicus*. While these teeth are distinguishable from members of *Microtus* by the presence of the "Pitymian rhombus" of $_1M_1$ and a similar confluence of anterior fields in $_2M_2$, they are not distinguishable to species level in this case (Brunet-Lecomte *et al.* 1987; Brunet-Lecomte *et al.* 1985). Metrical analysis was performed to attempt division of these two teeth to species level.

As indicated in Figures 6.5 and 6.6, no clear bimodality exists in the length of $_1M_1$ or $_2M_2$, combining all levels of El Juyo. As indicated in Table 6.9, $_1M_1$ of *P. lusitanicus* averages smaller than that of *P. pyrenaicus*. The former species was present in lower amounts as indicated by the $^3M^3$ of each (discussed below). The $_1M_1$ and $_2M_2$ measurements of *P. lusitanicus* therefore are most likely to fall towards the lower ends of Figures 6.5 and 6.6.

The *Pitymys* $_1M_1$ length sample mean for El Juyo is much lower than other modern populations of both *P. lusitanicus*

Table 6.10: Maximum Length (mm) of $^3M^3$ of *Pitymys lusitanicus* from El Juyo.

Level	N	Mean	S.D.	Range
8	1	1.84	*	*
4R	1	1.83	*	*
4S	10	1.58	0.21	1.40-1.97
Total	12	1.62	0.21	1.40-1.97

Table 6.12: Maximum Length (mm) of $^3M^3$ of *Pitymys pyrenaicus* from El Juyo.

Level	N	Mean	S.D.	Range
9	1	1.76	*	*
8	12	1.73	0.11	1.51-1.88
6	13	1.69	0.18	1.45-1.98
4S	47	1.69	0.13	1.38-1.93
4	16	1.64	0.12	1.42-1.81
Total	89	1.69	0.13	1.38-1.98

Table 6.11: Comparison of $^3M^3$ Length (mm) of *Pitymys lusitanicus*.

Site	n=	Mean	S.D.	Range	*t*	p=
El Juyo (Total)	12	1.62	0.21	1.40-1.97	*	*
Spain (Modern)[1]	49	1.52	0.10	1.31-1.75	1.64	0.13

[1]El Juyo Quarry Site-2 (Chapter 4)

Table 6.13: Comparison of $^3M^3$ Length (mm) of *Pitymys pyrenaicus*.

Site	n=	Mean	S.D.	Range	*t*	p=
El Juyo (Total)	89	1.69	0.13	1.38-1.98	*	*
Spain/France (Modern)[1]	8	1.78	0.11	1.67-1.92	-2.35	**0.043**
El Juyo (*P. lusitanicus*)	12	1.62	0.21	1.40-1.97	1.06	0.31

[1]Niethammer and Krapp (1982:444)

and *P. pyrenaicus* from Spain and France, despite the probable dominance of the latter, larger species in the site sample (Table 6.9). No negative correlation between body size and temperature (Bergmann's Rule) can therefore be supported in this case.

Pitymys lusitanicus

In the case of $^3M^3$, identification is possible to species level for the *Pitymys* taxa present in El Juyo (Brunet-Lecomte *et al.* 1994; Heim de Balsac and De Beaufort 1967; Kratochvil 1970). All metrical data presented in Table 6.10 and Figure 6.7 are for *P. lusitanicus* only. The length of this molar (Figure 6.1.1) was not significantly different from that of a modern Spanish population derived from barn owl pellets in the vicinity of El Juyo (t=1.64, p=0.13), though the sample mean was slightly larger for the former (Table 6.11). Since the complexity of the occlusal pattern of $^3M^3$ varies greatly among microtines, variations in molar length may be more related to intergradations between patterns of occlusal form than overall body size changes.

Pitymys pyrenaicus

As with *P. lusitanicus* above, this species may be differentiated from the other microtines present or potentially present in El Juyo by $^3M^3$ (Brunet-Lecomte *et al.* 1994; Heim de Balsac and De Beaufort 1969; Kratochvil 1970). All metrical data presented in Table 6.12 and Figure 6.8 therefore are for *P. pyrenaicus* only. No significant differences exist between the samples from individual levels of El Juyo. The length of this molar (Figure 6.1.1) was significantly smaller than that of a modern sample from Spain and France (t = -2.35, p = 0.043). The length of this molar was not significantly different from the $^3M^3$ of *P. lusitanicus* from all levels of El Juyo, though sample size for the latter tooth was small (Table 6.13). Separation therefore was upon morphological grounds alone. As with the $^3M^3$ of *P. lusitanicus*, metrical variation in this molar may be more due to slight changes in occlusal pattern than to changes in body size.

Microtus agrestis/arvalis

These two species are not reliably distinguishable by dental criteria, except in the case of $^2M^2$ discussed below. Some subtle differences have been identified among the other molars (Dienski 1969), but these are partially subjective in their application and do not necessarily hold for disparate populations. Other measurement criteria are generally not applicable to archaeological samples (Da Luz Madureira 1983). The measurements (Figure 6.6.1) presented in Table 6.14 for $_1M_1$ include individuals from both microtine species, which may aid in separation between these two species. Since 99 molars ($^2M^2$) from El Juyo are attributable to *M. agrestis* and 393 $_1M_1$ are attributable to *M. agrestis/arvalis*, the expected proportion of *M. agrestis* among these equivocal dental remains is about 25%.

As indicated in Table 6.15, the $_1M_1$ of *M. agrestis* tends to be larger than the $_1M_1$ of *M. arvalis*, though much overlap occurs. Much variability can be seen among modern European

Table 6.14: Maximum Length (mm) of $_1M_1$ of *Microtus agrestis/arvalis* from El Juyo

Level	N	Mean	S.D.	Range
10	1	2.47	*	*
9	4	2.66	0.17	2.42-2.79
8	9	2.69	0.15	2.41-2.86
7	3	2.57	0.17	2.40-2.74
6	18	2.68	0.14	2.41-2.93
4R	2	2.57	0.31	2.35-2.79
4S	95	2.69	0.16	2.26-3.07
4	52	2.66	0.21	2.27-3.23
2/4	2	2.71	0.14	2.61-2.81
Total	186	2.67	0.17	2.26-3.23

Table 6.16: Maximum Length of $^2M^2$ (mm) of *Microtus agrestis* from El Juyo.

Level	N	Mean	S.D.	Range
9	1	1.59	*	*
8	5	1.60	0.08	1.52-1.69
7	1	1.55	*	*
6	2	1.65	0.08	1.59-1.70
4R	1	1.59	*	*
4S	29	1.60	0.07	1.45-1.72
4	7	1.65	0.07	1.57-1.76
2	1	1.56	*	*
Total	47	1.61	0.07	1.45-1.76

Table 6.15: Comparison of $_1M_1$ Length (mm) of *Microtus agrestis/arvalis.*

Site	n=	Mean	S.D.	Range	*t*	p=
El Juyo (Total)	186	2.67	0.17	2.26-3.23	*	*
M. agrestis (Spain-Modern)[1]	19	2.75	0.11	2.57-3.03	-2.31	**0.029**
M. agrestis (Poland-Modern)[2]	269	2.8	0.2			
M. arvalis (Poland-Modern)[2]	103	2.6	0.2			
M. agrestis (Poland-Modern)[3]	30	3.05	0.13	2.83-3.32		
M. arvalis (Poland-Modern)[3]	30	2.67	0.10	2.48-2.84		
M. agrestis (Switzerland-Modern)[4]	19	3.04	0.16	2.8-3.3	-9.19	**<0.0001**
M. arvalis (Germany-Modern)[4]	19	2.63	0.13	2.4-2.95	1.35	0.19

[1]El Juyo Quarry Site-2 (Chapter 4)
[2]Fedyk and Ruprecht (1971)
[3]Nadachowski (1982)
[4]Niethammer and Krapp (1982)

Table 6.17: Comparison of $^2M^2$ Length (mm)of *Microtus agrestis.*

Site	n=	Mean	S.D.	Range	*t*	p=
El Juyo (Total)	47	1.61	0.07	1.45-1.76	*	*
Spain (Modern)[1]	15	1.62	0.04	1.53-1.70	-1.11	0.27

[1]El Juyo Quarry Site-2 (Chapter 4)

populations of *M. agrestis*, with sample means ranging from 2.75 mm to 3.05 mm. *M. arvalis* shows lower variation, with sample means ranging from 2.6 mm to 2.67 mm. The smaller sample mean from El Juyo (2.67 mm) reinforces the conclusion that the majority of individuals represented are *M. arvalis* (expected to contribute about 75% of the total individuals). Variations in occlusal pattern are also known from modern populations (Janossy and Schmidt 1975; Krystufek 1986). No bimodal distribution exists in the length of $_1M_1$ (Figure 6.9), but individuals from the larger size classes are more likely to belong to *M. agrestis.*

Microtus agrestis

Identification to species level is possible for this microtine with the $^2M^2$, which has an additional fifth field on its posterior end that is absent in the other taxa present or possibly present in El Juyo. Separation in this case is possible even from the otherwise highly similar *M. arvalis* (Dienski 1969). As discussed above, 99 total $^2M^2$ from El Juyo can be identified to this species, 47 of which were complete enough for measurement (Figure 6.1.1). Very little variation was detected in the length of this molar (Table 6.16 and Figure 6.10). As indicated in Table 6.17, no significant difference was detected between the total sample from El Juyo and a modern sample from nearby the site derived from barn owl pellets ($t = -1.11$, $p = 0.27$).

Microtus nivalis

This microtine species can be distinguished from the others present by the anterior portion of $_1M_1$. Differences among the other molars are not consistent enough to allow confident separation. Since the presence of this alpine species has important paleoenvironmental consequences (Chapters 5 and 8), the most conservative approach was taken. Measurement follows Figure 6.1.1.

Little statistical comment can be made upon the remains of this species, since only 3 measurable $_1M_1$ have been identi-

Table 6.18: Comparison of $_1M_1$ Length (mm) of *Microtus nivalis.*

Site	n=	Mean	S.D.	Range	*t*	p=
El Juyo (Total)	3	2.95	0.33	2.61-3.27	*	*
Grotte de la Garenne (Late Pleistocene)[1]	34	2.77	0.11	2.48-3.00	0.93	0.45
Switzerland/Italy (Modern)[2]	20	2.93	0.18	2.6-3.2	0.09	0.94
Aitzbitarte (Upper Paleolithic)[3]	10	2.91	0.14	2.7-3.1		
Ekain (Upper Paleolithic)[4]	3	3.03	0.06	3.0-3.1	-0.45	0.70

[1]Chaline (1972:364)
[2]Niethammer and Krapp (1982:265)
[3]Altuna (1972:231)
[4]Zabala (1984:328)

fied from El Juyo. These have been summarized in Table 6.18. This molar has a larger average length than that of *M. agrestis*, *M. arvalis* or *M. oeconomus*. No statistical difference was detected for this small sample from modern or archaeological populations. This uniformity is surprising, given the fragmented Pleistocene relict distribution of this species.

Pliomys lenki
A single molar from Level 4S was identified to this species. Maximum anterior-posterior length (Figure 6.1.1) of this M^2 is 2.11 mm. Identification was possible by the prismatic occlusal pattern, the presence of roots that develop through the individual's lifespan and the absence of cementum between the molar prisms. The occlusal pattern is similar to that found in *Clethrionomys glareolus*, but the $^2M^2$ of this species has cementum between the molar prisms. The $^2M^2$ of *C. glareolus* is also significantly smaller. This molar averaged 1.40 mm (n = 30, s.d. = 0.08, range: 1.19-1.50 mm) in a modern population from Finland (Niethammer and Krapp 1982: 111). The single specimen from El Juyo is significantly larger (t = -46.00, $p<0.0001$). Two $^2M^2$ from Pleistocene Italy measured 1.9 mm and 2.05 mm. This molar averaged 1.92 mm (n = 7, range: 1.87-1.98 mm) for a sample from Pleistocene Poland (Bartolomei *et al.* 1975). Metrical analysis therefore supports the designation of this specimen as belonging to *Pliomys lenki*. As such this identification is the youngest known for the Cantabrian region, the next latest survival of this species being the Upper Solutrean occupation of Amalda (Pemán 1990).

***Apodemus* spp.**
The large molars identified to *Apodemus* ($_1M_1$, $_2M_2$, $^1M^1$, and $^2M^2$) were measured in order to attempt separation between two species possibly present: *A. sylvaticus* (wood mouse) and *A. flavicollis* (yellow-necked mouse). Both murids are common in Europe and can be found in the same general habitat areas. Each is commonly trapped in the environs of El Juyo today (Chapter 4). Their molars are not separable morphologically, especially if advanced stages of wear have been obtained. While inhabiting the same climatic zone, the two species are found in separate local habitats (Chapter 3). Identification to species level therefore affects paleoenvironmental interpretation of immediate site surroundings. No significant differences were detected for the four measured molars (Figure 6.1.2) among the levels of El Juyo, though sample size is generally small (Table 6.19.1-.4).

The biometrical data indicates that both species may have been present at El Juyo. Data from modern and archaeological populations of *A. flavicollis* and *A. sylvaticus* indicate that the former species has slightly larger molars. Sample mean variation among European populations makes species assignment of the El Juyo fauna equivocal. The El Juyo $_1M_1$ total mean is 1.81 mm. Means for *A. flavicollis* range from 1.82 mm to 2.00 mm, with the upper end of this scale more typical. Sample means for *A. sylvaticus* range from 1.64 mm to 1.73 mm, placing the El Juyo sample midway between the normal extremes for these two species. The El Juyo sample therefore most likely represents a mixture of both species, with the larger individuals belonging to *A. flavicollis*. The same results, however, could be obtained if the local population of *A. sylvaticus* was larger during the colder conditions of the Pleistocene. The distribution of these measurements (Figure 6.11) does not show bimodality, though most molars cluster at the lower end of the scale.

Table 6.19.1: Maximum Length (mm) of $_1M_1$ of *Apodemus* from El Juyo..

Level	N	Mean	S.D.	Range
8	2	1.81	0.13	1.71-1.90
6	1	1.84	*	*
4S	18	1.79	0.10	1.66-2.00
4	5	1.89	0.05	1.82-1.95
Total	26	1.81	0.10	1.66-2.00

Table 6.19.2: Maximum Length (mm) of $_2M_2$ of *Apodemus* from El Juyo.

Level	N	Mean	S.D.	Range
8	1	1.26	*	*
4S	8	1.25	0.05	1.19-1.26
4	1	1.17	*	*
3	1	1.38	*	*
Total	11	1.26	0.06	1.17-1.38

Table 6.19.3: Maximum Length (mm) of $^{1}M^{1}$ of *Apodemus* from El Juyo.

Level	N	Mean	S.D.	Range
4S	15	1.88	0.12	1.59-2.10
4	1	1.96	*	*
Total	16	1.88	0.12	1.59-2.10

Table 6.19.4: Maximum Length (mm) of $^{2}M^{2}$ of *Apodemus* from El Juyo.

Level	N	Mean	S.D.	Range
4S	9	1.26	0.10	1.11-1.41
4	2	1.28	0.01	1.27-1.28
Total	11	1.26	0.09	1.11-1.41

Table 6.20.1: Comparison of $_{1}M_{1}$ Length (mm) of *Apodemus.*

Site	n=	Mean	S.D.	Range	*t*	p=
El Juyo (Total)	26	1.81	0.10	1.66-2.00	*	*
A. sylvaticus (Poland-Modern)[1]	15	1.64	0.04	1.57-1.71		
A. sylvaticus (Poland-Late Pleistocene)[1]	15	1.66	0.30	1.56-1.95		
A. flavicollis (Poland-Modern)[1]	15	1.82	0.12	1.62-2.05		
A. flavicollis (Poland-Holocene)[1]	30	1.94	0.09	1.69-2.18		
A. flavicollis (Poland-Late Glacial)[1]	11	2.00	0.06	1.91-2.08		
A. flavicollis (Germany-Modern)[2]	30	1.90	0.08	1.73-2.05	-4.00	**0.0002**
A. sylvaticus (Germany-Modern)[2]	30	1.73	0.05	1.60-1.80	3.76	**0.0006**
A. flavicollis (Spain-Modern)[3]	7	1.83	0.06	1.74-1.92	-0.67	0.51
A. sylvaticus (Spain-Modern)[3]	12	1.76	0.05	1.68-1.84	1.99	0.095

Table 6.20.2: Comparison of $_{2}M_{2}$ Length (mm) of *Apodemus.*

Site	n=	Mean	S.D.	Range	*t*	p=
El Juyo (Total)	11	1.26	0.06	1.17-1.38	*	*
A. flavicollis (Spain-Modern)[3]	7	1.27	0.07	1.17-1.37	-0.48	0.64
A. sylvaticus (Spain-Modern)[3]	12	1.21	0.05	1.11-1.28	2.07	0.052

Table 6.20.3: Comparison of $^{1}M^{1}$ Length (mm) of *Apodemus.*

Site	n=	Mean	S.D.	Range	*t*	p=
El Juyo (Total)	14	1.88	0.13	1.59-2.10	*	*
A. flavicollis (Poland-Modern)[1]	15	1.96	0.10	1.76-2.10		
A. flavicollis (Poland-Holocene)[1]	20	2.11	0.11	1.68-2.26		
A. flavicollis (Poland-Late Glacial)[1]	9	2.09	0.10	1.93-2.23		
A. flavicollis (Germany-Modern)[2]	30	1.99	0.07	1.85-2.12	-3.42	**0.0027**
A. sylvaticus (Germany-Modern)[2]	30	1.80	0.06	1.64-1.95	2.53	**0.021**
A. flavicollis (Spain-Modern)[3]	7	1.92	0.08	1.76-2.02	-0.89	0.38
A. sylvaticus (Spain-Modern)[3]	12	1.83	0.05	1.72-1.98	1.44	0.16

Table 6.20.4: Comparison of $^{2}M^{2}$ Length (mm) of *Apodemus.*

Site	n=	Mean	S.D.	Range	*t*	p=
El Juyo (Total)	11	1.26	0.09	1.11-1.41	*	*
A. flavicollis (Spain-Modern)[3]	7	1.25	0.07	1.14-1.32	0.20	0.84
A. sylvaticus (Spain-Modern)[3]	12	1.21	0.06	1.11-1.32	1.60	0.13

[1]Nadachowski (1982)
[2]Niethammer and Krapp (1978)
[3]Modern Cantabrian sample from trapping (Chapter 4)

Results for $_{2}M_{2}$ (Table 6.20.2) indicate the El Juyo sample mean is more in accordance with modern Spanish samples of *A. flavicollis* than *A. sylvaticus*. A modern sample of the former species was not significantly different from the El Juyo sample mean (t = -0.48, p = 0.64), while the modern sample for *A. sylvaticus* was far closer to statistically significant difference (t = 2.07, p = 0.052). Sample size for this tooth is very small (total n=11), and no bimodality is exhibited in the distribution of its length measurements (Figure 6.12).

The pattern for $_{1}M_{1}$ is repeated for the $^{1}M^{1}$ of *Apodemus.* Total sample mean for El Juyo is 1.88 mm. This molar has a mean of 1.80 mm for a modern population of *A. sylvaticus,* which was not significantly different from the El Juyo sample (t = 1.98, p = 0.068). Sample means for *A. flavicollis* range from 1.96 mm to 2.11 mm (Table 6.20.3). *A. sylvaticus* probably makes up the majority if not all of the El Juyo sample, though the presence of *A. flavicollis* cannot be discounted. The distribution of this sample (Figure 6.13) is bimodal, but the size is very small (n = 14 measurable $^1M^1$). The individuals at the upper end of this range are more likely to belong to *A. flavicollis*. One cannot discount the possibility that *A. sylvaticus* was larger during the Late Pleistocene, so that the slightly higher sample mean of El Juyo is due to changes in this species and not due to the presence of some *A. flavicollis*.

Results (Table 6.20.4 and Figure 6.14) for the $^2M^2$ are similar to those derived for the $_2M_2$: the total El Juyo sample is closer to that derived from modern Spanish samples of *A. flavicollis* than *A. sylvaticus*. The results of the larger molars ($_1M_1$ and $^1M^1$) should be considered more conclusive. These indicate probable assignment of the El Juyo remains to *A. sylvaticus*, though the presence of *A. flavicollis* cannot be discounted.

Sorex coronatus

Remains of this species are separable from those of *S. araneus* by morphological differences in the mandible (Castro Bernardez 1986; Hausser and Jammot 1974; Turni and Müller 1996). In each case where these indicators were present, the remains could be assigned confidently to *S. coronatus*. Since the modern ranges of these two species are mutually exclusive with slight overlap in contact zones (Castién and Gosálbez 1995a; Corbet and Ovenden 1982; Neet and Hausser 1990; López-Fuster *et al.* 1988), it is unlikely that any of the remains present are those of *S. araneus*. Thus, the metrical data in Table 6.21.1-.4 reflect *S. coronatus* alone. The measurements for all shrew species (Figure 6.1.3) follow Buchalczyk and Raczynski (1961) and Buhler (1964). No significant differences were detected among the levels of El Juyo.

Comparison with metrical data from a modern (barn owl pellet-derived) population near El Juyo shows statistically significant difference in one case only. Mandibular height for El Juyo was significantly smaller (t = -4.71, p = 0.0005) than for the modern Spanish population. There was no significant difference (Table 6.22.1-.4) between the El Juyo and modern populations for length of the $_1M_1$, $_2M_2$ and $_3M_3$ for *S. coronatus*. The difference in mandibular ramus height may therefore represent real microevolutionary change in cranial morphology of this species' local presence in Spain. No bimodality of distribution for these measurements was detected (Figures 6.14-6.17).

Sorex minutus

The tooth and mandibular portions of this smallest member of the genus *Sorex* are clearly distinguishable from those of the larger members. The metrical data presented in Table 6.23.1-.4 is representative therefore of this species only and reconfirms identification. Data have been grouped for all levels due to the low sample sizes for these elements. Comparison with a modern sample mean from Eastern Europe indicates the mean mandibular ramus height (Figure 6.1.3) of the El Juyo sample is higher than expected, but the ranges of these two samples are generally similar. A smaller sample from modern France and Germany is significantly larger than the El Juyo sample, indicating that much variation exists in this species across its large range. This size variation may relate to the presence of competitor shrew species, where *S. minutus* assumes a smaller mean size in response to resource pressure by larger competition.

Table 6.21.1: Maximum Ramus Height (mm) of *Sorex coronatus* from El Juyo.

Level	N	Mean	S.D.	Range
8	2	4.12	0.34	3.88-4.36
6	5	4.25	0.17	4.00-4.44
4S	2	4.23	0.16	4.11-4.34
4	2	4.20	0.08	4.14-4.25
Total	11	4.21	0.17	3.88-4.44

Table 6.21.2: Maximum Length (mm) of $_1M_1$ of *Sorex coronatus* from El Juyo.

Level	N	Mean	S.D.	Range
8	1	1.67	*	*
6	8	1.46	0.05	1.39-1.55
4S	5	1.47	0.03	1.42-1.50
4	4	1.40	0.05	1.34-1.45
Total	18	1.46	0.07	1.34-1.67

Table 6.21.3: Maximum Length (mm) of $_2M_2$ of *Sorex coronatus* from El Juyo..

Level	N	Mean	S.D.	Range
8	2	1.30	0.04	1.27-1.32
6	9	1.29	0.08	1.16-1.42
4S	5	1.26	0.04	1.22-1.33
4	3	1.25	0.02	1.24-1.27
Total	19	1.28	0.06	1.16-1.42

Table 6.21.4: Maximum Length (mm) of $_3M_3$ of *Sorex coronatus* from El Juyo.

Level	N	Mean	S.D.	Range
8	2	1.07	0.01	1.06-1.08
6	5	1.01	0.03	0.98-1.07
4S	2	0.98	0.11	0.90-1.05
4	1	1.04	*	*
Total	10	1.02	0.05	0.90-1.08

Neomys fodiens

Metrical analysis (Figure 6.1.3) of this species is limited by the small sample size of measurable elements (n = 2), all of which come from Level 4S. These data are presented in Table 6.24.1-.3. Even with this small sample size, no significant difference was found in ramus height with a modern sample from Germany (t = 0.21, p = 0.84).

Table 6.22.1: Comparison of Ramus Height (mm) of *Sorex coronatus.*

Site	n=	Mean	S.D.	Range	*t*	p=
El Juyo	11	4.21	0.17	3.88-4.44	*	*
Spain (Modern)[1]	64	4.47	0.13	4.13-4.78	-4.71	**0.0005**

Table 6.22.2: Comparison of $_1M_1$ Length (mm) of *Sorex coronatus.*

Site	n=	Mean	S.D.	Range	*t*	p=
El Juyo (Total)	18	1.46	0.07	1.34-1.67	*	*
Spain (Modern)[1]	62	1.42	0.05	1.30-1.51	2.14	.044

Table 6.22.3: Comparison of $_2M_2$ Length (mm) of *Sorex coronatus.*

Site	n=	Mean	S.D.	Range	*t*	p=
El Juyo (Total)	19	1.28	0.06	1.16-1.42	*	*
Spain (Modern)[1]	59	1.29	0.04	1.20-1.39	-.62	0.54

Table 6.22.4: Comparison of $_3M_3$ Length (mm) of *Sorex coronatus.*

Site	n=	Mean	S.D.	Range	*t*	p=
El Juyo (Total)	10	1.02	0.05	0.90-1.08	*	*
Spain (Modern)[1]	48	1.02	0.04	0.91-1.10	-0.22	.83

[1]El Juyo Quarry Site-2 (Chapter 4)

Table 6.23.1: Comparison of Ramus Height (mm) of *Sorex minutus.*

Site	n=	Mean	S.D.	Range	*t*	p=
El Juyo (Total)	8	3.07	0.17	2.71-3.26	*	*
Eastern Europe (Modern)[1]	286	2.91	0.20	2.7-3.1		
France/Germany (Modern)[2]	20	3.32	0.32	2.8-4.1	-2.63	**0.015**

Table 6.23.2: Maximum Length (mm) of $_1M_1$ of *Sorex minutus* from El Juyo.

Site	n=	Mean	S.D.	Range	*t*	p=
El Juyo (Total)	5	1.29	0.07	1.24-1.41	*	*

Table 6.23.3: Maximum Length (mm) of $_2M_2$ of *Sorex minutus* from El Juyo.

Site	n=	Mean	S.D.	Range	*t*	p=
El Juyo (Total)	7	1.11	0.04	1.04-1.15	*	*

Table 6.23.4: Maximum Length (mm) of $_3M_3$ of *Sorex minutus* from El Juyo.

Site	n=	Mean	S.D.	Range	*t*	p=
El Juyo (Total)	7	0.93	0.03	0.89-0.98	*	*

[1]Ruprecht (1971)
[2]Niethammer and Krapp (1990:193)

Table 6.24.1: Comparison of Ramus Height (mm) of *Neomys fodiens.*

Site	n=	Mean	S.D.	Range	*t*	p=
El Juyo (Total)	2	4.68	0.03	4.66-4.70	*	*
Eastern Europe (Modern)[1]	216	4.92	0.25	4.5-5.3		
Germany (Modern)[2]	14	4.67	0.14	4.45-4.9	0.21	0.84

Table 6.24.2: Maximum Length (mm) of $_1M_1$ of *Neomys fodiens* from El Juyo.

Site	n=	Mean	S.D.	Range	*t*	p=
El Juyo (Total)	2	1.81	0.01	1.80-1.81	*	*

Table 6.24.3: Maximum Length (mm) of $_2M_2$ of *Neomys fodiens* from El Juyo.

Site	n=	Mean	S.D.	Range	*t*	p=
El Juyo (Total)	2	1.62	0.04	1.59-1.64	*	*

[1]Ruprecht (1971)
[2]Niethammer and Krapp (1990:335)

***Talpa* spp.**
The remains of two species of this genus are possibly present in El Juyo: *T. europaea* (common mole) and *T. occidentalis* (blind mole). The modern range of these species overlaps in part in northern Spain. Separation of these remains is desirable using metrical analysis, since *T. occidentalis* has a range extending north from the western Mediterranean zone only in this relatively mild portion of the Atlantic zone. *T. europaea* ranges across temperate Europe into Asia (Chapter 3), so the two species mostly occupy regions with very different temperature regimes. Differentiation between their remains therefore affects paleoenvironmental interpretation.

Separation of these two species (Figure 6.1.4) can be obtained using maximum humerus length (Altuna 1972). This extremely robust element can be significantly larger in the generally larger *T. europaea*, though overlap does occur among populations. The presence of the other species may affect overall size for these species, since measurements for *T. europaea* in Spain (thus overlapping with the range of *T. occidentalis*) tend to be larger than in other portions of the former species' range. This size difference may be due to some kind of resource partitioning that reduces direct competition among these two species of mole.

Niethammer and Krapp (1990:116) report that the humerus length for three populations of *T. europaea* from Germany ranged somewhat low: 13.1-14.5 mm (n = 15), 13.6-15.9 mm (n = 18) and 14.5-16.6 mm (n = 14). A population of *T. europaea* from Spain ranged from 15.2-18.2 mm (n = 16). Two *T. europaea* humeri from modern owl pellets near the site of El Juyo (Chapter 4) measured 15.23 mm and 15.02 mm. The humeri lengths from Spanish populations of *T. europaea* do not overlap with this measurement for *T. occidentalis*. Niethammer and Krapp (1990:153) report average humerus length of 13.9 mm (n = 7, range: 13.6-14.6 mm) and 12.7 mm (n = 15, range: 12.1-13.7 mm) for two samples. These measurements do overlap with those of *T. europaea* for non-Spanish populations.

Humerus lengths <14.8 mm most likely belong to *T. occidentalis*, while those >14.8 mm most likely belong to *T. europaea*. Some leeway must be included in these determinations, since mean adult mass for these two species may have differed during the cold periods of the Pleistocene. A low percentage of total *Talpa* humeri remained in an intact, measurable state. This situation may be due to a high proportion of juveniles in the population, with their unfused limb elements succumbing to more destruction. The following measurements (Table 6.25) were taken for the El Juyo material and assigned to these two species.

The presence of both species at El Juyo is indicated. Additional, unmeasured humeri appeared to belong to both species. For example, the lengths of two damaged humeri fell just at the lower end of the range for *T. europaea* and probably belong to that species. Many other fragments were on a scale indicative of their probable assignment to *T. occidentalis*. The other skeletal remains of *Talpa* (teeth, cranial, other postcranial) may therefore belong to either species. Sample size of measurable humeri is not large enough to indicate shifting abundances of the two species per level. The presence of either species in each level therefore cannot be discounted.

Table 6.25: Humerus Length (mm) of *Talpa* from El Juyo.

Talpa europaea	*Talpa occidentalis*
15.90	12.13
16.10	13.24
17.13	13.44

Erinaceus europaeus
The remains of *Erinaceus europaeus* (hedgehog) were examined metrically (Figure 6.1.5) for comparison with remains from other sites and modern specimens. No other member of this genus is found in Spanish Pleistocene sites, and all remains are assignable to this single species. (The presence of *E. algirus* in modern Spain and France is attributable to recent deliberate or accidental introduction by humans.) Very few remains were measurable, the majority of identified teeth being incisors, canines and small premolars for which no comparable data has been found by the author. Measurements were therefore restricted to typically analyzed elements: mandible length, $_1M_1$ length and $^1M^1$ length.

Analysis of these elements (Table 6.26.1-.3) indicates a generally large size for this Late Pleistocene population of *E. europaeus*. Length of the single measurable mandible fell above the range of a modern European sample. $_1M_1$ length was higher than the modern average but not significantly so (t = -1.67, p = 0.17). $^1M^1$ length was significantly larger (t = -5.08, p = 0.015) than for the modern sample. A larger sample from El Juyo may continue the pattern of above average size, which in turn may correlate with colder conditions.

Oryctolagus cuniculus
A single tooth was identified to this species. This molar ($_2M$) from Level 4S has an anterior-posterior maximum length of 2.24 mm. Only one other remain assignable to this species was found in the same level: a fragment of proximal unfused femur. No other statistical comment can be made. This species is found in many Cantabrian Upper Paleolithic sites and

generally in very small amounts compared with other fauna, both macro- and micro- (Chapter 9).

Mustela nivalis

Metrical data from the small sample of measurable teeth (Figure 6.1.6) identified to this mustelid species are presented in Table 6.27.1-.3. Size for this species can vary greatly across its Holarctic range (Chapter 3), so some overlap with the generally larger *M. erminea* may occur. The measurements listed in follow the guidelines of Niethammer and Krapp (1993:XII). These dimensions indicate clearly that the remains are only attributable to *Mustela nivalis*. No statistically significant difference was found between the El Juyo sample of $_1M_1$ and a large modern sample from Germany (t = 0.31, p = 0.78), nor from a small sample from the Magdalenian level of Aitzbitarte (t = 1.93, p = 0.10). A larger sample from El Juyo may allow statistically significant separation.

Table 6.26.1: Comparison of Mandibular Length (mm) of *Erinaceus europaeus.*

Site	n=	Mean	S.D.	Range	*t*	p=
El Juyo (Total)	1	46.81	*	*	*	*
Europe (Modern)[1]	20	42.17	2.03	39.6-46.1	10.24	**<0.0001**

Table 6.26.2: Comparison of $_1M_1$ Length (mm) of *Erinaceus europaeus.*

Site	n=	Mean	S.D.	Range	*t*	p=
El Juyo (Total)	5	6.49	0.58	5.84-7.25	*	*
Europe (Modern)[1]	20	6.05	0.20	5.7-6.4	-1.67	0.17

Table 6.26.3: Comparison of $^1M^1$ Length (mm) of *Erinaceus europaeus.*

Site	n=	Mean	S.D.	Range	*t*	p=
El Juyo (Total)	4	5.92	0.33	5.60-6.34	*	*
Europe (Modern)[1]	20	5.04	0.27	4.6-5.5	-5.08	**0.015**

[1]Niethammer and Krapp (1990:30)

Table 6.27.1: Comparison of $_1M_1$ Length (mm) of *Mustela nivalis..*

Site	n=	Mean	S.D.	Range	*t*	p=
El Juyo (Total)	3	4.35	0.39	3.90-4.60	*	*
Germany (Modern)[1]	92	4.28	0.32	3.6-5.0	0.31	0.78
Aitzbitarte (Magdalenian)[2]	7	3.7	0.67	3.0-4.4	1.93	0.10

Table 6.27.2: Maximum Length (mm) of $_4P_4$ of *Mustela nivalis* from El Juyo.

Site	n=	Mean	S.D.	Range	*t*	p=
El Juyo (Total)	2	1.96	0.09	1.89-2.02	*	*

Table 6.27.3: Comparison of $^4P^4$ Length (mm) of *Mustela nivalis.*

Site	n=	Mean	S.D.	Range	*t*	p=
El Juyo (Total)	1	3.39	*	*	*	*
Germany (Modern)[1]	93	3.73	0.31	3.1-4.4		

[1]Niethammer and Krapp (1993:582-584)
[2]Altuna (1972:276)

Figure 6.2: Distribution of *Arvicola* $_1M_1$ Length (mm) from El Juyo.

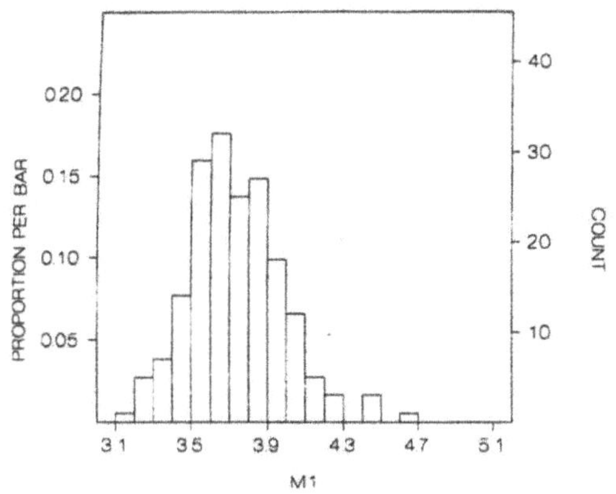

Figure 6.3: Distribution of *Arvicola* $^3M^3$ Length (mm) from El Juyo.

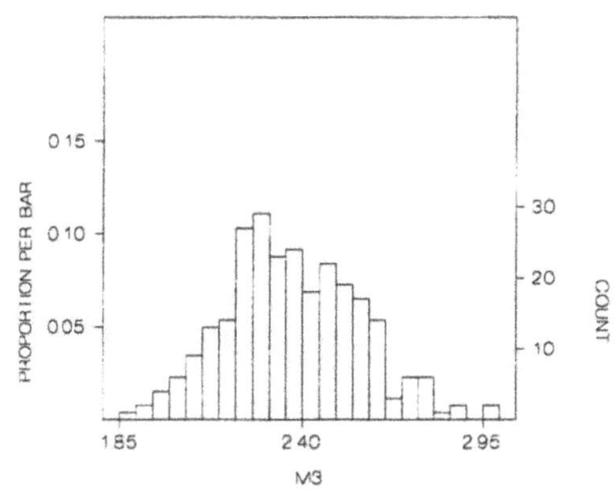

Figure 6.4: Distribution of *Microtus oeconomus* $_1M_1$ Length (mm) from El Juyo.

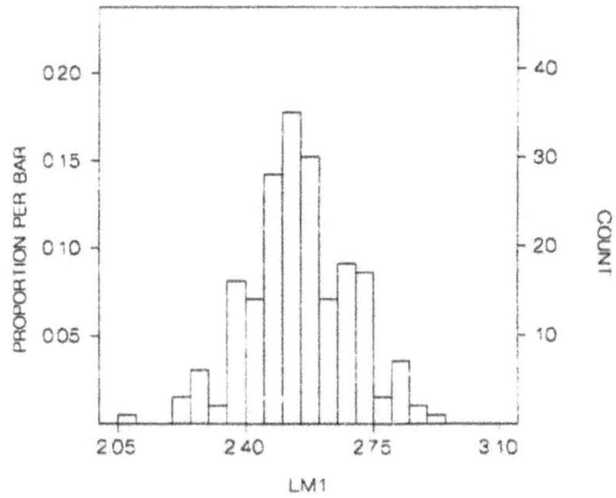

Figure 6.5: Distribution of *Pitymys* $_1M_1$ Length (mm) from El Juyo.

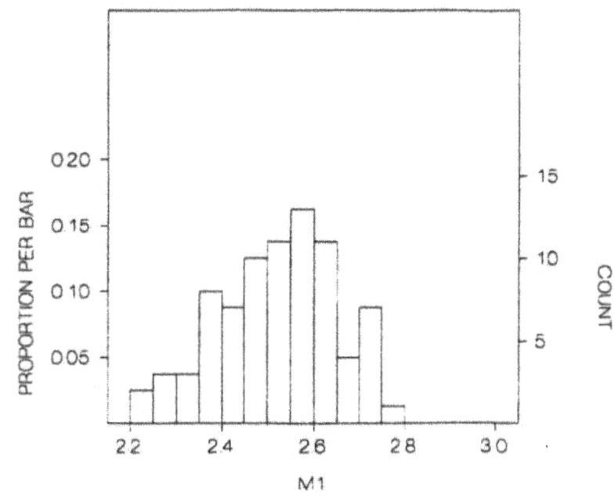

Figure 6.6: Distribution of *Pitymys* $_2M_2$ Length (mm) from El Juyo.

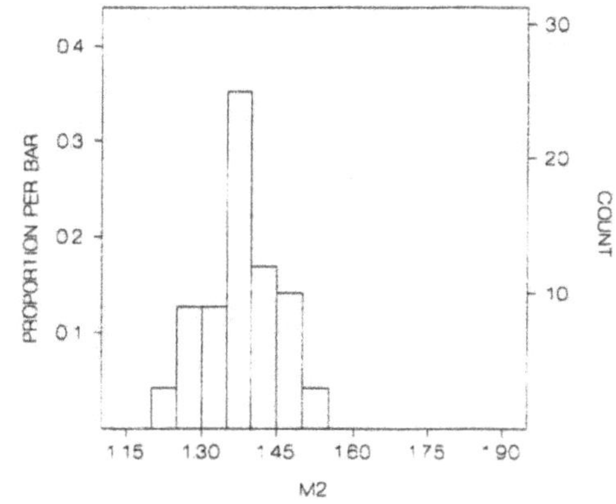

Figure 6.7: Distribution of *Pitymys lusitanicus* $^3M^3$ Length (mm) from El Juyo.

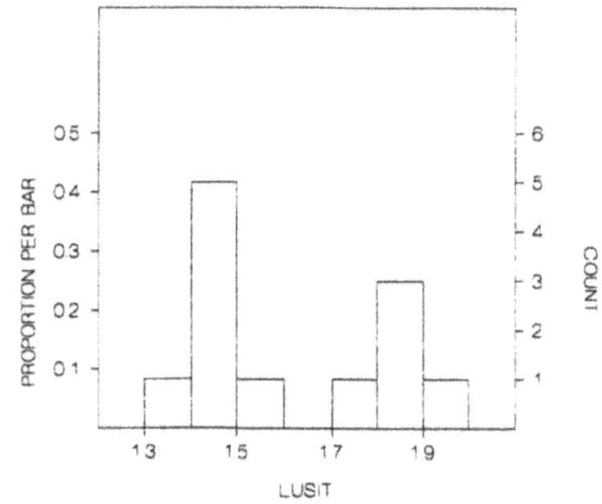

Figure 6.8: Distribution of *Pitymys pyrenaicus* $^3M^3$ Length (mm) from El Juyo.

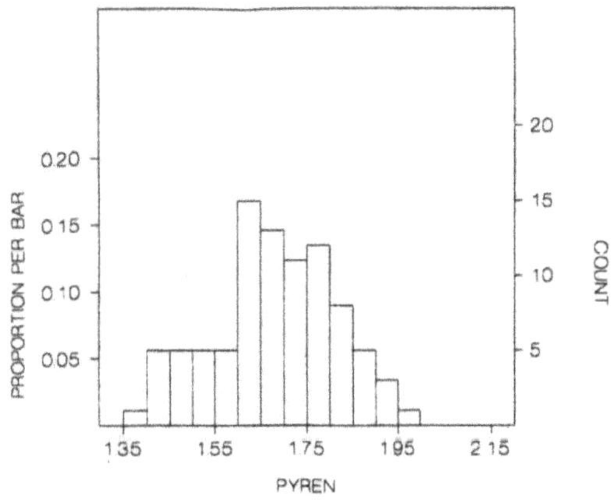

Figure 6.9: Distribution of *Microtus agrestis/arvalis* $_1M_1$ Length (mm) from El Juyo.

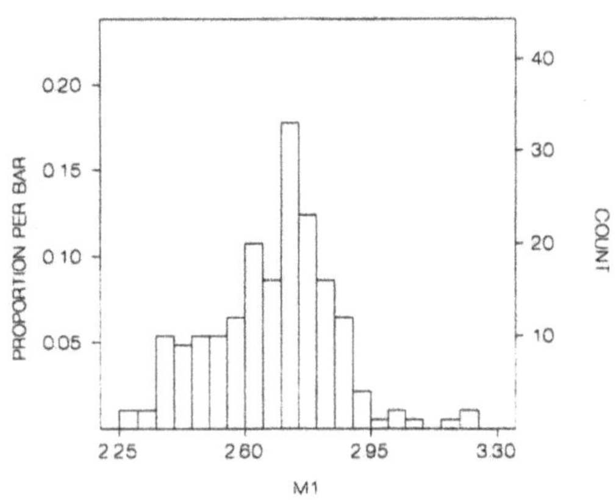

Figure 6.10: Distribution of *Microtus agrestis* $^2M^2$ Length (mm) from El Juyo.

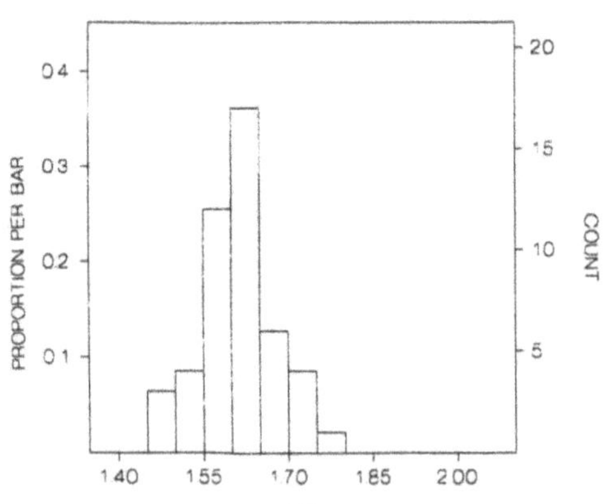

Figure 6.11: Distribution of *Apodemus* $_1M_1$ Length (mm) from El Juyo.

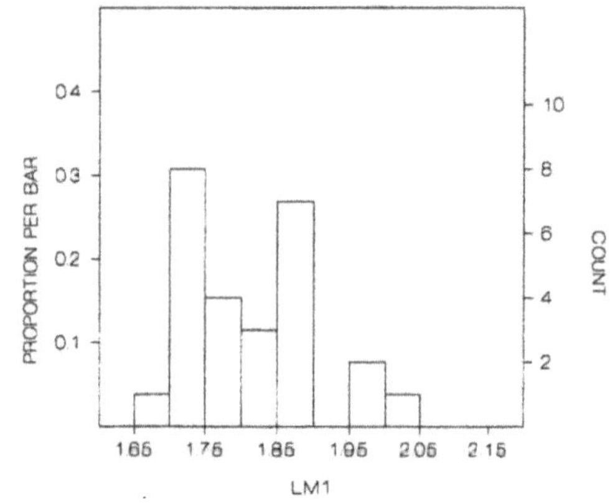

Figure 6.12: Distribution of *Apodemus* $_2M_2$ Length (mm) from El Juyo.

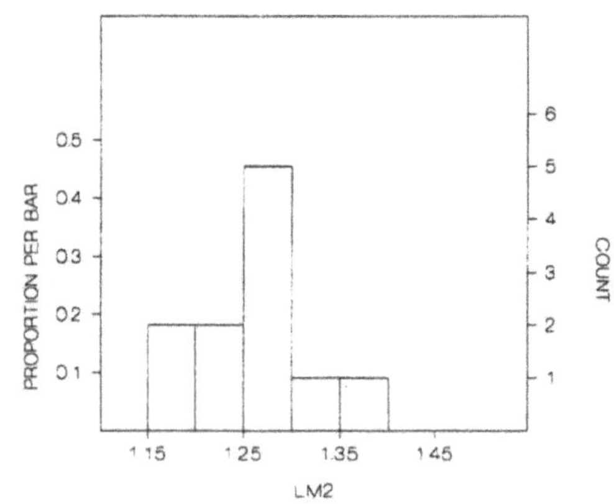

Figure 6.13: Distribution of *Apodemus* $^1M^1$ Length (mm) from El Juyo.

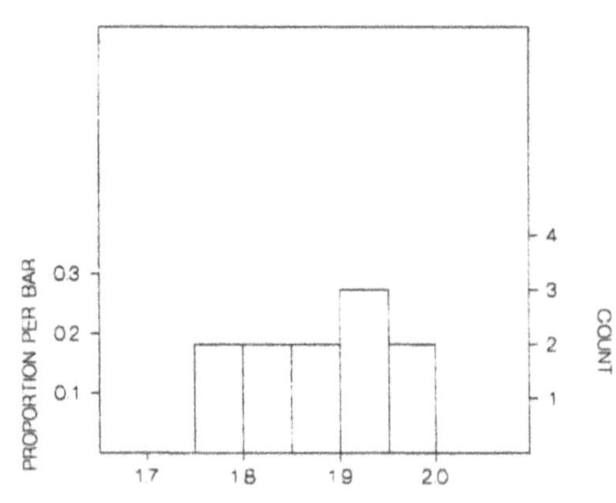

Figure 6.14: Distribution of *Apodemus* $^2M^2$ Length (mm) from El Juyo.

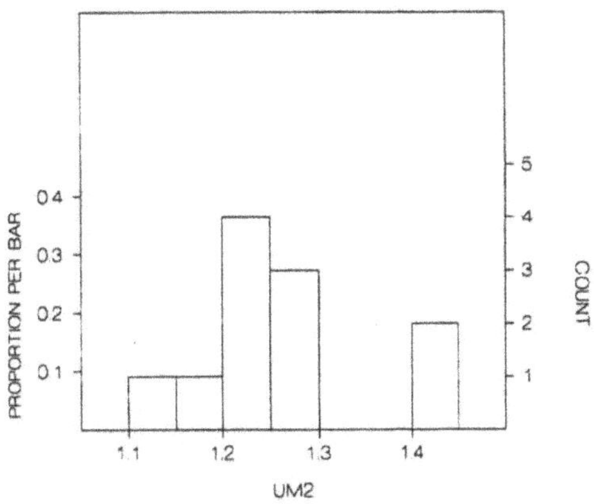

Figure 6.15: Distribution of *Sorex coronatus* Ramus Height (mm) from El Juyo

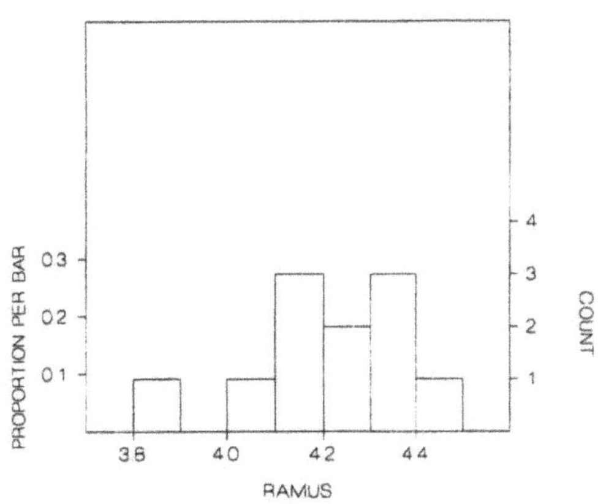

Figure 6.16: Distribution of *Sorex coronatus* $_1M_1$ Length from El Juyo.

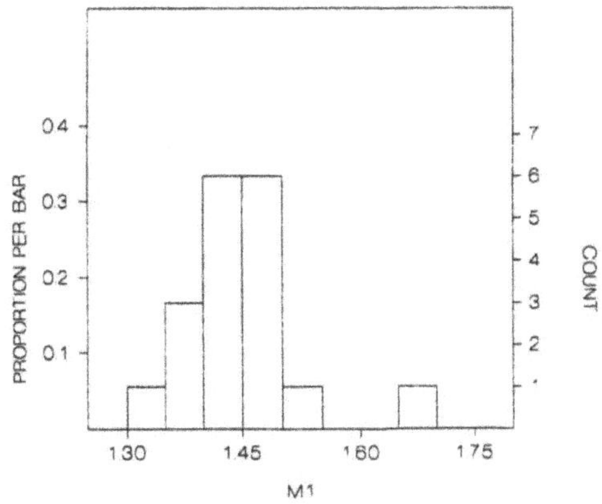

Figure 6.17: Distribution of *Sorex coronatus* $_2M_2$ Length (mm) from El Juyo.

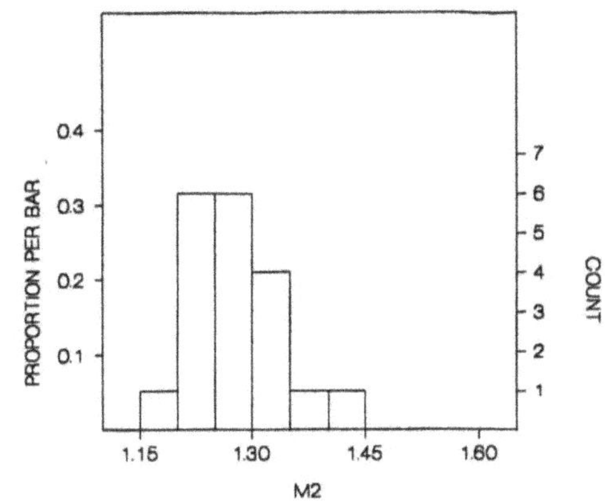

Figure 6.18: Distribution of *Sorex coronatus* $_3M_3$ Length from El Juyo.

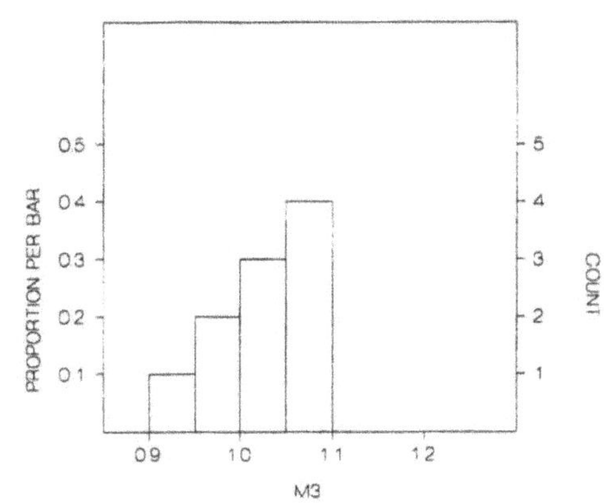

CHAPTER 7
OTHER CLIMATIC INDICATORS FROM EL JUYO

This chapter presents the many other sources of paleoenvironmental data from El Juyo. These are correlation with global ocean level chronology, large terrestrial mammals (red deer, bison, horse, etc.), aquatic resources (fish and molluscs), sedimentology, palynology, and macrobotanical remains (carbonized seeds and other plant remains). The limitations and biases of these methods are also discussed. These multiple sources along with the microfaunal data are synthesized in Chapter 8 for El Juyo and the Cantabrian region to give a more complete paleoenvironmental reconstruction of the Lower Magdalenian and the surrounding periods.

GLOBAL SEA LEVEL CHANGES

A framework for research into Pleistocene environments is best built around changes in global sea level and serves both chronological and environmental considerations. The former is indicated by the simultaneous application to all points on the globe: fluctuations in ocean level in one part of the European coast or elsewhere in the world occurred at precisely the same time. The latter is reflected in what causes these changes in ocean level: the volume of terrestrial fresh water contained in fluctuating ice caps. Lowering of sea levels are thereby taken to indicate a concomitant deterioration of climate. This latter consideration has much more inherent variation, since local climate during a general glaciation of the Northern Hemisphere presents a huge range of extremes. Proximity to continental or alpine glaciers, proximity to coast, latitude, altitude, humidity patterns, and the effects of ocean currents all may affect local conditions. Hence, other data are necessary beyond that of isotopic stage/relative ocean level in order to reconstruct regional paleoenvironments.

El Juyo was occupied during the close of Oxygen Isotope Stage 2, during which time sea level was gradually rising from the low point at the last glacial maximum (Figure 7.1). Global sea levels ca. 22,000-18,000 BP were at least 85 m lower than present-day levels (CLIMAP 1976). Other estimates place sea level from 90 m to 130 m lower than present (see Bailey 1983; Butzer 1983). Along with level, ocean circulation and temperature were very different from the temperate and humid climate found along the northern Spanish coast (CLIMAP 1976; McIntyre *et al.* 1976; Ruddiman and McIntyre 1976; Shackleton and Opdyke 1976; van Donk 1976). Deep sea core foraminifera indicate that North Atlantic surface ocean water temperature at 42^0 North latitude (approximately the Cantabrian coast) averaged 10^0 C colder at 18,000 B.P. than at present. Such a drop indicates polar water conditions with attendant pack ice (McIntyre *et al.* 1976). Since the southward advance of the polar ice cap into the North Atlantic may have deflected the Gulf Stream abruptly southward (Ruddiman and McIntyre 1976), this source of local climatic warming may have been switched on or off over relatively short time periods. This effect is particularly relevant to Cantabrian Spain, which owes in large part its current warm-temperate climate to its coastal proximity and the warm waters of the Gulf Stream. Cantabrian Spain is at the same latitude as New England, yet the two regions are in no way comparable in terms of temperature, snowfall, and year-round vegetation supported.

One may thus posit that increasing glaciation in Europe might manifest itself in a gradual deterioration of local Cantabrian climate, followed by an abrupt plunge into worse conditions as the Gulf stream becomes deflected southward past the Portuguese coast. Air temperatures at glacial maximum for the coast near El Juyo had a February average of around -1.6^0 C and an August average of 10.4^0 C. The nearest modern analog for these weather conditions is the treeless coast of Iceland at 66^0 N latitude (Butzer 1986:215).

Terminal moraines left by alpine glaciers during the last glacial maximum in the Cantabrian Cordillera reached as low as 550 m above sea level (Clark 1981:7). During the harshest climatic conditions of El Juyo's later occupation, global sea level data agrees with the interpretation from microfauna that the terrain around the site had much less forest cover and much more open grassland/heathland than modern, unaltered conditions allow.

SEDIMENTOLOGY

Sedimentological analysis of El Juyo deposits indicates a return to colder conditions during the time of deposition of the upper habitation levels (Freeman *et al.* 1988:11-12), which follows an earlier period of drier, more temperate conditions in Levels 13-7. At the time of deposition of Level 6, humidity had increased such that the water table of the cave rose to this level and caused the deposition of silt, especially in the dugout structure. Both Level 5 (a sterile flowstone layer) and Level 4 show an increase in *éboulis*, taken to indicate an increase in cold regime due to spalling of the cave ceiling. Level 3 was formed by the collapse of the main entrance followed by further sealing of the vestibule deposits with a layer of travertine, suggesting a climate regime of high humidity followed by a period of cold, dry conditions.

Assessments of cold conditions are primarily based upon the degree of cryoclastic spalling giving a relative indication of increasing in cold conditions. Any paleoenvironmental reconstruction, however, is based upon extrapolation from paleotemperature regimes and analogy with present-day cold climates (Butzer 1986). Human activity has a generally untested potential to alter sediment deposition in ways that may be mistaken for changes in environment. Increases in fluvial deposition of fine particulates (clay, silt) is normally interpreted as indication of expanded runoff due to an increase in humidity. Human trampling of the vegetation in front of a cave site during an extended occupation season may cause a similar sediment influx. Denuded slopes are far more unstable than vegetated ones, which hold the upper soil layers in place with a mesh of roots and provide protection from the

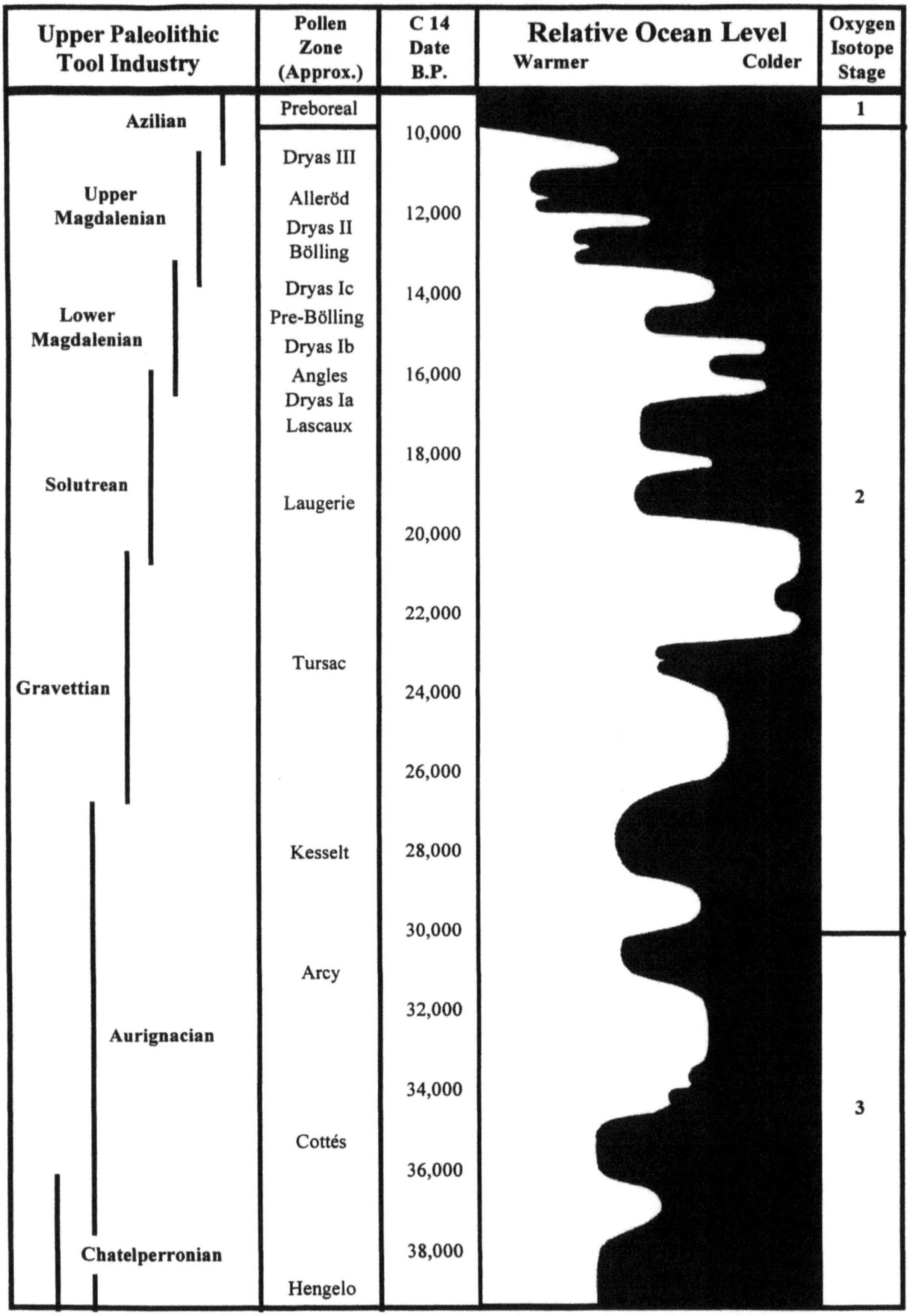

Figure 7.1: Relative Sea Level During the Upper Paleolithic (adapted from Sánchez Goñi 1993:144).

direct impact of rain with a foliage canopy and a layer of humus. This situation is particularly true for a depositional setting like that of El Juyo, where the substantial cave entrance is situated at the base of a large, funnel-shaped sink hole that would channel runoff naturally in its direction. An increase in the degree of cryoclastic spalling is also generally interpreted as an indication of increasing cold conditions (Farrand 1975). Larger and more frequent chipping away of the cave ceiling is seen as the direct result of colder average temperatures. This theory too lacks adequate testing. Spalling occurs most frequently on rock surfaces that are naturally fissured. The parent rock of the cave itself (including how fossiliferous it is) may therefore variably affect spalling rates. Spalling also occurs with greatest ease where the freezing point is passed the most frequently. Water seeps into microfissures and is frozen, causing it to expand into the crack and force it farther apart. Subsequent thawing will then allow the water to seep deeper into the microfissure and for the process to continue as the freeze/thaw cycle continues. Spalling should therefore be greatest where the freezing point is passed back and forth the most times, not where temperature is coldest. Humans have the potential to alter this freeze-thaw regime of a cave or rockshelter ceiling. The erection of barriers to air flow may alter the natural moderation of cave temperature. The use of fire in a cave has the potential to pass the ceiling through the freeze/thaw point as frequently as once a day, especially in the daylight zone where El Juyo's occupation debris was accumulated. These effects need greater understanding through experimentation. Sedimentology thus is problematic as a faithful reflection of climate change.

Paleoclimatic interpretation from sedimentology is in partial disagreement with that derived from microfaunal data. The former indicates a relatively dry and temperate climate through Levels 13-7 (Freeman *et al* 1988). The microfauna sample of these levels, while small, indicates high humidity accompanying relatively warmer temperatures. This high humidity continues through the upper levels (6-4). The two sources of data agree in the interpretation of deterioration during the deposition of Level 4, with its increase in *éboulis* and the appearance of an alpine species in the coastal lowlands (*Microtus nivalis*).

LARGE FAUNA

The large fauna of El Juyo and other Cantabrian sites have been affected by cultural biases in their selective harvest, making their relative proportions generally unsatisfactory for paleoenvironmental reconstruction. They are useful as indicator species, however, and complement other sources of data. The presence of reindeer (*Rangifer tarandus*), a tundra and boreal forest dwelling species would be of particular importance in Cantabrian Spain where they are rarely recorded during the Pleistocene. Similarly, the presence of open range species such as bison (*Bison bonasus*) and horse (*Equus caballus*) indicates the existence of grassland, though not necessarily in close proximity to a site. Both boar (*Sus scrofa*) and roe deer (*Capreolus capreolus*) indicate the presence of deciduous forest. The principal prey species at El Juyo (and in general throughout the Upper Paleolithic in Cantabrian Spain) are red deer (*Cervus elaphus*), which thrive across a range of latitudes and environments throughout the Holarctic and are indicative of no one habitat. A mosaic environment of parkland and open deciduous forest does allow the greatest abundance for this species (Corbet and Ovenden 1982:202), as this pattern increases microhabitat edge effects and the attendant increase in browsable vegetation. The presence of alpine ibex (*Capra pyrenaica*) in El Juyo in significant numbers says little about the paleoenvironment. This prey species was undoubtedly manually transported from its upland habitat after its acquisition and may be found in the Cantabrian Cordillera today within a relatively short distance from El Juyo. Larger species of microfauna (e.g. lagomorphs or small mustelids), while at the upper limits of owl prey, may also have been food or fur resources for humans and must be critically evaluated for evidence of their inclusion in the site having been culturally affected. Large fauna must therefore be considered of secondary importance in paleoenvironmental interpretation.

Table 7.1: Large Fauna from El Juyo.

	Level 4 NISP/MNI	Level 6/7 NISP/MNI	Level 8/9 NISP/MNI
red deer *Cervus elaphus*	3925/44	2627/78	972/15
roe deer *Capreolus capreolus*	178/4	85/4	78/5
bison *Bison bonasus*	94/4	32/2	25/1
horse *Equus caballus*	79/3	31/1	34/3
ibex *Capra pyrenaica*	40/2	47/4	2/1
boar *Sus scrofa*	1/1	1/1	
lion *Panthera leo*			2/1
wolf *Canis lupus*	1/1		

(Freeman *et al.* 1988:29)

Excavation at El Juyo is ongoing, so the faunal counts reported in Table 7.1 will change substantially in reports that follow. One additional large species now detected in very small amounts is brown bear (*Ursus arctos*). The general proportions of represented species, however, have not changed with the accumulation of additional (unpublished) faunal data. Hence the basic pattern of this earlier summation has not been altered, but in fact reinforced. The faunal profile is dominated by red deer, which in each level are as abundant in both NISP and MNI as all the other large mammal species combined. Found in generally similar amounts are horse, roe deer, bison and ibex, with trace amounts of boar, lion and wolf. Large mammal hunting at El Juyo was therefore centered upon red deer, which into historical times was the most abundant large terrestrial species available in Cantabrian Spain. This situation is very different from that of Late Pleistocene France, where the more gregarious reindeer was the primary large terrestrial mammal hunted during times of deteriorated climate.

The deer harvested in Levels 4 and 6/7 show catastrophic age profiles (Klein *et al.* 1981), indicating that whole herds may have been taken all at once. The implication for hunting behavior is that of cooperative drive and surround techniques to allow efficient, mass harvest of one primary species. This trend had already begun in Cantabrian Spain by the Solutrean and had accelerated into the Lower and Upper Magdalenian periods (Straus 1977, 1992; Freeman 1981). This pattern of exploitation is also reflected in the osseous projectile technology of this period (Pokines 1993, in press). Subsequent excavation of more material from Levels 8/9 may show that a similar pattern existed there regarding the mass harvest of red deer. These levels so far have been notable for their abundance of molluscan remains and approach the density of a true midden in this regard. Human choice in faunal acquisition has thus played a large role in determining the proportions of large mammals and molluscs present. As such their relative abundances may not accurately reflect the paleoenvironment.

The large fauna profile does generally agree with that derived from microfauna. These species are compatible with a mostly open heathland with patches of forest and grassland. Proximity to the alpine zone is supported by the presence of ibex in small amounts. However, no significant shift in the relative abundances of these large fauna species is noted over the course of Lower Magdalenian habitation.

AQUATIC FAUNA

While plentiful at El Juyo and most Cantabrian Late Upper Paleolithic and Mesolithic sites (Krupa 1994), molluscan remains are of lesser paleoenvironmental significance than their abundance would suggest. The two primary harvest species, periwinkles (*Littorina littorea*) and limpets (*Patella vulgata*) are tolerant of a broad temperature range and are common along much of the European Atlantic coast. Species endemic to cold water are found in small numbers at El Juyo, but these may have been washed up on the shore from a greater depth and collected for their aesthetic or curiosity value (as were the many fossils found at El Juyo).

Molluscan data (Table 7.2) accumulated thus far from El Juyo has been summarized by Krupa (1994) and continues to be analyzed as part of the ongoing investigation at this site. Incoming molluscan data are very unlikely to alter the basic pattern already established by the thousands of remains already identified. Of these species, three are indicators of a cold water environment: *Buccinum undatum* (Level 4), *Trivia arctica* (Levels 4 and 6), and *Cyprina islandica* (Level 6). The simultaneous presence in level 6 of two warm-water species, *Nassarius mutabilis* and *Dentalium vulgare*, indicates the problematic nature of this data source. The presence of the latter two species could be explained by their transport from the Mediterranean coast (Krupa 1994:41-42).

Most marine food harvest at El Juyo was selectively focused upon *P. vulgata* and *L. littorea*. These two species were utilized with increasing intensity during the Late Upper Paleolithic in Cantabrian Spain (Straus 1992; etc.), culminating in the shell middens of the (Mesolithic) Asturian period (Clark 1971). A similar pattern of intensive exploitation of these resources was discovered at the Late Upper Paleolithic levels of Altamira (Altuna and Straus 1976) and La Lloseta (Clark 1971). These same mollusc species are also the predominant forms at Tito Bustillo (Moure Romanillo and Cano Herrera 1986), El Castillo (Cabrera 1984), Cueto de la Mina (Vega del Sella 1916), Otero (González Echegaray *et al.* 1966), and La Riera (Straus and Clark 1986). Their utilization parallels that of red deer, in that over the same time period their intense harvest was intensified beyond that of more generalized, random use (Chapter 10). In some of these sites we see "wild harvest" (Freeman 1973, 1975) or intensive utilization of highly productive key resources that were very abundant in the local Pleistocene environment.

Table 7.2: Molluscan Remains from El Juyo

Taxon	Level 4	Level 6	Level 8
Patella vulgata	8483	6588	6677
Littorina littorea	4182	1357	2387
L. obtusata	48	36	12
L. neritodes	1		1
L. saxitalis	1	1	2
Gibulla sp.	1		
Aporrhais pespelicani		1	
Buccinum undatum	2		
Buccinum sp.	2		
Turritella communis		1	
Trivia arctica	3	2	
Nassarius mutabilis		1	
Nucella lapillus	2		
Chlamys varia			1
Chlamys sp.			1
Dentalium novemcostatum	2	1	
D. vulgare		1	
Pecten maximus	1	2	
Teredo navalis		1	
Cardium edule	1		
Cyprina islandica		1	
Macrocallista chione		1	
Mytilus sp.	2		
Skenea planorbis	1		
Janthina janthina			1
Helix sp.	9	3	
Azeca goodalli	1		
Azeca sp.	1		

(Following Krupa 1994:35)

Riverine resources were also important at El Juyo (Krupa 1996; Pokines and Krupa 1997). The site has currently yielded over 500 intact vertebral centra and many times this amount of vertebral fragments of large fish species, primarily Atlantic salmon (*Salmo salar*). These remains await full identification. Far more startling are the additional remains of fragmentary fish ribs and spines yielded by the fine screening of all sediments from this site. Nearly 45,000 of these fragments, averaging about 1 cm in length, have been recovered to date from all levels and areas. These remains are extremely fragile (Lubinski 1996), far more so than terrestrial microfauna recovered from the same levels, and therefore must represent only a small fraction of all fish remains introduced to the site. A complete lack of the far more durable otoliths and only the tiny representation of cranial elements (primarily isolated teeth) from the Magdalenian levels indicates that initial processing probably took place away from the site. Fish exploitation is also indicated by the presence of arc-shaped points that when lashed laterally to a shaft form a projecting, rear-facing barb. This point design was later replaced by multiply barbed harpoon points during the Upper Magdalenian and Azilian periods (Pokines and Krupa 1997). Many other Cantabrian Late Upper Paleolithic levels have yielded fish remains, though in much smaller amounts, probably due to differences among excavation recovery systems. These include most notably La Riera, the remnant stratigraphy of which has yielded 81 identified *Salmo* remains from its Late Upper Paleolithic sequence (Straus and Clark 1986).

Paleoenvironmental data derived from these fish remains is limited. Atlantic salmon were undoubtedly taken during their entry to freshwater before the spawning season (November to January) when these fish have returned to their birthing streams to mate (Netboy 1980:38). Unlike Pacific salmon (*Oncorhynchus* spp.), Atlantic salmon do not necessarily die after spawning and may return to the ocean, only to return to their birthing stream following spawning season. High mortality makes more than two spawning runs a very rare occurrence. Harvest most likely took place along the Río Pas, the nearest substantial river to El Juyo which today still yields very small, controlled harvests of salmon (Diputación Regional de Cantabria 1994). Spain and northern Portugal are at the modern southernmost extreme of Atlantic salmon's European range, which may have extended farther south during the colder episodes of the Pleistocene. The presence of salmon in large amounts therefore is not a useful indicator of past climate, as these fish thrived (in historic times) across a range of latitude that today extends as far as northern Norway.

POLLEN PROFILES AND TAPHONOMY

The interpretation of pollen and spores derived from archaeological sites and other depositional settings must take into consideration the multiple taphonomic processes undergone by these microscopic remains. With a notable recent exception (Sanchez Goñi 1993), this question has been scarcely dealt with in the assessment of pollen profiles from Cantabrian cave sites. The problems of pollen taphonomy are extensive enough even to inhibit their use as indicator species and even more greatly inhibits the use of their numerical proportions as direct environmental gages (though see Hevly 1981 for a rebuttal).

Differences in pollination syndrome are a major cause for bias in a site pollen diagram (Dunwiddie 1987; Faegri and Iversen 1989; Prentice 1986). Different species have evolved multiple methods for insuring that propagation occurs. This means that their pollen has very different properties and abundances. Bee, fly, humming bird, and lepidopteran pollinated species tend to produce lesser amounts of large, sticky pollen designed to be picked up by the pollinator and brushed off on the stigma of the same or another flower of the same species. Wind pollinated plants, by contrast, produce relatively huge amounts of buoyant pollen broadcast in the eventuality that some of it alights on the stigmas of the proper species. Long distance, aerial transport is a natural function of the pollen's evolutionary design. Direction of prevailing winds is therefore an additional factor in pollen profile interpretation (Candau and González Minero 1992), which is of great relevance to the altitudinally zonated, riverine bisected, and coastal Cantabrian region. One may generally expect that 59% of the pollen introduced into a site is local in origin (0-20 m), neighborhood pollen (20-500 m) constituting 10% and regional pollen (500 m-10 km) constituting 30%. Long-distance pollen (>10 km) generally accounts for about 1% (Sanchez Goñi 1993:24).

Ironically, the relative indestructibility of pollen and spore grains itself contributes to their taphonomic problems. Traverse (1990) noted previous studies where pollen was found to have been transported as far as 1400 km downstream. His own study of the Trinity River (Texas, U.S.A.) showed that palynomorphs from water samples included species reworked from weathered Paleogene rocks. One taxon of reworked pollen (*Engelhardtia*) showed up in almost every sample taken over the course of years. One therefore runs the risk, even in cave deposits, of ancient or simply distant pollen being introduced by flowing water and air currents.

Differential destruction is another serious problem. Grain corrosion (to the point of non-identifiability or destruction) is due to microbial attack, chemical oxidation, heating, or mechanical abrasion (Havinga 1967). While fire or sufficient mechanical buffeting will destroy any pollen grain, relative durability does vary by species. The former two destructive factors have a much greater potential to bias relative proportions of palynomorphs. While such factors as the physical structure of the exine (its degree of surface area subject to attack, its distinctiveness of shape, etc.) and the physical and chemical structure of the medine affect survivability, the sporopollenine content of the exine is by far the most important factor. This substance is extremely resistant to breakdown. Hence, variations in its concentration affect greatly the potential for palynomorph grains to survive corrosion to the point of identifiability. Havinga (1967:87-90) found that pollen grains of *Alnus*, *Corylus* and *Myrica* buried in river clay and leaf mould for 20 months underwent major corrosion. Only 4.2% (in river clay) and 1.8% (leaf mould) of *Alnus* survived without corrosion. These figures, respectively, for *Corylus* were 1.8% and 1.4% and for *Myrica* were 1.8% and 0.9%. Only rarely were grains completely destroyed. In other tested soil types no significant decline was observed, nor were significant declines noted in river clay and leaf mould for the 16 other pollen and spore species tested. Kwiatowski and Lubliner-Mianowska (1957, cited in Havinga 1967) also found a significant negative correlation between oxidation susceptibility and the sporopollenine content of pollen exines. Differential corrosion of palynomorphs therefore has a huge potential to alter their relative proportions in a site profile.

Palynomorph grains also suffer from differential transport. In a laboratory study, Holmes (1990) found differential buoyancy caused by varying amounts of trapped air bubbles in the surface structure of the exine of pollen grains. Already saturated grains introduced to a water flume showed sorting based upon size differences. At 25 cm/sec velocity, all pollen grains remained in suspension. Differential deposition with varying velocity was noted. Holmes also found in a field study of a shallow pond in England with a single inflow stream that differential deposition occurred. Overall pollen size range was 16 to 80 microns. The largest pollen types (Filicales, Gramineae) were deposited closest to the stream mouth. Mid-size pollen (*Corylus*) were deposited in greater amounts away from the stream mouth, and the smallest taxa (Urticaeae) were deposited in the greatest amounts in more distant sediments. Sediment type may also affect redeposition, in that sand-sized grains generally do not permanently trap pollen grains and allow them to be winnowed out and resuspended. Shape of pollen grains may also affect their differential transport, such as in the case of saccate shaped *Pinus* pollen grains.

Davis (1968) also found that seasonal water circulation causes redeposition of pollen grains. Shallow water sediment was stirred up and redeposited more than deep water sediment, the net result of repeated suspension being the movement of material from shallow to deep water. Davis further found that 80% of pollen input from sediment traps was out of season and therefore redeposited.

Sorting by grain size and hence correlation with sediment grain size has been reconfirmed by other researchers. Goodwin (1988) found that size sorting occurred in sampled Holocene glacio-lacustrine sediments. The percentage of nonarboreal pollen was greatest in fine deltaic sands, and arboreal pollen was preferentially concentrated in lacustrine silt and clay. Spruce (*Picea*) was most abundant in coarse silts, and alder (*Alnus*) was most abundant in fine silts and clays. Similarly, Fall (1987) found size sorting of the pollen deposited in a canyon stream (4 Corners area, Arizona, U.S.A). *Pinus* strongly correlated with clay sized particles from alluvial samples, suggesting that these were carried in suspension and settled out at low velocity or turbidity. *Quercus* and *Populus* were likewise correlated. Larger grained taxa such as Artemisia and Chenopodiaceae correlated positively with sand. Corroded (and therefore transport-damaged) grains also correlated with sand. *Juniperus*, *Ephedra* and *Populus* showed positive correlation with silt deposits.

Artificial introductions may also alter the picture of the natural pollen rain determined for a site. While the introduction by humans of wild plants to a campsite is of archaeological relevance of itself (Behre 1986; Hevly 1981), the danger of obscuring the local pollen rain pattern remains. Plants brought in as bedding, construction, food, or pharmaceutical material may leave artificially high levels of pollen behind. Similarly, animals entering a cave site of their own accord can differentially introduce insect-pollinated pollen types from the local environment. Humans may also alter the air circulation patterns in caves through the construction of tent or wall structures designed to prevent drafts or water dripping. Their fires may also alter air circulation patterns on a seasonal or circadian cycle, as well as destroy pollen already deposited. Human and animal feces deposited in a cave will also increase the overall levels of edible species' pollen. Small-scale human habitat alteration in the form of vegetation trampling immediately in front of a cave may also affect pollen fallout, given the proportion of pollen that originates in the immediate environment. These factors all have the potential to alter relative pollen levels within a site and hence the environmental picture derived from them.

As discussed above, water transport has the potential to rework and redeposit pollen already in a cave. In a depositional setting such as El Juyo, where multiple layers of travertine have been laid down directly over archaeological deposits and where episodes of flooding have occurred, the assumption that pollen grains may be associated directly and unambiguously with their level of correspondence is extremely tenuous. Similarly, the manner in which pollen samples are taken and processed is a potential source of bias. Sediment samples are taken typically from one freshly-exposed column from one location in a site. The assumption that the pollen taken from a few cubic centimeters within a level is representative of that entire level remains to be adequately tested. Furthermore, Havinga (1967) reports that ultrasound preparation of laboratory samples can damage or destroy pollen at certain frequencies. Some analyses are also faulted by non-statistically significant levels of pollen identification (see Sanchez Goñi 1993). Relative identifiability of species and the variability of climatic tolerances within higher taxonomic levels (if a given grain may not be identified to species level, which is a common occurrence) may also bias interpretation for or against certain taxa.

Some researchers attempt to control for some of these biases by developing pollen signatures for given types of biomes. This tactic involves sampling the local pollen rain from a variety of locations, on the assumption that variances in transport and production will cancel themselves out when comparing fossil and modern pollen profiles. Variation in survivability may also be compensated for by experimentation with pollen corrosion.

Palynological analysis has been undertaken for El Juyo (Boyer-Klein and Ar. Leroi-Gourhan 1987; Ar. Leroi-Gourhan 1994; Sanchez Goñi 1993), as well as for a suite of Cantabrian sites. These include Otero, Cueva Morín, and La Riera (Ar. Leroi-Gourhan 1966, 1978 and 1986, respectively); Tito Bustillo, Chufín, Rascaño and Eralla (Boyer-Klein 1976, 1984, 1981, 1985); Los Azules (López 1981); Ekain and Amalda (Dupré 1984, 1990); and Lezetxiki, Labeko and Urtiaga (Sanchez Goñi 1993). The information presented here concerning El Juyo will follow the most recently updated interpretation (Ar. Leroi-Gourhan 1994) of the El Juyo pollen. The analyzed samples were taken along the 3.5 m section of the cleaned test trench of the 1955-57 excavations (Janssens and González Echegaray 1958). Analysis of 39 of these yielded 5651 identified pollen grains from Levels 12 through 2. Levels 12 through 7 of El Juyo are interpreted as having been deposited during the Pre-Bölling (relatively temperate) phase, while levels 6 through 4 are placed within the Dryas I (dry and very cold) phase. Level 7 has a carbon date of 14,440 +/- 180 BP and level 4 has been dated at 13,920 +/- 210 BP.

In general, non-arboreal pollen dominates the spectrum, even in the lower levels of the site, and includes large proportions of Graminae (grasses) and Ericaceae (heather). Arboreal pollen oscillates around 25% of the total per sample through Levels 12 through 7, then drops off to lower levels in Levels 6 through 4. After a hiatus, arboreal pollen for (Bronze Age and more recent) levels 2 and 3 has returned to higher levels around 35%. Levels of *Pinus* (pine) pollen show a slight tapering from level 12 through 7, with values in the neighborhood of 1% to 10%. Other tree taxa represented include fir (*Abies*), oak (*Quercus*), walnut (*Juglans*), elm (*Ulmus*), ash (*Fraxinus*), birch (*Betula*), alder (*Alnus*) and hazel (*Corylus*). Forest cover is also indicated by the frequent inclusion in sediments of carbonized wood, primarily from the taxa *Pinus*, *Salix* and *Populus* (Freeman *et al.* 1988:12). Patches of mixed forest are thus indicated, with pine and oak dominating.

Levels 12 through 7 are dominated in particular by Ericaceae, which Ar. Leroi-Gourhan (1994:123) interprets as occupying most of the terrain, indicating a coastal heath vegetation with interspersed trees in more sheltered locations. Proportion of these species oscillates around 43% of pollen totals. Climatic stability is indicated for an extended period of time encompassing these lower levels. Temperate conditions are indicated by the presence of thermophiles such as walnut and oak.

Climatic deterioration is indicated in Levels 6 through 4 by the decrease in arboreal pollen, the sharp drop in Ericaceae pollen (from a peak of 67% down to 2%), and the sharp increase in Graminae to levels of over 50% in some samples. In particular, Cicoriaceae (chicory family) and Poaceae (meadow grasses) jump in importance. The interpretation for the Levels (6-4) of the Dryas I phase is for an open steppe environment with much colder and dryer conditions than previously. True tundra is not indicated. Oaks have nearly disappeared from this open landscape, as have any other thermophilic species.

The pollen profile of El Juyo, along with those of all Cantabrian Paleolithic sites, has been reinterpreted recently by Sanchez Goñi (1993). She finds the association in Levels 12 through 7 of large amounts of herbaceous pollen (50% Ericaceae, 10% Cichoriaceae and 10% Poaceae) along with many taxa typical of interglacial periods (*Ulmus*, *Fraxinus*, *Hedera* [ivy], and *Juglans*) to be problematic. The assignment of these lower levels to the Pre-Bölling climatic phase is thus in question. Since this phase has a 900 year duration (Ar. Leroi-Gourhan 1989), there is not sufficient time for these thermophilic taxa to spread to this region to establish a mixed community. Furthermore, the carbon dates for these lower levels correspond in continental Europe to the Dryas I phase. Since El Juyo is adjacent to the Cantabrian Sea, it is unlikely to have been spared as a refuge zone from the rigorous climatic conditions related to the presence of polar waters in the nearby ocean. It is possible that the pollen of these thermophilic species was percolated downwards. This interpretation is reinforced by their appearance in low numbers along the extent of the pollen profile.

This interpretation, however, is difficult to reconcile with the presence of travertine layers in El Juyo that effectively sealed lower deposits from this type of downward pollen intrusion from upper levels of the site. Since the entrance to El Juyo is located at the base of a large sinkhole feature, the potential does exist for the area immediately surrounding the cave to serve as a refuge area for plant species otherwise unable to survive in a harsher, colder climate in the coastal zone. The multiple river valleys transecting Cantabrian Spain also may have functioned in this role. This interpretation is supported by the microfauna of El Juyo and other Cantabrian Spanish Late Pleistocene sites. While these microfaunas (Chapters 5 and 8) overwhelmingly support the interpretation of a mostly open, very humid, and herbaceous plant-dominated terrain, at most sites (including El Juyo) forest-associated species show up in low levels. Their presence suggests the existence of some areas of forest cover, at least in the local sheltered surroundings near some sites. The presence of thermophile tree species in low amounts as indicated by palynology is thus supported by the microfaunal data and is not in opposition to the interpretation of the overall picture of climate and terrain indicated here.

Levels 4 through 6 (previously assigned to Dryas I) have an almost entirely herbaceous pollen profile (50% Cichoriaceae, 20% Ericaceae, 10% Poaceae, etc.) indicating an open countryside typical of a Dryas episode. Carbon dating of these levels also fits with established Dryas I chronology. Thus, these levels are the only ones which may be interpreted botanically and climatically, and their original pollen phase attribution is tenable (Sanchez Goñi 1993:114-116).

Other factors other than the possible percolation of pollen grains may affect the interpretation of the El Juyo sample, as were discussed earlier in this chapter. Though a total of 106 sediment samples were taken, only 39 were analyzed palynologically. Sample size could be increased dramatically. Additionally, cursory examination of sediment samples from areas in the site away from the main pollen coupe indicates that significant spatial discrepancies in pollen deposition occur (Freeman *et al.* 1988:12). Sediment samples secured from excavated lot (33 cm x 33 cm x about 5 cm of depth) are available for future analysis to address this issue. The cave of El Juyo is situated at the bottom of a large sinkhole of sufficient size to shelter plant species otherwise unable to survive climatically rigorous conditions, possibly explaining apparent mixtures of cold and warm flora. These palynological analyses also fail to address the question of the proportions of the landscape covered by the various plant species detected, a question which was addressed in Chapter 5. Palynology can offer a necessary framework for paleoenvironmental reconstruction that due to inherent problems of taphonomy must be reinforced with multiple sources of supporting data.

MACROBOTANICAL REMAINS

A further source of evidence utilized in conjunction with the microfauna are the botanical remains recovered from El Juyo. Excellent preservation combined with comprehensive retrieval practices has allowed unprecedented recovery of carbonized and uncarbonized plant parts and seeds (Crowe 1987; Freeman *et al.* 1988; Ibañez-Angulo 1991; Stettler 1995). This data suffers from some of the same biases in paleoenvironmental interpretation as pollen. Passage through a cultural filter in terms of site introduction and relative abundance means that variation cannot be used to infer relative environmental abundance. Differential potential of seeds for transport (by evolutionary design for their dispersal) also will affect their presence in the site: seeds are designed to travel from their parent plant(s) in some manner in order to increase the combined chances for successfully passing on their genome. Seeds also have differential susceptibility to destruction. Human use patterns will also affect which seeds are placed near enough to hearths to allow their carbonization and hence greatly enhanced chances for surviving destruction from decay. Local plant taxa represented through botanical remains serve as a check as to the local origin of pollen introduced to the site and as indicator species in themselves.

Table 7.3: Macrobotanical Remains from El Juyo.

Taxon	**English Name***
Agrostis	bent
Allium	ramsom, etc.
Alyssum	alyssum
Asteraceae	asters, etc.
Betula	birch
Bidens	sticktights
Carex	sedge
Caryophyllaceae	pinks, etc.
Chenopodium	goosefoot
Compositae	daisy, etc.
Corylus	hazelnut
Cyperaceae	galingales, etc.
Equisetum	horsetail
Erica	heather
Galium	bedstraw, etc.
Gramineae	grasses
Holcus lanatus	Yorkshire-fog
Ilex opaca	holly
Lens	lentil
Lepidium	(a grass)
Medicago	medick
Nardus	mat-grass/nard, spikenard
Papaver	poppy
Picris	oxtongue
Pinus	pine
Plantago	plantain
Poa	meadow-grass
Polygonum	bistort
Populus	poplar
Quercus	oak
Rosaceae	roses, etc.
Rubus	blackberry
Rumex	dock
Salix	willow
Sambucus	elder
Senecio	ragwort
Silene	campion, etc.
Statice	thrift, sea-ping
Trifolium	clover
Ulex	gorse
Viola	violet

*Following Polunin and Walters (1985) and Willis (1973).

The macrobotanical remains from El Juyo are listed in Table 7.3, as summarized recently by Stettler (1995:41-47). These remains are currently under study, and enough seeds have not been identified per level to allow statistical climatic comparisons between them. Some observations that are available thus far include relative abundances of these taxa. *Holcus*, *Senecio*, *Viola*, *Agrostis*, *Silene*, *Polygonum* and *Sambucus* are common. Additionally, *Bidens*, *Equisetum* and *Ulex* are well represented. *Bidens* (sticktights) are animal-dispersed taxa that cling to fur and clothing, and therefore their abundance in El Juyo is probably due to accidental transport. *Equisetum* (horsetail) is represented primarily through the remains of stems and hence has a greater amount of plant material (compared with the seeds of other taxa) available for carbonization/preservation. *Ulex* (gorse) is abundantly represented through its thorns, which are perhaps more survivable than other macrobotanical remains. Most charcoal from El Juyo comes from *Pinus*, *Salix* and *Populus* (Freeman *et al.* 1988:12). Since firewood is a heavy, daily necessity, one may hypothesize that the most abundant and local species were burned preferentially.

Some preliminary inferences by level have been drawn for El Juyo. Level 8 has a proportional abundance of marsh plants (Freeman *et al.* 1988:13). These include Cyperaceae, *Polygonum*, *Plantago* and *Bidens*, as well as *Statice*, a maritime form (Stettler 1995:46). Levels 6 and 7 have yielded macrobotanical remains of plants mostly from disturbed contexts, but Stettler (1995:47) notes that these taxa are common for all levels. These taxa include chenopods, Asteraceae and *Polygonum*. Such species commonly would be found in the trampled zone outside of a human-occupied cave site, so their inclusion throughout the site deposits is understandable. Level 4 shows an apparent shift in species composition, with a decrease in marsh plants and the appearance of plants from the subalpine zone. These subalpine taxa include *Betula*, *Salix*, *Medicago*, and Caryophyllaceae. These preliminary results are in agreement with the microfaunal data. Marsh plants in the lower levels (up to Level 6) reinforce the picture of humid climate indicated by the predominantly humid meadow and water edge species. Such a marshy environment would have been particularly suited to *Arvicola terrestris* and *Microtus oeconomus*, two of the dominant species. The climatic deterioration for Level 4 already indicated by the first appearance of the alpine species *Microtus nivalis* in El Juyo (and the continued importance of *Microtus oeconomus*) is reinforced by the concurrent appearance of alpine macrobotanical remains.

OTHER INDICATORS OF PALEOENVIRONMENT

Other indicators of paleoenvironment await further analysis from El Juyo. The fine fraction has yielded a very large sample of amphibian remains. While these have not yet been identified, their large number indicates significant humidity and reinforces the interpretation of wetlands very near to El Juyo. These species are small enough to have been introduced by barn owl predation. Smaller amounts of reptile remains also await identification. El Juyo has also yielded a small sample of avian remains, which may when identified provide further indication of paleoenvironment. Since many bird species are migratory, however, these may be of lesser utility. Preliminary analysis indicates that most of these remains are from smaller species (passerines) with a few larger species present. The majority could therefore have been introduced by barn owls, with perhaps the few larger species resulting from human predation.

SUMMARY

These multiple sources of paleoenvironmental data from El Juyo when considered alone give an incomplete picture of the waning Pleistocene landscape. Sea core foraminifera (CLIMAP 1976; McIntyre *et al.* 1976; Ruddiman and McIntyre 1976; Shackleton and Opdyke 1976; van Donk 1976) suggest that the North Atlantic surface ocean water temperatures at 42^0 North latitude (approximately the Cantabrian coast) averaged 10^0 C colder at 18,000 B.P. than at present, indicating polar water conditions with attendant pack ice. By 14,000 B.P., ocean levels were again lowering as continental ice caps expanded again after a period of retreat. Ocean level was fluctuating somewhere intermediate to the -100 m glacial maximum lowpoint and modern sea level that was reached shortly after the beginning of the Holocene. Surface ocean water temperatures and accompanying climate were therefore somewhere intermediate between the period of glacial maximum and the significantly warmer temperate/Atlantic climate of the Holocene. Local climate in north coastal Spain need not have followed with precision global changes in sea level, since much room for variation exists especially in shallow coastal areas affected by insolation and warming currents during some portion of the cycle of glacial advance and retreat. Changes in air circulation and weather patterns also allow great variation in local climate, hence the need to examine other more direct evidence of Cantabrian Pleistocene conditions.

Sedimentological analysis of the El Juyo deposits suggests that colder conditions returned during the time of deposition of the upper habitation Levels 6-4 (Freeman *et al.* 1988:11-12), which follow the earlier period of relatively dry, temperate conditions in Levels 7-13. Humidity (or precipitation concentration) had increased by the time of deposition of Level 6 such that the water table of the cave reached this level and deposited clay. Both Level 5 (a sterile flowstone layer) and Level 4 show an increase in *éboulis*, which may indicate an increase in cold regime due to spalling of the cave ceiling, and partial closure of the entry is required for the formation of Level 5. Level 3 was formed by the collapse of the main entrance and complete isolation of the vestibule in a typical cave environment, followed by further sealing of the vestibule deposits with a thick layer of travertine. This sequence suggests a climate regime of high humidity followed by a period of cold, dry conditions.

Neither large mammalian, marine molluscan, nor fish faunas give any definite indications of paleoenvironment. This situation is due to the wide environmental tolerances of the species involved, almost all of which are found today (or were extirpated locally in historical times) in Cantabrian Spain. El Juyo's large mammal fauna (in descending order of importance) consist of red deer, roe deer, bison, horse, ibex, boar, lion and wolf. These species now are or formerly were found throughout Europe. Red deer thrive on mixed closed/open habitats with their edge effects and resultant browsing foliage (Clutton-Brock *et al.* 1982). Both roe deer and boar, found in much smaller amounts than red deer, prefer forest habitats. Horses and bison are primarily open grassland species and each is found in small amounts throughout the site levels. Ibex are an upland species that must have been transported to El Juyo. The large fauna therefore indicate that a range of habitats were present in unknown proportions. Molluscan fauna were similarly inconclusive, since the two primary harvest species *Patella vulgata* and *Littorina littorea* are found throughout a wide range of Atlantic coastlines. Warm and cold indicator species, retrieved in very small amounts, provide conflicting or inconclusive evidence. The many salmonid remains recovered are also equivocal indicators of paleoenvironment, since the rivers of Holocene Spain and Portugal are at the southern extreme of their spawning range.

Palynological analyses offer a finer resolution of the paleoenvironment of El Juyo, but suffer from the limitations inherent to this type of research. Previous interpretation (Boyer-Klein and Ar. Leroi-Gourhan 1987; Ar. Leroi-Gourhan 1994) indicated that the lower levels (12-7) of El Juyo can be correlated with the relatively temperate Pre-Bölling climatic phase. This interpretation is based upon the presence of moderate levels of tree and steppe pollen and the dominance of Ericaceae (heather) pollen. The upper levels of El Juyo's Lower Magdalenian occupation are assigned to the Dryas I climatic phase, whereby a major deterioration of conditions is indicated by the dominance of steppic pollens at the expense of tree and shrubs. Sanchez Goñi (1993) reinterprets this pollen data as reflective still of the cold, dry conditions of Dryas I in the upper levels of the site, with the lower Levels 12-7 not assignable to the Pre-Bölling, but is equally open to criticism. The lower levels therefore may possibly have been deposited during Dryas conditions, which would better correlate with radiocarbon dates from the site. However, as with other sites in Cantabrian Spain, radiocarbon dating presents certain difficulties, and the dates are possibly in error.

The macrobotanical remains from El Juyo reinforce these mixed paleoenvironmental signals, with a varied group of species represented. Level 8 has yielded a proportional abundance of marsh-favoring and maritime plants, indicating wetter conditions during the formation of this stratum and/or more intensive exploitation of the nearby coast and estuaries. Subalpine taxa appear during the deposition of Level 4, reinforcing that colder conditions were obtained in the waning stages of occupation at El Juyo. Throughout all levels, taxa favoring disturbed habitats are abundant. The presence of these taxa indicates the selective nature of the macrobotanical remains introduced to the site both by deliberate and accidental human activities. As such, a cultural filter has in large part selected the relative abundances of these remains and lessened their value in paleoenvironmental reconstruction.

Using these many data sources in conjunction with the microfauna sample, the paleoenvironmental picture of El Juyo is far clearer. Conditions throughout deposition were intermediate between those of the glacial maximum and the warm temperate Holocene period. Large mammal indicator species agree with the interpretation that open conditions were intermixed with at least some areas of forest cover and grassland among predominating heath. The relative amounts of these vegetation types are best reflected by the microfauna, because these have not passed through a cultural filter in their acquisition. High humidity is reflected by the microfauna through all levels. Moist habitat microfauna are accompanied by macrobotanical remains of marsh plants in occupations below Level 4. Sedimentology, palynology and macrobotanical analysis all agree with the microfaunal data that further deterioration of climate occurred during the deposition of Level 4, indicating that cooling and drying of the climate occurred.

Conflicts still exist both within and between these data sets. However, a more refined picture of the Pleistocene landscape and the relative importance of vegetational types and the climate that they reflect is provided by the analysis of all these sources of data together.

CHAPTER 8
MICROFAUNA FROM CANTABRIAN REGION SITES

This chapter details the smaller microfaunal remains (Rodentia, Insectivora, and Carnivora) from Cantabrian sites and the interpretation of paleoenvironment from their relative proportions. Some site levels from the Mesolithic period and other later Holocene environments are presented for purposes of ecological comparison over the deglaciation transition. Some sites from just outside the Cantabrian region (Pyrenean France, Navarra province in Spain) are included for similar reasons. The remains of larger microfauna (Carnivora and Lagomorpha) are discussed elsewhere (Chapter 9). This analytical separation is not directly due to the differences in taxonomy of these species, but to the probable manner of introduction into sites of each size category. The smaller taxa usually entered sites as the bones in egested nocturnal raptor pellets, while the larger taxa usually entered as a direct result of human hunting for food or furs.

The Cantabrian region has many Paleolithic sites whose excavations have yielded significant amounts of terrestrial mammalian microfaunal remains (Figure 8.1). The smaller microfaunal species (Rodentia, Insectivora and Chiroptera) from these sites are summarized below according to province (Guipúzcoa, Vizcaya, Cantabria, Asturias). For many older excavations, only the presence of a species can be indicated. In these cases the presence of a species, particularly those from northern or alpine zones, still may serve as a valuable indicator of climate. Generally, a sharp increase in the recovery of microfauna is noted corresponding to the recentness of site excavation, both between sites and when comparing the results of earlier and later excavations at the same site. This discrepancy may be taken as an indirect report upon the history of screening deposits from Cantabrian sites, rather than absence of these smallest taxa from certain levels. The assumption that in most cases these species had existed at a site completely excavated in the past using less precise methods is logically surmisable, though unprovable unless backdirt has more recently been screened. Strong evidence for this failure of recovery is found at El Castillo, where there are no extant curated microfaunal remains from the Aurignacian deposits labeled as "Rodent Level" by earlier excavators (Cabrera 1984), although subsequent curational loss is possible alternative explanation. In all cases the excavation era/methods will be considered as a part of an assemblage's taphonomic history and will be considered in the interpretation of paleoenvironmental data. Only a few sites have yielded sufficient microfauna for full interpretation: Amalda (Pemán 1990), Erralla (Pemán 1985), Aitzbitarte (Altuna 1972) and Cueto de la Mina (Castro Bernardez 1986).

Further complications arise in the interpretation of microfaunal remains from other sites. Differences in counting method used by investigators can greatly affect the relative proportions of reported taxa. While the present study uses identified teeth to compare relative taxa abundance, other investigators may base their final counts upon all remains including postcranial (Eastham 1995) or upon varied single elements per taxa (Pemán 1985). Most studies of modern owl pellets base counts upon the MNI derived from crania or demimandibles. Variations may also exist in the taphonomic history of these assemblages. While most large accumulations in these settings result from barn owl activity, other possible accumulators would bias taxa proportions in more and different ways. These uncertainties limit the reinterpretation of site faunas.

PYRENEES/NAVARRA

Abauntz

Level/Tool Industry	Associated ^{14}C Dates
Eneolithic	4240+/-140 BP
Neolithic	5390+/-120 BP
Ancient Neolithic	6910+/-450 BP
Epipaleolithic	9530+/-399 "BO"
Lower Magdalenian	15800+/-350 BP

Abauntz is located peripherally to the Cantabrian region in the province of Navarra on the west edge of the Pyrenees (Altuna and Marriezkurrena 1982).

Microfaunal remains from this site consist only of larger taxa. The remains of *Lepus europaeus*, *Martes martes*, *Vulpes vulpes* and *Felis sylvestris* have been recovered in small amounts (Chapter 9). Only a few remains of *Erinaceus europaeus* (n = 3) from the Eneolithic level have been recovered.

Dufaure

Level/Tool Industry	Associated ^{14}C Dates
3 Azilian	9,600+/-290 BP
	9,750+/-110
	9,810+/-100
	10,310+/-270
4 sup Final Magdalenian	10,910+/-220
4 m-s Final Magdalenian	11,750+/-300
	12,030+/-280
4 m-i Final Magdalenian	
4 inf Final Magdalenian	12,260+/-400
5 Middle Magdalenian	12,690+/-230
	12,990+/-270
	14,570+/-390
6 Magdalenian	14,020+/-340
	14,590+/-100
	14,640+/-230

Dufaure is located less than 40 km from the modern coastline in the province of Basses-Pyrénées (France). Its altitude is 127 m above the modern sea level in generally flat coastal plain that was broadened significantly during periods of ocean lowering. This *abri* (rockshelter) is adjacent to the

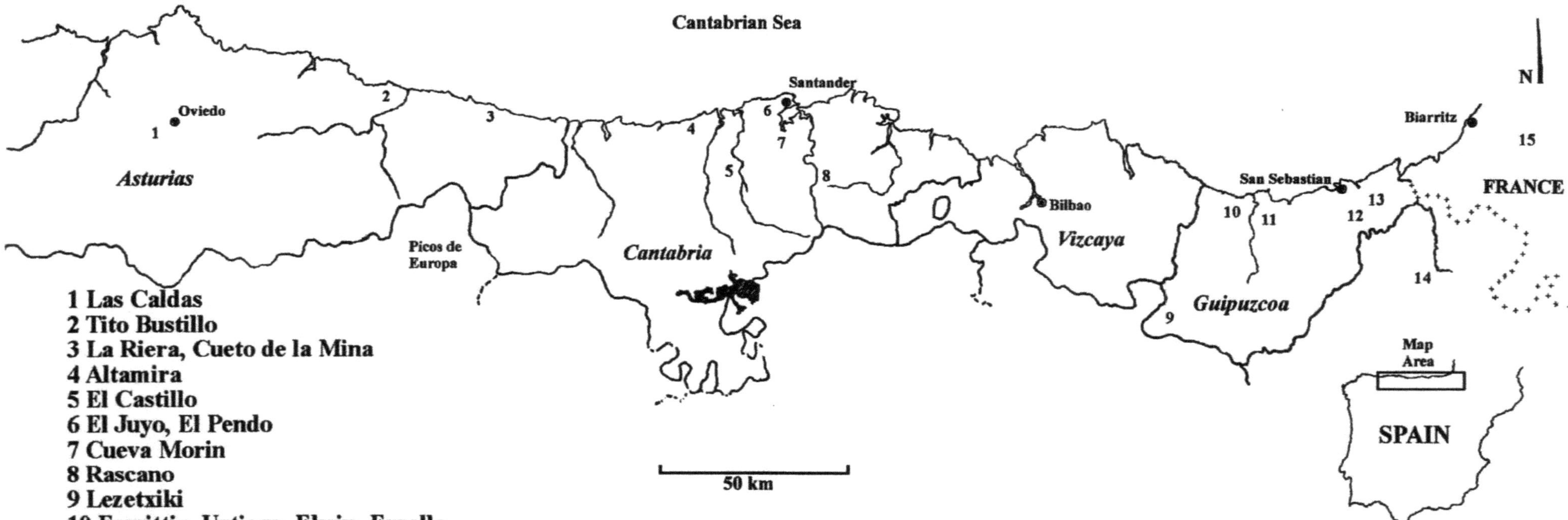

Figure 8.1: Cantabrian Microfaunal Sites Discussed in the Text

sites of Duruthy, le Grand Pastou and le Petit Pastou. The Gave d'Oleron, a tributary to the l'Adour River, is located less than a kilometer to the south. The Gave d'Oleron Valley is an important communication route between the coastal lowlands and the alpine zones of the Pyrenees which rise steeply to the south. The *abri* itself is 11 m wide and a maximum of 4 m deep (Straus 1995).

Dufaure was first discovered by R. Pottier and L. Lartet in 1873 during the course of their excavations at le Grand Pastou. This was the last of this cluster of sites to be excavated, with the first excavations being carried out there by P. Dubalen and H. Breuil in 1900. These sites received no additional attention for the next half century, until R. Arambourou in 1957 began new work on the terrace below Duruthy that proceeded nearly every following year until 1975. Excavations in 1968 and 1969 were shifted temporarily to the terrace beneath le Grand Pastou. Work at Dufaure itself began anew with the excavations directed by L. Straus from 1980 to 1984 (Straus 1995). The faunal collections from this site therefore come from meticulous modern excavations. The levels at the site span Azilian, Final Magdalenian, Middle Magdalenian and Lower Magdalenian occupations. The microfaunal data from this site span the time of general climatic amelioration at the end of the Pleistocene.

GUIPÚZCOA

Aitzbitarte IV
Level/Tool Industry

Azilian
Magdalenian
Solutrean
Aurignacian

Aitzbitarte IV is located 220 m a.s.l. with a modern distance of 7 km from the Cantabrian coastline (Altuna 1972), with coordinates of 43°15'50" N and 1°53'20". Nearby foothills climb to over 500 m, and the site overlooks the Oyarzun River valley to the north. The entry to this cave is 16 m wide and continues inward to bifurcate into two deep chambers. The excavated deposits are from this vestibule area, with Aurignacian, Solutrean, Magdalenian and Azilian levels. First excavation of this site was by M. del Valle, Conde de Lersundi in 1892. Excavations were continued by P.M. de Soraluce from 1896 to 1901, and from 1902 to 1906 included G. de Repáraz and E. Rotondo Nicolau. This site was visited by pioneering prehistorians Hugo Obermaier and J. Bouyssonie in 1909 and by Henri Breuil in 1917. Modern excavations were carried out under the direction of J. M. de Barandiarán from 1960 to 1964, which yielded the materials examined by Altuna (1972) and summarized below.

Amalda

Level/Tool Industry	Associated ^{14}C Dates
II Late Roman	
IIIA-B Chalcolithic	
IV Upper Solutrean (middle)	17,580+/-440
(superior)	16,200+/-380
	16,090+/-240
V Perigordian VII (central)	17,880+/-390
(base)	19,000+/-340
VI Perigordian V (central)	27,400+/-1,000
	27,400+/-1,100
VII Mousterian	
VIII Sterile	
X Sterile	

Amalda is located 205 m a.s.l. and about 8 km from the modern shoreline. The cave is not far from the water in the Alzolaras River valley, a tributary of the Urola River. The site of Erralla lies about 3 km upstream. Amalda is within easy reach of the uplands, with hills over 1000 m just a few kilometers to the south. The cave itself has a wide (12 m wide and 7 m high) entrance opening on a somewhat narrower front vestibule about 50 m deep. Mousterian, Perigordian and Upper Solutrean levels have been excavated here. Quite recent excavation was carried out from 1979 through 1984 under the direction of J. Altuna (Altuna *et al.* 1990).

Ermittia
Level/Tool Industry

(1) Neolithic
(2) Azilian
(3) Magdalenian
(4) Sterile
(5) Solutrean
(6) Sterile

Ermittia is located 125 m a.s.l. in the low coastal hills only 3 km from the modern coastline, with coordinates of 43°16'35" N and 2°21'30" W. The cave lies less than half a kilometer from the Deva River, though nearby foothills climb to over 600 m. The cave has a very narrow entrance only a few meters wide, which leads to a larger inner chamber. Excavations have yielded Solutrean, Magdalenian, Azilian and Neolithic levels. These were carried out by J. M. de Barandiarán and T. de Aranzadi from 1924 to 1926 in this narrow vestibule area, though later sampling in 1965 by J. M. Barandiarán and J. Altuna indicated the presence of archaeological deposits in the inner chamber (Altuna 1972).

Lezetxiki
Level/Tool Industry

Ia Upper Magdalenian
Ib Sterile
II Solutrean/Gravettian
IIIa Upper Aurignacian
IIIb Sterile
IVa Aurignacian
IVb Sterile
IVc Protoaurignacian
Va Sterile
Vb Mousterian
VI-VII Mousterian
VIII

Lezetxiki is located up the Deva River valley in the Cantabrian foothills 375 m a.s.l., at 43°05'20" N and 2°31'55" W. The modern coastline is 30 km away. This inland location places this site within easy access to the upland zone, with nearby hills reaching over 1000 m, yet proximity to the river valley allows easy communication with the coast. The cave consists of a long, narrow chamber intermittently open along its eastern face. Barandiarán first sampled this site in 1928 and carried out excavations from 1956 to 1968 (Altuna 1972). Mousterian and Aurignacian levels have been excavated at Lezetxiki.

Urtiaga

Level/Tool Industry	Associated ^{14}C Dates
A Mixed	
B Bronze Age	
C Azilian	8,700+/-170
D Final Magdalenian	10,280+/-190
E Sterile	
F Final Magdalenian (base)	17,050+/-140
G, I Sterile	

Urtiaga is located along the Cantabrian coast about 3 km east of Ermittia and 1.5 km from the present day shore. Its elevation is 130 m a.s.l., though the proximity of low hills to the coast place elevations of 600 m within easy distance. The cave itself is a very narrow-mouthed (about 1 m) tunnel that, while deep, does not widen beyond 4 meters in its main chamber. Its occupation during the Paleolithic as a semi-permanent camping site it thus restricted, with use as temporary hunting or emergency shelter by small groups more likely. Excavation was carried out in two phases, from 1928 to 1936 by T. de Aranzadi and J. M. Barandiarán and 1954, 1955 and 1959 by the latter researcher. These were carried out in the narrow entranceway and yielded materials from Final Magdalenian, Azilian and Bronze Age occupations (Altuna 1972).

Marizulo

Level/Tool Industry	Associated ^{14}C Dates
I Bronze Age I	5,285+/-65
II Neolithic	
III Final Mesolithic	

Marizulo (43°13'10" N and 1°58'10" W) is found 260 m a.s.l. and 12 km from the modern coastline. This site lies on the edge of the Cantabrian foothills, with heights over 800 m less than 2 km to the southeast. The Río Leizaran, a tributary of the Río Oria, is just over 2 km to the west. Marizulo itself is a very small cave, the entrance of which consists multiple narrow interlinked chambers no more than 2 meters wide. Its smallness perhaps explains why it apparently was never used as a habitation site during the Paleolithic, though it is possible that it was used sporadically as emergency shelter. Artifacts from Final Mesolithic, Neolithic and Bronze Age levels have been found there. This site was first sampled under the direction of J. M. Merino in 1962 and excavated from 1962 through 1967 (Altuna 1972).

Ekain

Level/Tool Industry	Associated ^{14}C Dates
I Sterile	
II Azilian	
III Azilian	12,750+/-250 B.P.
IV Azilian	9,460+/-185 B.P.
V Azilian	13,350+/-250 B.P.
VIa Final Magdalenian	
VIb Upper Magdalenian	12,050+/-190 B.P.
VIIa Lower Magdalenian	
VIIb Lower Magdalenian	16,030+/-240 B.P.
VIIc Lower Magdalenian	15,970+/-240 B.P.
VIId Lower Magdalenian	15,400+/-240 B.P.
VIIe Lower Magdalenian	
VIIf Lower Magdalenian	13,950+/-330 B.P.
	16,250+/-250 B.P.
VIIIa-b Interstadial Würm III-IV	20,900+/-450 B.P.
IXa	
IXb Aurignacian?	>30,600 B.P.
Xa Chatelperronian?	
Xb	
XI-XII Sterile	

Ekain (43°14'10" N,1°24'41" W) is located 90 m a.s.l. and less than 7 km from the modern shoreline. The cave is adjacent to a small tributary of the Río Urola, the valley of which allows easy passage between the coast and the rapidly rising highlands of the interior. Foothills of the Cantabrian Cordillera rise to over 800 m within 3 km south of the site. Site placement, as was typical among Upper Paleolithic Cantabrian occupations, allowed access to a range of habitat zones within an easy walking radius. The cave itself bifurcates to form inner galleries, with only a small inhabited vestibule area in some places only 2 m wide. While small and low-lying, the cave entrance has suitable shelf areas for raptors to roost, though these would have been within easy disturbance access to humans. The site is also significant for its parietal art, depicting mostly horses but also (in descending order) bison, ibex, red deer, bear, fish, and possible rhinoceros (Altuna and Merino 1984).

Excavations at Ekain began in 1969 under the direction of J. M. de Barandiarán and continued from 1971-1975. Direction for the final three campaigns was by Jesús Altuna. All sediments were screened through 2 mm mesh. This size is sufficient to retrieve most microfaunal remains, though isolated teeth of smaller species (shrews in particular) may be lost.

Erralla

Level/Tool Industry	Associated ^{14}C Dates
I Sterile	
II Final Magdalenian	
III Final Magdalenian	12,310+/-190
IV Sterile	14,570+/-300
	15,800+/-230
V Lower Magdalenian	15,740+/-240
	16,200+/-240
	10,580+/-270
	16,270+/-240
VI Sterile	
VI' Sterile	

Erralla (43°12'32" N, 1°30'20" W) is located 230 m a.s.l., about 9 km from the modern shoreline (Altuna 1985a). The cave is adjacent to the Alzolaras River, a tributary of the Urola. The site of Amalda lies about 3 km downstream. Altitudes of over 1000 m are reached less than 3 km to the south in the Cantabrian foothills. This location is thus intermediate between the coastal and highland zones and within easy travel of both. The cave itself has two entrances opening to a vestibule area approximately 6 m by 18 m. Suitability for roosting by raptors is indicated by the very large microfauna sample derived from this site, from both sterile and archaeological levels. Erralla has yielded Lower and Final Magdalenian occupations. Excavations at this site are very recent and were directed by J. Altuna in 1976 and 1977.

CANTABRIA

Altamira

Level/Tool Industry	Associated ^{14}C Dates
2 Lower Magdalenian	15,500+/-700
	14,520+/-260
	14,480+/-250
3 Solutrean	18,540+/-320

Altamira is located 153 m a.s.l. near the top of a small hill overlooking the coastal lowlands. The modern shoreline is 4 km distant. The cave itself originally opened onto a large vestibule, with substantial tunnels leading inwards to multiple chambers including the areas of paintings and engravings. Both Solutrean and Lower Magdalenian levels have been excavated there, though the major artwork episodes have been assigned a Magdalenian age (More Romanillo *et al.* 1996; Valladas *et al.* 1992).

The excavations at Altamira are among the earliest Paleolithic investigations in Cantabrian Spain (Altuna and Straus 1976; Freeman 1987, 1988; Straus 1976-77, 1992). This cave was discovered only in 1868 and was excavated for the first time in 1876 by Marcelino Sanz de Sautuola, a local amateur natural scientist. He continued excavations in 1879 in the vestibule area, at which time (according perhaps to legend) his young daughter discovered just inside the vestibule the world-renowned bichrome ceiling composition of bison and deer. These finds were met with skepticism over the association of the parietal art with primitive artifacts by the reigning French prehistorians of the day, notably G. de Mortillet and E. Cartailhac. In 1881 the cave was visited by E. Harlé at their behest and declared a probable forgery. Over time and in the face of growing evidence at other newly discovered parietal art cave sites, Cartailhac later recanted (1902). Some other researchers excavated in the cave in the intervening years before more substantial work by H. Alcalde del Rio in 1902-1903, digging in the vestibule area and in the "Great Hall" underneath the bison bichromes. Altamira was subsequently excavated by H. Obermaier in 1924 and 1925. His work was carried out primarily in the vestibule area. Modern excavations under the overall direction of J. González Echegaray took place in 1980 and 1981 in the "kitchen midden" area of the vestibule (Freeman 1988).

Though the early collections from the site have suffered great curational losses (Altuna and Straus 1976), huge densities of artifacts and living debris were deposited there during the Solutrean and Lower Magdalenian periods. The artistic styles recorded on osseous implements there have been used (Conkey 1978, 1980) to suggest that Altamira functioned as a seasonal aggregation site for dispersed groups living in the local region. With its artwork and large size, Altamira may have functioned as a meeting place for group ritual activities during the Lower Magdalenian period.

El Castillo

Level/Tool Industry	Associated ^{14}C Dates
1a Modern	
1b Stalagmite	
2 Neolithic	
3 Stalagmite	
4 Azilian	
5a-b Stalagmite	
6 Magdalenian A	10,310+/-120
	12,390+/-130
7 Magdalenian A/B	
8 Magdalenian B	16,850+/-220
9 Sterile	
10 Solutrean	
11	
12a-b Aurignacian A	
13 Aurignacian A/B	
14a-14c Aurignacian B	
15 Aurignacian B/G	
16a Rodents I	
16b-c Aurignacian G	
16d Rodents II	
17 Sterile	
18a-b Aurignacian D	39,000+/-2000
	(ave. of 3)
	(Bischoff *et al.* 1991)
19 Stalagmite	
20a-b Mousterian A upper&lower	
21 Sterile	
22a-c Mousterian B	
23a-c Sterile	89,000+11,000/-10,000
	U-series isochron date
	on flowstone
	(Bischoff *et al.* 1991)
24 Acheulean	
25a-b,26 Below Acheulean	

El Castillo is located 175 m a.s.l. on a steep, conical hill which itself rises to 375 m. The site overlooks the Río Pas valley, which delves southward into the foothills of the Cantabrian Cordillera. Its location allows access to the upland and coastal lowland zones, and easy communication to both along the river valley. The cave now consists of a high-ceilinged open *abri* at the entrance and a cavern leading back into the hill many hundreds of meters, with spectacular chambers and flowstone formations. This and other caverns in the Castillo hill have multiple instances of parietal art located throughout, even deep within in some cases. Located in the same hill radially along the same approximate altitude are the caves of Lago, Las Chimeneas (González Echegaray

1953, 1974), La Flecha, Pasiega and Las Monedas. Three of these (Chimeneas, Pasiega and Monedas) are major parietal art sites additional to El Castillo itself. La Flecha has yielded Mousterian artifacts. In addition to its art, El Castillo has yielded levels covering virtually the entire Paleolithic sequence of Western Europe: Acheulean, Mousterian, Aurignacian, Solutrean, Magdalenian, Azilian and also Neolithic levels have been discovered there.

Just as the excavations at Altamira are an early example of serious Paleolithic investigation in Cantabrian Spain, the excavations of El Castillo chart much of its subsequent history into the modern period (Cabrera 1984; Straus 1992). H. Alcalde del Rio discovered the parietal art of El Castillo in 1903, and subsequently dug in the upper levels of the massive deposits at the open *abri* at the front of the site. The newly formed Institut de Paléontologie Humaine and its patron, Prince Albert I of Monaco, sponsored massive excavations in the 18 m deep sequence in the front of the cave. These were carried out from 1910 through 1914 under the direction of H. Obermaier and included the participation of such notables as P. Wernert and the Conde de la Vega del Sella. Unfortunately (and far too frequently) these results were only sparsely published and poorly curated, with much loss of irreplaceable information (Cabrera 1984). Excavation of remnant stratigraphy continues at El Castillo currently, under the direction of F. Bernaldo de Quirós and V. Cabrera.

Cueva Morín

Level/Tool Industry	Associated ^{14}C Dates
1 Azilian	
2 Final Magdalenian	
3 Solutrean	
4 Gravettian	
5 Gravettian/Aurignacian II (Sup.)	20,710+/-350
6 Aurignacian I (7/6)	32,415+/-901
7 Aurignacian I	29,515+/-865
	28,055+/-1,535
8 Aurignacian 0 (8a)	28,435+/-556
	28,155+/-757
	28,515+/-1,324
9 Aurignacian 0	
10 Chatelperronian	28,610+/-577
	36,950+/-6,777
12 Mousterian	

Cueva Morín is located less than 2 km south of the southern shore of the Bay of Santander. Its proximity to the glacial age shoreline would have been significantly greater (over 15 km), since this shallow bay would have been dry land. Site setting is within the narrow coastal lowlands. At Cueva Morín, nearly the entire later Paleolithic cultural stratigraphic sequence has been recovered: Mousterian, Chatelperronian, Aurignacian, Gravettian, Final Magdalenian, and Azilian levels have been discovered. Chatelperronian Level 10 is the most secure identification of this tool complex for Cantabrian Spain. The earlier excavations were carried out there from 1912 to 1919 in alternating seasons by J. Carballo and by the Conde de la Vega del Sella. Brief tests were made at the site in 1962 and the witness section was more extensively sampled in 1966. Re-excavation of the cave in 1968 and 1969 was under the direction of J. González Echegaray and L. G. Freeman (1971, 1973, 1978). The ^{14}C dates for levels downward from 8a are suspected by the excavators to be unreliably young. These excavations covered most of the main vestibule area, which itself runs about 8 m by 8m, though the cave continues inwards from there.

El Pendo

Level/Tool Industry

I Azilian
II Final Magdalenian
IIab Final Magdalenian
IIcd Final Magdalenian
III Late Aurignacian
IV Late Aurignacian
V, Va Gravettian
Vb, VI Evolved Aurignacian
VII Aurignacian I
VIII Chatelperronian
VIIIa,b Archaic Aurignacian
VIIIc Sterile
VIIId Mousterian
IX Mousterian
XI-XIII Mousterian
XIV Mousterian
XV Mousterian
XVI a-c Mousterian

El Pendo is located less than 3 km inland from El Juyo, on the coastal lowlands about 10 km from the modern coast. The nearest significant river is the Río Pas, about 3 km to the west. The cave itself is expansive, with the main vestibule chamber about 30 m high, up to 20 m wide and many times that amount deep. This site has yielded a long stratified sequence with Mousterian, Aurignacian, Chatelperronian, Gravettian, Final Magdalenian and Azilian levels. It also contains parietal art in the form of bird engravings, a rarity in Paleolithic art.

The excavation history of El Pendo is long and complex. El Pendo was first excavated in 1878 by M. Sanz de Sautuola and again in 1880 with J. Vilanova y Piera. Some engravings in the site were described by H. Alcalde del Río in 1907. J. Carballo visited the site in 1908 accompanied by H. Obermaier, and it was also revisited the Alcalde del Río and the Abbé Breuil in that same year. J. Carballo returned to the cave in 1910 accompanied by W. Beatty and carried out some archaeological testing of the deposits. R. Shallcras, O. Cendrero, Gómez Llueca and C. Pérez Bustamente also sampled the deposits in 1914 and subsequently published some of these findings (Cendrero 1915). J. Carballo carried out several brief campaigns in the 1920's, 30's and 40's. These included visits by the American School of Prehistoric Research in 1930. Major excavations were undertaken under the direction of J. Martínez Santa Olalla (then director of the Seminario de Historia Primitiva de la Universidad de Madrid) with A. Cheinier, B. Saez Martín, V. Ruis Argilés, J. González Echegaray and C. Borden in 1953 in collaboration with the Diputación Provincial and A. García Lorenzo. Direction in 1954 and 1955 was by Ruis Argilés in collaboration with Francesca Minelono. Excavation in 1956 and 1957 was directed by Martínez Santa-Olalla and André Leroi-Gourhan. Publication of the results from the excavations of the 1950's was delayed many years (González Echegaray 1980). Current sampling of the remnant deposits is being undertaken under the direction of R. Montes and V. Cabrera.

Rascaño

Level/Tool Industry	Associated ^{14}C Dates
1 Azilian	10,486+/-90
	10,558+/-244
2 Upper Magdalenian	12,896+/-137
	12,282+/-164
3 Lower Magdalenian	15,173+/-160
4 Lower Magdalenian	15,988+/-193
5 Initial Magdalenian	16,433+/-131
6 Nearly Sterile	
7 Nearly Sterile	27,240+950-810
8 Nearly Sterile	
9 Aurignacian?	
10 Sterile	

The site of Rascaño (43°17'38" N, 3°41'44" W) is located 275 m a.s.l. at a distance from the modern shoreline of over 25 km. The cave is nearly adjacent to the Río Miera and lies on the edge of the foothills climbing to the Cantabrian Cordillera a few kilometers to the south. Rascaño is thus situated to allow access to the adjacent highland zone and easy communication with the coast along the river valley. Utilization of this site occurred possibly during the Aurignacian period, and clearly during the Lower Magdalenian, Upper Magdalenian and Azilian periods. This location was probably chosen for use as a specialized ibex (*Capra pyrenaica*) hunting camp, as is indicated by the abundance of this species' remains in the site deposits. The first excavation of this site occurred in 1912 under the direction of J. R. Gómez Riaño and J. Carballo in conjunction with the Comisión de Investigaciones Paleontológicas y Prehistóricas. Some excavation continued in subsequent years under the direction of P. Sierra and Gómez Riaño. Some investigations were also carried out there by Hugo Obermaier and the Conde de la Vega del Sella in 1921, and again in 1923 by Obermaier with J. and F. Riaño. Modern excavation of Rascaño waited until 1974 and was conducted under the direction of J. González Echegaray and I. Barandiarán Maestu (1981).

ASTURIAS

Las Caldas

Level/Tool Industry
1 Postglacial
2 Middle Magdalenian
3-6 Final Solutrean
7-9 Upper Solutrean
10-13 Sterile
14 Middle Solutrean-B
15 Sterile
16-18 Middle Solutrean-A
19 Sterile

Las Caldas (43°20'10" N, 2°14'05" W) is located 160 m a.s.l. near the capital of Asturias (Oviedo). This site is located about 25 km from the modern shoreline and similarly distant from the Cantabrian Cordillera to the south. Nearby foothills do not reach above 450 m, but within 8 km of the site peaks up to 1712 m (Pico Gamonal) are reached. The Río Nalon runs just south the entrances, the valley of which allows easy communication to the interior and coast. The cave itself consists of two separate, narrow chambers running parallel into a hillside. Each of these caves is no more than 5 m wide and usually much narrower, and extends about 35 m inwards (Corchón 1981).

Las Caldas was first mapped in 1958 by local speleologists R. Querol and J. Carrasco. Looting pits subsequently destroyed a large portion of vestibule deposits, which had been reduced previously by natural water flow. The intact deposits were first excavated from 1971 to 1973 under the direction of M. S. Corchón, which included at least some screening of site sediments. The faunal remains from Las Caldas were heavily fragmented and included very few microfaunal species.

La Riera

Level/Tool Industry	Associated ^{14}C Dates
26 Upper Mag./Azilian	
24 Upper Magdalenian	10,890+/-430
21-23 Upper Magdalenian	10,340+/-560
	12,620+/-300
20 Lower Magdalenian	9,090+/-570
	12,360+/-670
	17,160+/-440
19/20 Lower Magdalenian	
19 Lower Magdalenian	15,230+/-300
	15,520+/-350
	16,420+/-430
18 Lower Magdalenian	
17 Solutrean	16,900+/-200
	17,070+/-230
16 Solutrean	18,200+/-610
15 Solutrean	15,600+/-570
	17,225+/-350
14 Solutrean	15,690+/-310
13 Solutrean	
11 Solutrean	
10 Solutrean	19,820+/-390
9 Solutrean	
8 Solutrean	15,860+/-330
	20,690+/-810
7 Solutrean	
5 Solutrean	
4 Solutrean	20,970+/-620

La Riera is located at 43°25'31" N and about 4°52' W. The cave entrance is 30 m a.s.l., and is 1.5 km from the Río Calabres to the west and 1.75 km south of the modern shore. La Riera is situated in the coastal lowlands of Asturias, though foothills of over 600 m altitude can be reached just 4 km to the south. Its location made for easy coastal resource exploitation, even during times of lowered sea level. The cave itself has an entrance about 5 m wide and a main vestibule extending over 15 m inwards. Excavations here began with the Conde de la Vega del Sella in 1917 to 1918. A few examples of parietal art were discovered in the main chamber by M. Mallo, M. Pérez and J.A. Alvarez in 1968. In 1969, G. Clark tested the deposits in front of the cave, which was repeated in 1972 by J. M. Gómez-Tabanera and M. Pérez with two small pits. The cave also suffered many depredations by looters in the many years since its original excavation and the most recent project directed by L. G. Straus and G. Clark from 1976 through 1979 (Straus and Clark 1986; Straus *et al.* 1981). This site has yielded levels from the Solutrean, Lower and Upper Magdalenian, and Azilian periods as well as some Asturian period *conchero* deposits.

Cueto de la Mina

Level/Tool Industry

a Asturian
b Upper Magdalenian
c Middle Magdalenian
d Middle Magdalenian
e Lower Magdalenian
f Upper Solutrean
g Sterile
h Middle Solutrean
I Sterile
j Upper Aurignacian

The site of Cueto de la Mina is found in another vestibule opening of the same cavern complex that houses the site of La Riera. Cueto de la Mina is 35 m a.s.l., and about 1.5 km from the Río Calabres to the west and 1.75 km south of the modern shoreline. It is situated in the narrow coastal lowlands of Asturias, with foothills of over 600 m altitude just 4 km to the south. Its location made for easy coastal resource exploitation, even during times of lowered sea level. The occupied vestibule area is about 10 m by 10 m. Aurignacian, Solutrean, Lower and Upper Magdalenian, and Asturian deposits have been discovered here. This site was first discovered and excavated by Conde de la Vega del Sella in 1914 and tested by F. Jordá in 1953 (Castro Bernardez 1986).

Tito Bustillo

Level/Tool Industry	Associated ^{14}C Dates
1a/b Upper Magdalenian	14,220+/-180
	14,250+/-300
	15,180+/-300
	15,400+/-300
1b-c Upper Magdalenian	
1c Upper Magdalenian	13,520+/-300
	13,870+/-300

The cave of Tito Bustillo is located nearly at sea level, 500 m from the mouth of the Sella River and 750 m from the modern coastline. Tito Bustillo was discovered only in 1968, this delay caused by the sealing of its original entrance. The cave itself is quite massive, extending in one main tunnel tens of meters tall for hundreds of meters inward, with spectacular limestone formations and multiple side caverns. The site of La Lloseta was discovered at another entrance to this cavern complex. Tito Bustillo also has extensive parietal art in the form of multiple geometrics along the main cavern, engravings in a small side chamber, and the Reindeer Panel containing multiple superimposed engravings and polychrome images of reindeer and horse deep within the cavern. Occupation debris has been detected near the old main entrance and at the foot of the Reindeer Panel. All the levels excavated to date have been Upper Magdalenian, the oldest dated occurrence of this tool complex in Cantabrian Spain.

The first tests of its archaeological deposits were carried out by M. A. García Guinea in 1970, who due to the absence of antler harpoons and the presence of other types assigned this site to the Lower Magdalenian. More extensive excavations in 1972 and 1974-75 under the direction of J.A. Moure Romanillo unearthed harpoons, leading to reassignment to the Upper Magdalenian (Moure Romanillo 1975; Moure Romanillo and Cano Herrera 1976, 1978, 1979).

MOUSTERIAN ENVIRONMENTS

Very little can be said about varying Mousterian climate from the microfaunal data, since no large samples have yet been recovered. Only three sites (Amalda, Morín and Lezetxiki) have yielded microfauna from their Mousterian levels, and each has a small sample size (Table 8.1, end of chapter). In addition, the Mousterian period covers a long period from perhaps 130,000 B.P. to 40,000 B.P. and spans much of Oxygen Isotope Stages 5, 4 and 3: from full interglacial to interstadial conditions. Most paleoenvironmental information from this period comes from correlation with sea level changes, palynology and macromammalian fauna.

Amalda Levels X and VIII (Sterile overlying Mousterian) and VII (Mousterian) have yielded small microfaunal samples (Pemán 1990). The microtines present in Level X are *Microtus arvalis* (n = 133), *M. oeconomus* (n = 44), *Pitymys pyrenaicus* (n = 24), *M. nivalis* (n = 2) and *Arvicola* cf. *terrestris* (n = 1), which indicate a cold, treeless environment with open areas of grassland and perhaps some moist heath. Levels VIII and VII are less informative, with small amounts of *Arvicola* cf. *terrestris*, *Pliomys lenki* and *Sorex araneus* present. Cueva Morín has yielded only a single element of *Arvicola* and one additional unidentified rodent remain from its Mousterian Level 15 (Altuna 1978). Mousterian age microfauna from Lezetxiki Levels VII, Vi and Vb is similarly uninformative, with *Microtus agrestis/arvalis*, *Arvicola*, the extinct microtine *Pliomys lenki*, and *Sorex* present in small numbers.

EARLY UPPER PALEOLITHIC ENVIRONMENTS

While a larger body of microfaunal data exists from this shorter period, generally small sample sizes limit interpretation to the presence/absence of indicator species (Table 8.2). Microfauna from Early Upper Paleolithic levels has been reported from Amalda (Pemán 1984), Aitzbitarte (Altuna 1972), El Castillo (Cabrera 1984), Ekain (Zabala 1984), Lezetxiki (Altuna 1972) and Cueva Morín (Altuna 1978). Culture stratigraphic units cover Aurignacian and Chatelperronian (40,000 to 27,000 B.P.) and Gravettian/Upper Perigordian (27,000 to 20,500 B.P.) levels. Oxygen Isotope Stages span late 3 and early 2: from interstadial conditions to glacial maximum.

The microfaunal sequence from Lezetxiki (Levels IVc, IVb, IVa, IIIb and IIIa) spans Protoaurignacian to Upper Aurignacian deposits (Altuna 1972). Sample size is small, but the microfauna indicates a mostly treeless environment based upon the dominance of microtines. The uncertainly attributed Levels Xa (Chatelperronian) and IXb (Aurignacian) of Ekain are similarly dominated by microtines, indicating treeless conditions (Zabala 1984). Only *Arvicola* and *Talpa europaea* (both relatively large and easily identifiable) have been reported from the Superficial and Basal Aurignacian G Levels from El Castillo (Cabrera 1984). These two taxa have also been identified from the Gravettian levels of Cueva Morín (Altuna 1978).

More complete information is available from the Aurignacian level of Aitzbitarte (Altuna 1972). Treeless conditions are indicated here as well, with microtines dominating the sample: *A. terrestris* (n = 59), *Microtus oeconomus* (n = 48), *M. agrestis/arvalis* (n = 32), *Pitymys* spp. (n = 4) and *M. nivalis* (n = 1) have been recovered along with *Sorex araneus* and *Talpa europaea*. Humid conditions are also indicated by the presence of *A. terrestris* and *M. oeconomus*, with some bare rocky areas at least within a predator's hunting radius as indicated by *M. nivalis*. Open, grassy areas are indicated by *M. agrestis/arvalis* and *Pitymys*. The habitat indicated by this microfauna is similar to El Juyo Level 4/4S, though Aitzbitarte exists in a slightly higher altitudinal zone (220 m a.s.l.) with nearby foothills climbing to over 500 m.

Cueto de la Mina has yielded a large microfaunal sample from its Aurignacian Level 7 (Castro Bernardez 1986). Some partial tree cover is indicated for the time of deposition of this assemblage by the presence of arboreal *Eliomys quercinus* and forest-favoring *Apodemus sylvaticus* and *A. flavicollis*. Other fauna indicate significantly colder and more open conditions than today. Significant downward shift of the alpine zone is indicated by the presence of *Galemys pyrenaicus* and *Microtus nivalis*. Significant open grassy areas are indicated by *M. agrestis/arvalis* and open water/humid meadow by *Arvicola, Sorex coronatus, S. minutus* and *M. oeconomus*.

Level V (Evolved Perigordian) from Amalda (Pemán 1990) has been radiocarbon dated to a time of interstadial ocean level (27,000+/-1,000 and 27,000+/-1,100 B.P.). The small microfaunal sample is indicative of open, grassy conditions with *Microtus agrestis* (n = 69) and *M. arvalis* (n = 112) predominating over *M. oeconomus* (n = 39), *M. nivalis* (n = 1) and *A.* cf. *terrestris* (n = 22). Level VI (Perigordian V) from Amalda has been radiocarbon dated to around the last glacial maximum (17,880+/-390 and 19,000+/-340 B.P.) and coincides with more extreme environment. The small microfaunal sample is even more dominated by *M. arvalis* (n = 96) and is indicative of the spread of drier, open grass environment.

SOLUTREAN ENVIRONMENTS

Of shorter duration than the Mousterian and Early Upper Paleolithic, the Solutrean in Cantabrian Spain (20,500 to 16,500 B.P.) spanning roughly the last glacial maximum (ca. 22,000-18,000 B.P.) is underrepresented by microfaunal data. Some is reported from the sites of Amalda (Pemán 1990), Aitzbitarte (Altuna 1972), Las Caldas (Corchón 1981), Ermittia (Altuna 1972), Morín (Altuna 1978) and La Riera (Altuna 1986). No large samples have been recovered, so paleoenvironmental inferences must be drawn by the presence/absence of indicator species. The small amounts of microfauna recovered may also be a partial result of an environment unfavorable to barn owls (very cold), which reduced the population of this accumulating species in Cantabrian Spain.

Cold and treeless conditions are indicated by the sample from the Upper Solutrean levels (IVb and IVa) from Amalda (Pemán 1990). These have been radiocarbon dated to after the glacial maximum: dates of 17,580+/-440, 16,200+/-300 and 16,090+/-240 B.P. indicate some overlap with the succeeding Lower Magdalenian period. The microfaunal samples from these two levels are dominated by open meadow microtines (*Microtus arvalis, M. agrestis, Pitymys pyrenaicus* and *P. lusitanicus*) indicative of drier and cooler conditions. *M. nivalis* is also found in high amounts for a coastal Cantabrian site (total n = 22) and indicates open rocky areas near Amalda. Humid indicator species are reduced, including *A.* cf. *terrestris* and *M. oeconomus*. Some tree cover is indicated by the presence of *Apodemus*.

The Solutrean level microfaunal sample of Aitzbitarte exhibits a definite cold character (Altuna 1972). While *A. terrestris* has the highest frequency among the microtines (n = 120), *M. oeconomus* is similarly abundant (n = 109) and far outweighs both *M. agrestis/arvalis* (n = 36) and *Pitymys* (n = 13). Alpine-dwelling *M. nivalis* is also present (n = 3), while no *Apodemus* has been recovered from this level. *Talpa* is the most abundant taxon (n = 616) due to its greater ease of identification from multiple durable elements. This microfauna assemblage indicates a humid, cold, open environment with perhaps sparse trees and open water nearby and is similar to El Juyo Level 4/4S.

The Middle Solutrean levels (15, 16, 17) of Las Caldas are noted only for the presence of *Apodemus* cf. *flavicollis, M. arvalis, Microtus* spp., *Arvicola sapidus* and *Pitymys* cf. *lusitanicus* (Corchón 1981). These species are also common to the modern Cantabrian fauna, though the microtines favor open environments and *Apodemus* favors woodland. Material from Solutrean Level 3 at Cueva Morín is similarly uninformative, with only a single remain each of *Arvicola* and *Talpa* reported (Altuna 1978). The Solutrean level at Ermittia (Altuna 1972) has yielded only *Microtus agrestis/arvalis, M. oeconomus, Arvicola terrestris* and *Talpa europaea*. The presence of *M. oeconomus* agrees with the cold phase as indicated by lowered ocean level. La Riera has yielded a small microfaunal sample from its Solutrean levels (4, 5, 7-11 and 14-17) that consists mostly of *Arvicola terrestris*. The presence of *Galemys pyrenaicus* in Level 7 is a strong indicator of downward altitudinal zone shift, since this species (Pyrenean desman) is today found only adjacent to water in the alpine zone above 300 m (Chapter 3), while La Riera is almost upon the modern coast.

MAGDALENIAN ENVIRONMENTS

The Magdalenian period has sufficient amounts of recovered microfaunal remains to allow paleoenvironmental interpretation beyond indicator species for portions of this time (Table 8.4). This culture stratigraphic phase spans 16,500 to 13,500 B.P. (Lower Magdalenian) and 13,500 to 11,000 B.P. (Upper Magdalenian) in Cantabrian Spain, with some overlap between these two major divisions. In addition to El Juyo (Chapter 5), microfaunal data has been reported for Magdalenian levels from Aitzbitarte (Altuna 1972), Altamira (Castro Bernardez 1986), Ekain (Zabala 1984), Ermittia (Altuna 1972), Erralla (Pemán 1985), El Pendo (Fuentes Vidarte 1980), Lezetxiki (Altuna 1972), Cueva Morín (Altuna 1978), Rascaño (Altuna 1981), La Riera (Altuna 1986), Tito Bustillo (Altuna 1976) and Urtiaga (Altuna 1972). This archaeological period covers the late Oxygen Isotope Stage 2 and the onset of Stage 1, representing sea level rise from the glacial maximum during the Solutrean period towards the high levels of the Holocene. Fluctuations between these two extremes complicate interpretation, since ocean level rise was not steady and went through repeated temporary lowerings before obtaining Holocene levels ca. 10,000 B.P. In addition, warming temperatures can be expected to precede an increase in ocean level, since melting of continental and alpine glaciers is not instantaneous. Similarly, the onset of cooling temperatures will be followed by lowering ocean level as ice accumulates on land.

Lower Magdalenian environments in Cantabrian Spain correspond to generally higher ocean levels and cold conditions somewhat ameliorated from the glacial maximum. Lower Magdalenian levels just postdating the glacial maximum have been excavated at the upland site of Rascaño (Altuna 1981). Initial Magdalenian Level 5 has been dated to 16,433+/-131 B.P., and Lower Magdalenian Levels 4 and 3 have yielded dates respectively of 15,988+/-193 B.P. and 15,173+/-160 B.P. (González Echegaray and Barandiarán 1981:213). The small microfaunal sample from Level 4b includes somewhat anomalously the presence of *Glis glis*, indicating some forest cover. Other species present reflect open and humid conditions. The Lower Magdalenian levels of La Riera (18, 19 and 20) also indicate open and humid conditions and include one alpine indicator (*Microtus nivalis*) at a site adjacent to the modern coastline (Altuna 1986). These levels have yielded comparable radiocarbon determinations of 16,420+/-430, 15,520+/-350 and 15,230+/-300 B.P. (Straus and Clark 1986).

The Lower Magdalenian Levels VIIa-f at Ekain show a similar pattern of open, cold and humid conditions (Zabala 1984). Microtines are dominated by *Arvicola* and *Microtus oeconomus* with *M. nivalis* present. *Pitymys* and *M. agrestis/arvalis* (indicative of dryer grasslands) are present in much smaller numbers. Altamira Lower Magdalenian Level 2+3 reflects a similarly open and humid environment, but with more indications of grassland and a forest species (*Apodemus sylvaticus*) (Castro Bernardez 1986). Greater tree cover is indicated by the Magdalenian (Level 6) and Middle Magdalenian (Level 5) from Dufaure in nearby France, where *Apodemus* is a significant portion of the microfaunal assemblage (Eastham 1995). A cold, open and humid environment is also indicated by the dominance of *Arvicola terrestris*, *Microtus ratticeps* (= *oeconomus*) and *M. agrestis*.

More complete information comes from the Lower Magdalenian Level V of Erralla, which overlies culturally-sterile Level IV (Pemán 1985). Level V has yielded dates of 15,740+/-240 and 16,200+/-240 B.P. (Altuna 1985b:26). This deposit corresponds with temporarily lowered ocean level and the Dryas I pollen zone, and hence cooler conditions of nearly those obtained at the last glacial maximum (ca. 22,000-18,000 B.P.).

As discussed in Chapter 5, the Lower Magdalenian levels of El Juyo show gradual climatic deterioration. Microfauna from Levels 9, 8, 7 (dated to 14,400+/-180 B.P.) and 6 show a cold, very humid and open environment. Palynological and macrobotanical analysis indicates that heath predominated, along with many wetland-preferring taxa. Some slight forest cover and areas of grassland are indicated. These levels correspond to the relatively temperate Pre-Bölling pollen zone. Microfauna from Level 4/4S (dated to 13,920+/-240 B.P.) shows deterioration of this environment corresponding to temporary lowering of ocean level and the Dryas I pollen zone. An increase in alpine species is noted among the microfaunal, pollen and macrobotanical remains which indicates significant lowering of alpine zones to within a few kilometers of this coastal lowland site.

The undesignated Magdalenian level at Ermittia (Altuna 1972) has too small a sample to allow interpretation, with only the larger taxa *A. terrestris* and *T. europaea* present. Both indicate open and humid conditions. The microfaunal sample of the undesignated Magdalenian level at Aitzbitarte (Altuna 1972) reflects a harsher environment and is hence most probably Lower Magdalenian in age. Cold, open and wet conditions are indicated by the dominance of *A. terrestris* (n = 195) and *M. oeconomus* (n = 144) and the presence of *M. nivalis* (n = 5), while no *Apodemus* have been recovered. Some dryer grassland is present based upon the presence in significant amounts of *M. agrestis/arvalis* (n = 65) and *Pitymys* (n = 18).

The microfaunal sample from Lower Magdalenian Level V at Erralla is large enough to allow closer statistical examination. This sample is interpreted by Pemán (1985) as having resulted from owl pellet egestion in the cave. While he does not specify which species, examination of the body size profile of the microfaunal prey indicates probable accumulation by a medium owl. The accumulation of such a large sample in a cave setting and the very light damage upon the individual elements rules out other medium sized owl species other than barn owls (Chapter 2). Analysis of this sample will proceed with the assumption that barn owls were its primary accumulators, though subsequent taphonomic examination may alter this view and necessitate adjustment of paleoenvironmental interpretation.

Analysis of Erralla Level V is presented in Table 8.5. This occupation has been radiocarbon dated to 15,740+/-240 B.P. and 16,200+/-240 B.P. (Altuna 1985b:26), predating the Lower Magdalenian occupation of El Juyo. This time corresponds with rising ocean levels just postdating the glacial maximum ca. 22,000-18,000 B.P., so climatically rigorous conditions were likely. This level has the largest total MNI (7469) for any Cantabrian Upper Paleolithic level published to date. This sample has been analyzed in the same manner as the El Juyo microfauna (Chapter 5): individual species MNI have been converted to total prey units, which in turn represent the total biomass derived from each habitat type and its proportional representation of landscape area.

Similar to the Lower Magdalenian occupation of El Juyo, habitat area (Figure 8.2) is dominated by humid meadow/water edge species (72.21%). This total includes a very high proportion of *Microtus oeconomus*, indicating significantly cold climate as well as humid climate. This open landscape also had significant areas (10.8%) of dry meadow (grasses). The individuals of the genus *Pitymys* were not identified to species. In this colder regime, these most likely belonged to *P. pyrenaicus*, a generalist species. Some portion of the *Pitymys* total may have belonged to *P. lusitanicus*, which would increase the total area of humid meadow present. Forest cover was minimal within the barn owl hunting radius of the site, amounting to well under 1% of the land area. That some forest was present is also indicated by the presence of *Glis glis*, the presence of which in this level is somewhat anomalous to the general environmental picture.

Any tree cover was found in sheltered areas on the landscape. Harsh conditions are reinforced by the presence of the alpine species *Microtus nivalis* and the only Cantabrian Upper Paleolithic occurrence of *Microtus gregalis* (steppe lemming). This latter species is found today in the tundra and steppe biomes of Asia (Chapter 3). These two species indicate significant altitude zonal shift downwards to within a radius of a few kilometers of this lowland site. The proportion of them must have been greater in the alpine zone of the Cantabrian Cordillera to the south during this occupation of Erralla. This site does not include any remains of the alpine species *Galemys pyrenaicus*.

Cantabrian Upper Magdalenian environments change in a fluctuating manner from relatively harsh interstadial to warmer and more forested conditions. Only inconclusive samples are available from Upper Magdalenian Level 2 at Rascaño (Altuna 1981) and Level I at Lezetxiki (Altuna 1972).

The Upper Magdalenian Levels 1a-c of Tito Bustillo have been dated to the very early Upper Magdalenian period and in fact overlap with the end of the Lower Magdalenian. These radiocarbon determinations range from 13,520+/-300 to (perhaps anomalous) 15,400+/-300 B.P. (Moure Romanillo 1976). A date just preceding 13,000 B.P. would place Tito Bustillo amidst the relatively temperate Bölling pollen zone. The microfaunal sample from this site (Altuna 1976) suggests that somewhat earlier dating is correct, placing the time of occupation during the end of the colder pollen phase Dryas I. While small, the sample has only a single remain of a forest species (*Apodemus*) and is dominated by open-environment microtines: *A. terrestris* (n = 61), *M. oeconomus* (n = 7), and *M. agrestis/arvalis* (n = 6). A cold-phase attribution for this occupation is further supported by the presence of a single remain of reindeer (*Rangifer tarandus*) from level 1b (Altuna 1976:153). This species rarely penetrated as far south as Cantabrian Spain and is associated with true tundra environments. Original interpretation of this site indicated a Lower Magdalenian industrial attribution, until the subsequent discovery of antler harpoon heads placed it by definition within the Upper Magdalenian (Moure Romanillo 1975; Moure Romanillo and Cano Herrera 1976, 1978, 1979). Tito Bustillo is adjacent to a major river and near the modern coastline. The early presence in it of harpoon heads (presumably for fishing) may be more due to site function than to changing artifact style. A similar microfaunal sample has been recovered from the Upper Magdalenian Level VIb of Ekain (Zabala 1984).

The later occurrences of the Upper Magdalenian show mixed amounts of environmental amelioration and may correspond to the Bölling (warmer), Dryas II (cooler), Alleröd (warmer) and Dryas III (cooler) pollen zones. Levels 24 and 21-23 of La Riera have been dated to the end of this phase (Straus and Clark 1986). This site retains alpine vole *Microtus nivalis* in Level 24 and (Altuna 1986). The sample size is too small to conclude more about the environment other than more probable assignment to a cooler phase. The Final Magdalenian Level VIa at Ekain has a very similar microfaunal sample (other than the relative reduction in *Arvicola*) as the Upper Magdalenian Level VIb at this site (Zabala 1984).

Microfaunal samples from the Final Magdalenian level of Urtiaga (Altuna 1972) and Level 2 of Cueva Morín (Altuna 1978) are uninformative, as each is dominated by larger microfauna as the probable result of excavation technique. *Apodemus sylvaticus* does show up in Final Magdalenian Levels IIab and II at El Pendo (Fuentes Vidarte 1980), indicating some return of forest cover.

The Final Magdalenian levels at Erralla are more illuminating due to their greater sample size. Level III has been dated to 12,310+/-190 B.P. (Altuna 1985b:26), coinciding with the Dryas II or Alleröd pollen phases. Assignment to the temporary warming conditions of the Alleröd is more likely based upon the microfaunal sample (Pemán 1985). The arboreal *Glis glis* shows up for the first time at Erralla in these levels, as does forest-dwelling *Clethrionomys glareolus*. *Apodemus* is also abundant, while *Microtus nivalis* declines from previously higher levels. The other alpine species (*M. gregalis*) does not recur. While *M. oeconomus* is still present, it no longer dominates the microfaunal sample. Other indicators of humid open cover have declined (*Arvicola*, *Talpa europaea* and *Sorex* spp.). The changed microfaunal profile indicates the cool, humid heath with sparse trees of the Lower Magdalenian have shifted towards dryer grassland with increasing tree cover and a retreat of the alpine zone away from the hunting radius of Erralla. In contrast, the nearby French site of Dufaure shows relatively little microfaunal turnover from its Magdalenian level (6) through its Middle (5) and Final Magdalenian levels (4 inf. through 4 sup.) (Eastham 1995).

AZILIAN ENVIRONMENTS

The Azilian archaeological period in Cantabrian Spain (11,000 to 9,000 B.P.) occurred during the Terminal Pleistocene/Initial Holocene and therefore spans the time of abrupt ocean level rise and general climatic amelioration. This period begins during the (cooler) Dryas III pollen zone and continues into Preboreal. Global sea level is recorded as early Oxygen Isotope Stage 1, where levels rapidly rose to the approximate modern level by 10,000 B.P. Microfaunal samples (Table 8.6) are less adequate than those for the Magdalenian period and come from the sites of Aitzbitarte (Altuna 1972), Ekain (Zabala 1984), Ermittia (Altuna 1972), El Pendo (Fuentes Vidarte 1980), Rascaño (Altuna 1981), La Riera (Altuna 1986), and Urtiaga (Altuna 1972).

The changing Cantabrian Azilian environment reflects the warming climate with a spread of forests, the cliserial shift northwards of cold-adapted species ranges and the retreat upwards of alpine biotic zones (Peñalba 1994). This transition is noted at La Riera, where the small microfaunal sample from Azilian levels 27-30 includes the arboreal dormouse species *Glis glis* for the first time (Altuna 1986). Dormice tend to be rare owl prey, since they dwell among tree crowns and are thus more difficult to hunt. Their presence in a site therefore strongly indicates forested conditions. A similar pattern is found at El Pendo (Fuentes Vidarte 1980). The microfaunal samples from the Azilian levels of Rascaño (Altuna 1981), Ermittia and Urtiaga (Altuna 1972), and Ekain (Zabala 1984) are uninformative as they yield primarily *Arvicola* and *Talpa* (perhaps not coincidentally the two most visible taxa among the smaller microfauna). The situation is similar for Azilian Level 3 from the nearby French site of Dufaure, though the arctic vole *Microtus oeconomus* persists locally (Eastham 1995).

Aitzbitarte has yielded a more substantial microfaunal sample from its Azilian level (Altuna 1972). *Apodemus* shows up for the first time in this site. Aitzbitarte has a late persistence of *Microtus oeconomus* and *M. nivalis*, though the former species decreases relative to other microtines in the Magdalenian level at this site.

POST-PALEOLITHIC ENVIRONMENTS

While not the focus of this research, data from post-paleolithic levels of sites also yielding paleolithic levels are presented to illustrate the environmental changes that ensued after the Pleistocene/Holocene transition ca. 10,000 B.P. Western European pollen zones proceeded through the Boreal (9,000-8,000 B.P.), Atlantic (8,000-5,000 B.P.) and Sub-Boreal (5,000-3,000 B.P.). Microfaunal data is presented here from the sites of Abauntz (Altuna and Mariezkurrena 1982), Amalda (Pemán 1990), Marizulo and Urtiaga (Altuna 1972). The microfauna from these levels reflects the general warming that had occurred and with it the return of forested conditions to the Cantabrian coast. The modern open Cantabrian environment (Chapter 4) is largely due to the effects of agriculture, pastoralism, forestry and urbanization and does not reflect subsequent climatic change.

Microfauna from these levels is listed in Table 8.7. The Final Mesolithic is represented at Marizulo (Altuna 1972), where the only smaller microfaunal species represented is *Glis glis*. The Eneolithic Level b2-b1 at Abauntz has yielded only the remains of larger microfauna including *Erinaceus europaeus*. This level has yielded a radiocarbon date of 4,240+/-140 B.P. The Neolithic is represented at Marizulo, where remains of ubiquitous *Arvicola terrestris* and *Talpa* are reported.

The Chalcolithic levels IIIb and IIIa at Amalda (Pemán 1990) have yielded some remains of arboreal *Glis glis* as well as forest-favoring *Apodemus* and *Clethrionomys glareolus*. Also common are *Microtus agrestis* and *Pitymys pyrenaicus*, which favor more open environments. Amalda also shows the late survival of a locally relict population of *Microtus oeconomus*. The presence of this species in a warmer climate in competition with *M. agrestis* and *M. arvalis* indicates the presence of a wetland/humid microhabitat near the site. Also present is *M. nivalis*, an alpine species which inhabits exposed rocky areas. The location of Amalda in the Cantabrian Cordillera foothills (with hills above 600 m within 2 km of the site, well within the hunting range of many owl species) explains the persistence of *M. nivalis* to this late date after the general climatic amelioration. The Bronze Age is represented by Level B at Urtiaga (Altuna 1972). Significant forest cover in the vicinity of this site is indicated by the presence (n = 10) of *Glis glis*, with *A. terrestris*, *Talpa europaea* and three taxa of bats also present.

The Late Roman period (Level II) sample from Amalda is more similar to the contemporary microfauna of Cantabrian Spain than to that of the Late Pleistocene or earlier Holocene. This level has high proportions (n = 156) of the forest taxon *Apodemus* and some (n = 11) *Clethrionomys glareolus* and (n = 3) *Glis glis*. *Arvicola* cf. *terrestris* has nearly disappeared from the microfaunal profile. While other humid meadow/water edge taxa are present (*Pitymys lusitanicus*, *Sorex araneus* and *Neomys*), dry meadow species (*Microtus agrestis* and *Crocidura* spp.) are found in greater amounts. Their great abundance indicates the significant local alteration of landscape for agriculture, and is reinforced by the arrival of the murids *Micromys minutus* and *Mus musculus*. The former species, while not a true commensal, is associated with the spread of agricultural fields. The latter species is of course associated with the spread of human habitations.

Table 8.5: Lower Magdalenian Level V of Erralla.

	MNI	%MNI	Prey Units	Total Biomass	%Biomass	Habitat Association	%Habitat Area
Microtus nivalis	129	1.7	1	129	1.4	Alpine	
Microtus gregalis	2	0.03	1	2	0.02	Tundra	1.42
Glis glis	X	*	[4]	*		Forest	
Apodemus	8	0.1	1	8	0.1	Forest	0.1
Talpa cf. *europaea*	274	3.7	5	1370	15.3	Humid Meadow	
Sorex araneus	1482	19.8	0.5	741	8.3	Humid Meadow	
Sorex minutus	5	0.07	0.2	1	0.01	Humid Meadow	
Neomys fodiens	29	0.4	0.75	21.75	0.2	Wetlands/Water	
Microtus oeconomus	2910	39.0	1	2910	32.5	Humid Meadow	
Arvicola	285	3.8	5	1425	15.9	Humid Mead/Water	72.21
Microtus agrestis-arvalis	968	13.0	1	968	10.8	Dry Meadow	10.8
Pitymys	1377	18.4	1	1377	15.4	Generalist?	15.4
Total	7469	100.0		8952.75	99.9		99.9

Figure 8.2: Paleoenvironment of Erralla Level V.

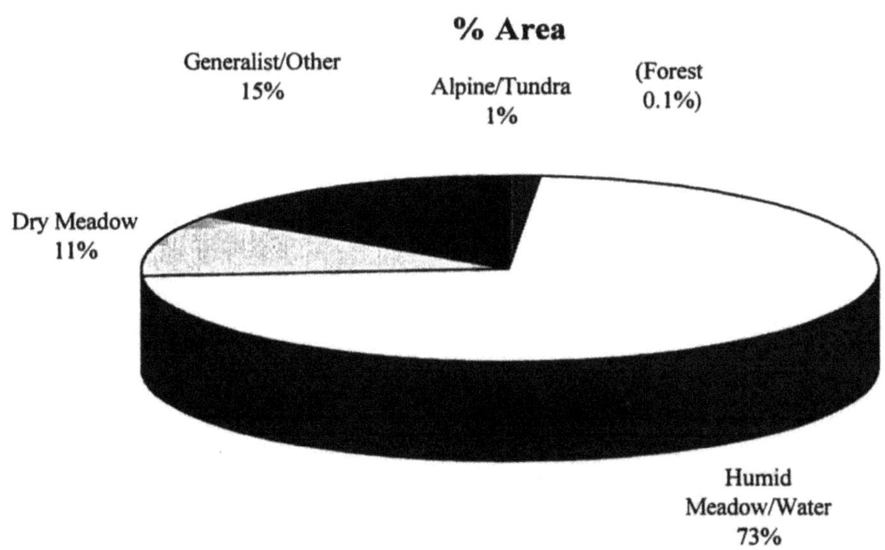

Table 8.1: Microfauna from Cantabrian Mousterian Levels.

	Amalda VII Mousterian	Amalda VIII Sterile	Amalda X Sterile	Lezetxiki Va Sterile	Lezetxiki Vb Mousterian	Lezetxiki VI Mousterian	Morín 15 Mousterian
Glis glis							
Eliomys quercinus							
Apodemus				1			
Apodemus sylvaticus							
Apodemus flavicollis							
Micromys minutus							
Mus musculus							
Rattus rattus							
Microtus							
Microtus arvalis			133				
Microtus agrestis							
Microtus agrestis-arvalis				4	1		
Microtus nivalis			2				
Microtus oeconomus			44				
Microtus gregalis							
Arvicola				3	1	1	1
Arvicola terrestris	3		1				
Arvicola sapidus							
Clethrionomys glareolus			X				
Pitymys				2			
Pitymys lusitanicus			1				
Pitymys pyrenaicus			24				
Pliomys lenki	2	X	1	2	1	1	
Talpa							
Talpa europaea			1				
Talpa occidentalis							
Sorex				1	1		
Sorex araneus	1		3				
Sorex coronatus							
Sorex minutus							
Neomys							
Neomys fodiens							
Neomys anomalus							
Crocidura			1				
Crocidura russula							
Erinaceus europaeus							
Galemys pyrenaicus							
Chiroptera							

Table 8.2: Microfauna from Cantabrian Early Upper Paleolithic Levels.

	Amalda V Evolved Perigordian	Amalda VI Perigordian V	Aitzbitarte Aurignacian	Castillo Sup. G. Aurignacian	Castillo Base G. Aurignacian	C. de la Mina 7 Aurignacian	Ekain VIIIa Interstadial	Ekain VIIIb Interstadial	Ekain IXa	Ekain IXb Aurignacian	Ekain Xa Chatelperronian
Glis glis											
Eliomys quercinus						1					
Apodemus		2									
Apodemus sylvaticus						2					
Apodemus flavicollis						3					
Micromys minutus											
Mus musculus											
Rattus rattus											
Microtus											
Microtus arvalis	96	112									
Microtus agrestis		69									
Microtus agrestis-arvalis			32			286		1		1	2
Microtus nivalis		1	1			7	2				
Microtus oeconomus	15	39	48			76	4				
Microtus gregalis											
Arvicola						120	44	20	4	27	9
Arvicola terrestris	6	22	59	X							
Arvicola sapidus											
Clethrionomys glareolus		X									
Pitymys			4			75	2			2	
Pitymys lusitanicus	X	X				14					
Pitymys pyrenaicus	X	X				3					
Pliomys lenki	1	4								1	
Talpa											
Talpa europaea	1	3	96		X	51	8	3			
Talpa occidentalis						160					
Sorex							1				
Sorex araneus	7	32	2								
Sorex coronatus						142					
Sorex minutus						9					
Neomys		2									
Neomys fodiens											
Neomys anomalus						25					
Crocidura											
Crocidura russula											
Erinaceus europaeus						2					
Galemys pyrenaicus						3					
Chiroptera											

Table 8.2: (Cont.)

	Lezetxiki IIIa Upper Aurignacian	Lezetxiki IIIb Sterile	Lezetxiki IVa Aurignacian	Lezetxiki IVb Sterile	Lezetxiki IVc Proto Aurignacian	Lezetxiki Va Sterile	Morín 4 Gravettian	Morín 5a Gravettian
Glis glis								
Eliomys quercinus								
Apodemus						1		
Apodemus sylvaticus								
Apodemus flavicollis								
Micromys minutus								
Mus musculus								
Rattus rattus								
Microtus				1				
Microtus arvalis								
Microtus agrestis								
Microtus agrestis-arvalis	2	1		3		4		
Microtus nivalis								
Microtus oeconomus								
Microtus gregalis								
Arvicola	10	1	1		3	3	1	1
Arvicola terrestris								
Arvicola sapidus								
Clethrionomys glareolus								
Pitymys						2		
Pitymys lusitanicus								
Pitymys pyrenaicus								
Pliomys lenki			1		1	2		
Talpa							1	
Talpa europaea	2							
Talpa occidentalis								
Sorex			1			1		
Sorex araneus								
Sorex coronatus								
Sorex minutus								
Neomys								
Neomys fodiens								
Neomys anomalus								
Crocidura								
Crocidura russula								
Erinaceus europaeus								
Galemys pyrenaicus								
Chiroptera				1				

Table 8.3: Microfauna from Cantabrian Solutrean Levels.

	La Riera 17	La Riera 16	La Riera 15	La Riera 14	La Riera 11	La Riera 10	La Riera 9	La Riera 8	La Riera 7	La Riera 5	La Riera 4	La Riera 3+2	La Riera 1
Glis glis													
Eliomys quercinus													
Apodemus													
Apodemus sylvaticus													
Apodemus flavicollis													
Micromys minutus													
Mus musculus													
Rattus rattus													
Microtus													
Microtus arvalis													
Microtus agrestis													
Microtus agrestis-arvalis									1				
Microtus nivalis													
Microtus oeconomus									1				
Microtus gregalis													
Arvicola													
Arvicola terrestris	1	6		3	1	2	2	5	1		3	1	3
Arvicola sapidus							1						
Clethrionomys glareolus													
Pitymys		1							1		1		
Pitymys lusitanicus													
Pitymys pyrenaicus													
Pliomys lenki													
Talpa	1	11	1					1	4	1	1		
Talpa europaea													
Talpa occidentalis													
Sorex													
Sorex araneus													
Sorex coronatus													
Sorex minutus													
Neomys													
Neomys fodiens													
Neomys anomalus													
Crocidura													
Crocidura russula													
Erinaceus europaeus												1	
Galemys pyrenaicus									1				
Chiroptera													

Table 8.3: (Cont.)

	Amalda IVa	Amalda IVb	Aitzbitarte	Las Caldas 15-17	Ermittia	Morín 3
Glis glis						
Eliomys quercinus						
Apodemus	14	1				
Apodemus sylvaticus						
Apodemus flavicollis				X		
Micromys minutus						
Mus musculus						
Rattus rattus						
Microtus				X		
Microtus arvalis	138	189		X		
Microtus agrestis	71	5				
Microtus agrestis-arvalis			36		3[2]	
Microtus nivalis	19	3	3			
Microtus oeconomus	64	30	109		2[2]	
Microtus gregalis						
Arvicola						1
Arvicola terrestris	20	11	120		27[4]	
Arvicola sapidus				X		
Clethrionomys glareolus						
Pitymys			13			
Pitymys lusitanicus	7			X		
Pitymys pyrenaicus	55	23				
Pliomys lenki	2					
Talpa						1
Talpa europaea	1	X	616		14[3]	
Talpa occidentalis						
Sorex						
Sorex araneus	61	18	8			
Sorex coronatus						
Sorex minutus	2	1				
Neomys	2	2				
Neomys fodiens						
Neomys anomalus						
Crocidura	5					
Crocidura russula						
Erinaceus europaeus						
Galemys pyrenaicus						
Chiroptera	1					

Table 8.4: Microfauna from Magdalenian Levels.

	Aitzbitarte	Altamira 2+3 Lower	Dufaure 4 Sup. Final	Dufaure 4 M-S Final	Dufaure 4 M-I Final	Dufaure 4 Inf. Final	Dufaure 5 Middle	Dufaure 6	Ekain VIa Final	Ekain VIb Upper	Ekain VIIa Lower	Ekain VIIb Lower	Ekain VIIc Lower	Ekain VIId Lower	Ekain VIIe Lower	Ekain VIIf Lower	Ermittia
Glis glis																	
Eliomys quercinus																	
Apodemus			8	60	36	8	84	26									
Apodemus sylvaticus		3				1											
Apodemus flavicollis																	
Micromys minutus																	
Mus musculus																	
Rattus rattus																	
Microtus			37	119	94	15	299	74									
Microtus arvalis																	
Microtus agrestis			3	23	12	4	30	1									
Microtus agrestis-arvalis	65	15							1	1			1		1	1	
Microtus nivalis	5										1	2			1		
Microtus oeconomus	144	29	5	19	14	1	39	3	2	3		9	2	2	3	3	
Microtus gregalis																	
Arvicola		69							47	147	7	190	88	24	32	65	
Arvicola terrestris	195		166	560	289	71	1125	240									8[3]
Arvicola sapidus																	
Clethrionomys glareolus		12															
Pitymys	18											1		1	1	1	
Pitymys lusitanicus																	
Pitymys pyrenaicus		3															
Pliomys lenki																	
Talpa																	
Talpa europaea	759	6	75	215	103	45	321	34	8	11	1	21	13	6	7	10	1[1]
Talpa occidentalis		19															
Sorex				1										1	1	2	
Sorex araneus	9																
Sorex coronatus		9															
Sorex minutus		4															
Neomys																	
Neomys fodiens							2										
Neomys anomalus																	
Crocidura																	
Crocidura russula	1																
Erinaceus europaeus		9					1										
Galemys pyrenaicus																	
Chiroptera																	

Table 8.4: (Cont.)

	Erralla I Sterile	Erralla II Final	Erralla III Final	Erralla IV Sterile	Erralla V Lower	Erralla VI Sterile	Erralla VI' Sterile	Lezetxiki I Upper	Morín 2 Final	El Pendo II Final	El Pendo IIab Final	El Pendo IIcd Final	Rascaño 2 Upper	Rascaño 3 Lower	Rascaño 4a Lower	Rascaño 4b Lower	Rascaño 5 Initial	Rascaño 7 Sterile
Glis glis		9	3		X											2		
Eliomys quercinus																		
Apodemus	1	12	8	1	8	2												
Apodemus sylvaticus										2	1							
Apodemus flavicollis																		
Micromys minutus																		
Mus musculus																		
Rattus rattus	1																	
Microtus																		
Microtus arvalis																		
Microtus agrestis											1	2						
Microtus agrestis-arvalis	6	13	21	1	968	318	1440	1					3	9				1
Microtus nivalis	1	1	7	1	129	20	31											
Microtus oeconomus	6	16	28	2	2910	748	2085	1					8		1		1	
Microtus gregalis					2													
Arvicola	3	2	2	1	285	9	24											
Arvicola terrestris										5	4	7		21	3	1	5	1
Arvicola sapidus																		1
Clethrionomys glareolus		1																
Pitymys	4	6	15		1377	261	745				3	2	1	11				
Pitymys lusitanicus																		
Pitymys pyrenaicus																		
Pliomys lenki																		
Talpa	1	X	5	3	274	13	17		2									
Talpa europaea											4	4	6	10	2			
Talpa occidentalis														1			1	
Sorex																		
Sorex araneus	0	5	7		1482	335	1931											
Sorex coronatus																		
Sorex minutus					5		31											
Neomys																		
Neomys fodiens		1	1		29	4	24											
Neomys anomalus																		
Crocidura									3									
Crocidura russula		4	1															
Erinaceus europaeus												1		1				
Galemys pyrenaicus																		
Chiroptera																		

Table 8.4: (Cont.)

	La Riera 24 Upper	La Riera 23,22,21 Upper	La Riera 20 Lower	La Riera 19 Lower	La Riera 18 Lower	Tito Bustillo 1a Upper	Tito Bustillo 1b Upper	Tito Bustillo 1c Upper	Urtiaga Final	Urtiaga Sterile	Urtiaga Final	Urtiaga Sterile	Urtiaga Sterile
Glis glis													
Eliomys quercinus													
Apodemus								1					
Apodemus sylvaticus													
Apodemus flavicollis													
Micromys minutus													
Mus musculus													
Rattus rattus													
Microtus													
Microtus arvalis													
Microtus agrestis													
Microtus agrestis-arvalis	2	1				1	1	4					
Microtus nivalis	1				1								
Microtus oeconomus						2	2	3					
Microtus gregalis													
Arvicola													
Arvicola terrestris	16		2	10	9	12	20	29	63	7	16		3
Arvicola sapidus													
Clethrionomys glareolus													
Pitymys													
Pitymys lusitanicus													
Pitymys pyrenaicus													
Pliomys lenki													
Talpa		1	2	3		1		1					
Talpa europaea									3		3		
Talpa occidentalis													
Sorex													
Sorex araneus													
Sorex coronatus													
Sorex minutus													
Neomys													
Neomys fodiens													
Neomys anomalus													
Crocidura													
Crocidura russula													
Erinaceus europaeus		1				1		1					
Galemys pyrenaicus													
Chiroptera													

Table 8.6: Microfauna from Cantabrian Azilian Levels.

	Aitzbitarte	Dufaure 3	Ekain II	Ekain III	Ekain IV	Ekain V	Ermittia	El Pendo I	Rascaño 1	La Riera 30.1-30	La Riera 29	La Riera 28	La Riera 27	La Riera 26	Urtiaga
	Azilian	Azilian	Azilian	Azilian	Azilian	Azilian	Azilian	Azilian	Azilian					Upper Mag/Azil	Azilian
Glis glis								1			1	6	1		
Eliomys quercinus															
Apodemus	1														
Apodemus sylvaticus		1													
Apodemus flavicollis															
Micromys minutus															
Mus musculus															
Rattus rattus															
Microtus		9													
Microtus arvalis															
Microtus agrestis															
Microtus agrestis-arvalis	25														
Microtus nivalis	3														
Microtus oeconomus	32	1											1		
Microtus gregalis															
Arvicola			4	26	12	87									
Arvicola terrestris	200	49					3[2]	12	1	4	2	5	6	7	92
Arvicola sapidus															
Clethrionomys glareolus															
Pitymys	5														
Pitymys lusitanicus															
Pitymys pyrenaicus															
Pliomys lenki															
Talpa													1		
Talpa europaea	321	18	1	7	15	6	4[2]								11
Talpa occidentalis															
Sorex															
Sorex araneus	5														
Sorex coronatus															
Sorex minutus															
Neomys															
Neomys fodiens															
Neomys anomalus															
Crocidura															
Crocidura russula															
Erinaceus europaeus								1							
Galemys pyrenaicus															
Chiroptera	6				2										1

Table 8.7: Microfauna from Cantabrian Post-Paleolithic Levels.

	Abauntz	Amalda	Amalda II	Amalda IIIa	Amalda IIIb	Marizulo	Marizulo	Marizulo	Urtiaga
	Eneolithic	Current	Late Roman	Chalcolithic	Chalcolithic	Bronze Age	Neolithic	Final Mesolithic	Bronze Age
Glis glis		2	3	X	2	4		2	10
Eliomys quercinus									
Apodemus		1219	156	26	20	1			
Apodemus sylvaticus									
Apodemus flavicollis									
Micromys minutus		4	7						
Mus musculus		9	4						
Rattus rattus		2							
Microtus									
Microtus arvalis			1		7				
Microtus agrestis		431	100	15	31				
Microtus agrestis-arvalis									
Microtus nivalis			1	1	2				
Microtus oeconomus			2	1	15				
Microtus gregalis									
Arvicola									
Arvicola terrestris		7	1	3	2	1	2		3
Arvicola sapidus									
Clethrionomys glareolus		84	11	1	7				
Pitymys									
Pitymys lusitanicus		522	65	3	7				
Pitymys pyrenaicus		196	14	12	30				
Pliomys lenki									
Talpa									
Talpa europaea		14	2	X	X	1			4
Talpa occidentalis									
Sorex									
Sorex araneus		506	126	19	24				
Sorex coronatus									
Sorex minutus		37	23	3	3				
Neomys		46	13	3	2				
Neomys fodiens									
Neomys anomalus									
Crocidura		976	122	18	8				
Crocidura russula									
Erinaceus europaeus	3[1]						2		
Galemys pyrenaicus									
Chiroptera		3	1			4			5

Key to Tables:
All numbers denote NISP, except where bracketed [= MNI]
"X" denotes presence

Sources:
Abauntz: (Altuna and Mariezkurrena 1982)
Aitzbitarte IV: (Altuna 1972)
Altamira: (Castro Bernardez 1986)
Amalda: (Pemán 1990)
Las Caldas: (Corchón Soledad 1981)
El Castillo: (Cabrera 1984)
Cueto de la Mina: (Castro Bernardez 1986)
Dufaure: (Eastham 1995)
Ekain: (Zabala 1984)
Ermittia: (Altuna 1972)
Erralla: (Pemán 1985)
Lezetxiki: (Altuna 1972)
Marizulo: (Altuna 1972)
Cueva Morín: (Altuna 1978)
El Pendo: (Fuentes Vidarte 1980)
Rascaño: (Altuna 1981)
La Riera: (Altuna 1986)
Tito Bustillo: (Altuna 1976)
Urtiaga: (Altuna 1972)

CHAPTER 9
EXPLOITATION OF FUR-BEARING MAMMALS

A related subcategory of small mammal was recovered in significant amounts from the fine screen deposits of El Juyo: the remains of medium sized fur-bearing species, most of which belong to Order Carnivora. While many of these species fall below the upper boundary of 5 kg used in delimiting microfauna, many do not (at least as adults). These larger species are included in discussion here, since their inclusion in the site shows the same pattern as the smaller remains within this group. Even the smaller of the taxa discussed here, while fitting under the category of micromammals, are generally larger than the typical rodent and insectivore fauna described in the previous chapters. As such, these taxa experience different taphonomic processes, including source of mortality, manner of transport to the site, pattern and degree of skeletal element destruction, and deposition within the site. While it is possible that some of these individuals (especially the smallest species such as *Mustela nivalis* and *M. erminea* or juveniles of the larger species) were introduced as the prey of owls, the pattern of skeletal element recovery suggests that these were brought into the cave by the direct action of humans. The depositional pattern of these fur-bearing mammals is also in direct contrast to that of hedgehog (*Erinaceus europaeus*) remains recovered from El Juyo (Chapter 2). While this species is of a similar size to some fur-bearers (about 1 kg adult mass), it is not a usable source of fur due to its protective spines. The pattern of fur-bearer representation and its implications will be discussed below.

El Juyo has yielded to date 72 foot elements, primarily phalanges and metapodials of carnivores in the size range of ermine (*Mustela erminea*), ferret (*M. putorius*), and marten (*Martes* spp.). The presence of these elements is in striking contrast to the virtual lack of *any* other elements, including teeth, from these taxa (Chapter 5). Non-carnivores of this size range (such as rabbit or hedgehog) do not show this same element representation discrepancy, nor do smaller, more fragile rodent and small insectivore taxa. Differential destruction of non-foot elements of the fur-bearing species therefore may be ruled out. The author attributes the preponderance of fur-bearing mammal foot elements to their selective introduction to the site by a human agency. Such a pattern is the logical result of foot bones left in a pelt that was skinned from an animal trapped or hunted away from the site, with the remaining carcass including all other skeletal elements discarded in the field. Mustelid meat has a strong, unpleasant flavor due to their strong scent glands (personal observation) and may have been therefore a less desirable food choice. These species were probably acquired wholly for their pelts, with no other archaeologically recoverable portion of them reaching the site.

Due to the nature of the evidence involved, the exploitation of animals for their pelts is difficult to detect archaeologically. First, it presupposes a pattern of faunal procurement and processing that, while a logical expedient, is by no means the only system possible for fur acquisition. Medium carnivores may have been used for their meat also (such as fox or lynx), and in such cases an entire carcass would be returned to and disposed of in a site. Entire small carnivoran carcasses might also have been introduced to a site prior to skinning, simply because the cost of transporting them was not deemed significant or was less than the effort of skinning them in the field. Mammal foot elements (carpals/tarsals, metapodials and phalanges) are relatively small and durable (Klein and Cruz-Uribe 1984) and have a relatively low utility in associated meat, contained marrow, or as a raw material for tool manufacture. They are thus less likely than many other elements (longbones, vertebrae, ribs) to be destroyed by postdepositional forces or to be damaged as a part of butchery, marrow extraction, or artifact manufacture. Their overrepresentation in some sites is due to these taphonomic factors. In addition, carnivores often den in caves alternately with any human occupants or may encroach upon human occupation in order to scavenge food. Carnivore remains may be present in a site therefore as a result of natural mortality (Kurtén 1972, 1976) and have little to do with human activity.

A pattern of fur exploitation has been previously proposed for Upper Paleolithic sites in Cantabrian Spain (Straus 1977, 1982). Species recovered from Cantabrian sites range across the entire carnivore spectrum from the smallest carnivoran (*Mustela nivalis*) through medium-sized mustelids, foxes, and larger species such as wolf and bear. These are summarized by Straus (1992). The majority of these remains may be attributed more economically to the utilization of caves by carnivores independent of human actions. Aitzbitarte IV in Basque country has yielded several remains of *Mustela nivalis*, *M. erminea*, and *M. putorius* in its Solutrean and Magdalenian levels (Altuna 1972). Bolinkoba, a Basque site, has also yielded the remains of Mustelids and red fox (*Vulpes vulpes*).

THE LOWER PALEOLITHIC

The Acheulean in Cantabrian Spain is poorly represented, so making meaningful statements about small faunal exploitation is difficult (Figure 9.1). Only the basal levels of El Castillo have yielded stratified deposits from this period (Cabrera 1984). Some equivocal open air sites also have been found, such as Bañugues (Rodríguez and Flor 1979), and there are some isolated surface finds. Faunal information is therefore scarce. Small mammal finds in other than geological settings consist only of the presence of *Marmota marmota* in El Castillo Level 25a (Cabrera 1984:134). Typical large fauna (*Canis lupus*, *Equus caballus*, *Ursus spelaeus* and *Cervus elaphus*) are also represented in generally much greater amounts. There is no direct evidence that *M. marmota* was introduced to the El Castillo cave by hominid action, with natural mortality or animal predation being likely alternatives.

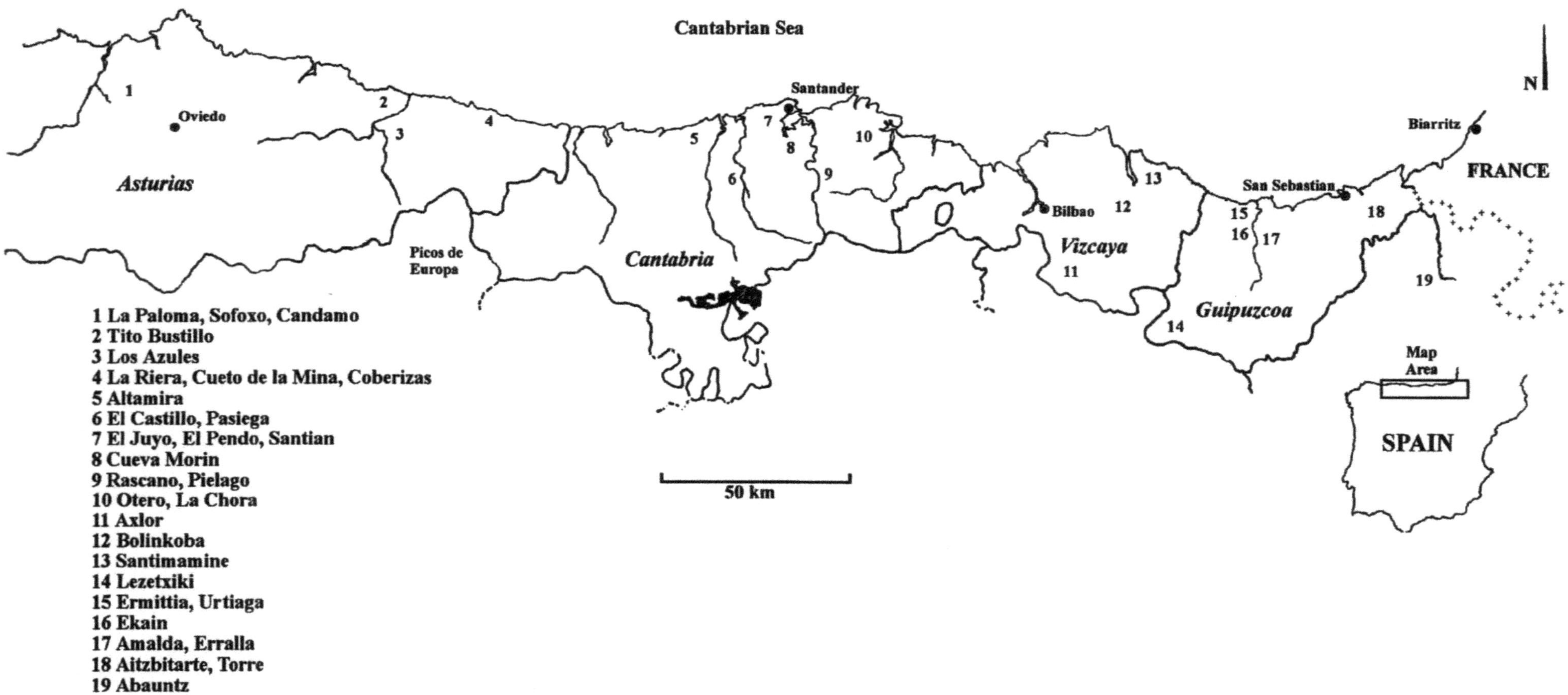

Figure 9.1: Cantabrian Middle and Upper Paleolithic Sites Yielding the Remains of Fur-bearing Species.

THE MIDDLE PALEOLITHIC

The Mousterian period in Cantabrian Spain (Table 9.1, end of chapter) is far better documented, and multiple stratified cave sites have been excavated. Large fauna exploitation by the Neanderthal inhabitants during this period exhibits a more eclectic acquisition pattern (Freeman 1973, 1981; Straus 1982, 1992) than is usually the case for the Upper Paleolithic. Typical large prey species such as red deer, horse, and bison were taken. No specialization upon red deer or ibex, such as evidence for mass drives of entire herds that occurs later (Straus 1977; Klein *et al.* 1981), was yet exhibited during this earlier period.

An eclectic pattern also holds for small mammal exploitation during the Middle Paleolithic in Spain and elsewhere in Europe (Stiner 1993, 1994). Some small mammals are found in small amounts in some levels. El Castillo Level 20 includes the presence of *Vulpes vulpes*, and Level VII at Amalda yielded 29 remains of this species (Pemán 1990). Level VI at Lezetxiki (Altuna 1972) yielded fox, plus *Meles meles*, *Felis lynx* and *Oryctolagus cuniculus* in small amounts, while both fox and *Felis silvestris* were found at Axlor (Straus 1982; Altuna 1989). This low recovery level of small fur-bearing mammals for the Mousterian may be due in some cases to the antiquity of the excavations yielding the data. Such is not the case for Amalda, which perhaps not coincidentally yielded the largest number of identified remains of any of these taxa. Contrariwise, the modern excavations of Cueva Morín with its multiple Mousterian levels (González Echegaray and Freeman 1978) did not yield any of these taxa, even though smaller rodent and insectivore species were recovered (Altuna 1978). This situation is maintained at El Pendo, whose less modern excavation (1953-57) did yield rodent, insectivore, and small fur-bearing mammal taxa from its Upper Paleolithic levels (Fuentes Vidarte 1980), but none of these taxa from its Mousterian levels. Lezetxiki also yielded a greater abundance and diversity of small fur-bearing mammals from its Upper Paleolithic levels (Altuna 1972). This variance within sites between their Middle and Upper Paleolithic levels would tend to indicate that these taxa were not exploited as intensively by Cantabrian Spain's Neanderthal inhabitants.

These generally low levels of remains are also in accordance with the interpretation that natural mortality in these taxa or the actions of non-human predators caused their introduction into these sites. There is no evidence to suggest that these taxa were exploited by hominids principally for their fur, other than their general suitability for this purpose. Other faunal niches that were un- or under-exploited during this period include game birds, molluscs, and fish. The lack of specialization in fur-trapping therefore fits the overall pattern of non-specialization visible among other resources represented archaeologically.

THE EARLY UPPER PALEOLITHIC

The Early Upper Paleolithic of Cantabrian Spain (Aurignacian, Chatelperronian, Perigordian/Gravettian), with the arrival of anatomically modern *Homo sapiens*, marks somewhat of a shift in faunal acquisition. This transition also marks major changes throughout the Old World in stone and osseous tool manufacture, landscape usage, and symbolic behavior (Chase and Dibble 1987; Harrold 1980; Klein 1989; Mellars 1990; Mellars and Stringer 1989). While the same suite of large fauna (red deer, horse, bison and ibex) were the primary prey species, some widening of the food base has been detected (Freeman 1973; Straus 1977). This shift is more apparent among small fur-bearing mammal taxa, where there is an increase in quantity and diversity of remains recovered from these levels.

Several taxa are recovered for the first time in Cantabrian Upper Paleolithic archaeological contexts (Table 9.2). Mustelids begin to assume some significance among these assemblages. *Mustela nivalis* is present in El Castillo (Aurignacian) Level 16 (Cabrera 1984), Lezetxiki (Aurignacian) Level IVa (Altuna 1972), Amalda (Perigordian) Level V (Altuna 1990) and the Aurignacian level of Aitzbitarte (Altuna 1972). *Mustela erminea* is represented in multiple Aurignacian levels at El Castillo and the Aurignacian of Aitzbitarte, as well as (Aurignacian) Level 7 of Rascaño (Altuna 1981). *Mustela putorius* also appears and is found at El Castillo, Lezetxiki and Amalda. Significantly, the Mousterian levels from each of these sites did not yield any reported remains from these taxa. Mustelids are further represented by 2 remains of *Martes* at Level Xa (Chatelperronian?) of Ekain (Altuna and Marriezkurrena 1984) and one element from Level IIIa (Aurignacian) at Lezetxiki (Altuna 1972), and the presence of *Meles meles* in this same level and at El Castillo. *Gulo gulo* is uniquely represented in a Cantabrian archaeological context at Lezetxiki Level II (Gravettian/Solutrean) with a single identified element. Since wolverine is boreal/tundra zone species, this level may be more likely of Solutrean affiliation given the coincidence of this tool complex and glacial maximum conditions.

While present in Mousterian levels, *Vulpes vulpes* remains become frequently represented, though in generally small amounts among these levels. Fox has been identified at Santimamiñe (Altuna 1972), Santian (Altuna 1972), Otero (González Echegaray *et al.* 1966), Morín (Altuna 1978), Gatzarria (Altuna 1972), Ekain (Altuna and Marriezkurrena 1984), El Castillo, Lezetxiki, Bolinkoba and Amalda. At this last site, 133 remains were recovered from (Perigordian V) Level VI (Altuna 1990). This level also yielded the only instance of arctic fox (*Alopex lagopus*) from a Cantabrian context. Lagomorphs (*Oryctolagus cuniculus* and *Lepus* spp.) also continue to show up in small amounts, such as at El Castillo, Morín, Rascaño and Amalda. Small- and medium-sized felids (*Felis silvestris* and *Felis lynx*) also show up more frequently, as at the sites of El Castillo, Gatzarria, Morín and Lezetxiki.

With the exception of the fox remains from Perigordian Levels V and VI at Amalda, these small fur-bearing mammal taxa have not been recovered in large amounts. This pattern may reflect the generally low utilization by Early Upper Paleolithic groups of this faunal niche. This pattern may also reflect less strenuous excavation/recovery techniques, which are biased against the recovery of these small remains even where these were present in larger numbers. This interpretation is strengthened by noting that the most recent excavations were those of Amalda carried out from 1979-1984 (Altuna *et al.* 1990). However, the modern excavations of Morín (González Echegaray and Freeman 1978) yielded only 10 remains among these taxa from its Early Upper Paleolithic levels (Altuna 1978). These deposits were wet-screened with a 2 mm mesh, with some samples floated for the recovery of fine remains. Low levels of small fauna remains may therefore accurately reflect low levels of exploitation in at least some cases. While the increased diversity and abundance of these taxa does suggest their exploitation by humans instead of other predators, their exploitation may have been a result of random encounter hunting and not of a true shift toward intensive utilization.

With one exception, their use specifically as fur-bearing mammals cannot be determined from the current data, given the low per part abundances. The fox sample from Amalda Level VI is large enough to examine meaningfully in terms of part representation. These remains, while including 32 (24.1%) 1st, 2nd and 3rd phalanges, also include a wide spread among all body parts with cranial elements, vertebrae and ribs, and long bones represented (Altuna 1990:174). Thus an overabundance of foot elements is only suggested in this case. However, the presence of mustelid remains (the meat of which is generally unpalatable to humans) in Early Upper Paleolithic levels does argue for fur exploitation during this period.

THE LATE UPPER PALEOLITHIC

The Late Upper Paleolithic of Cantabrian Spain, beginning with the Solutrean and continuing with the Magdalenian and Azilian periods marks a further shift in the spectrum of faunal exploitation. Solutrean levels in Cantabrian Spain (Table 9.3) show a pattern similar to that of the Early Upper Paleolithic, with generally low numbers of remains from many species. This pattern may in this case be more attributable to methods of archaeological recovery, given the antiquity of most of these excavations. Amalda again is an exception, the Upper Solutrean level (IV) of which yielded 27 *V. vulpes* remains (Altuna 1990). The recent project at La Riera Cave failed to yield large amounts of these taxa (Altuna 1986), but in this case only remnant stratigraphy in the rear of the cave was available for excavation (Straus and Clark 1986). *V. vulpes* continues to show up in small amounts at nearly all sites, along with felids and lagomorphs in some levels. Also significant is the Solutrean level of Aitzbitarte with the taxa *Mustela nivalis*, *M. erminea*, and *M. putorius* (Altuna 1972).

The Lower and Upper Magdalenian levels from Cantabrian Spain (Table 9.4) show a marked increase in the intensity of utilization of these taxa as well as definite indications that fur acquisition was a specific objective. This apparent change in faunal exploitation may also be explained in terms of the increasing scarcity of large carnivores, which had been in direct competition with humans for denning caves, prey species and occupational range. A hypothetical decline or local/total extinction throughout the Upper Paleolithic of species such as cave bear (*Ursus spelaeus*), cave lion (*Panthera leo*) or leopard (*P. pardus*) as well as the more commonplace wolf and European brown bear would (seemingly inevitably) accompany human population expansion in this or any other area. A shift to smaller fur-bearers may therefore have been in partial response to unavailability of larger species. These smaller species are today far more available in Europe than their larger counterparts (Corbet and Ovenden 1982) and in historical times were the ones that made up the bulk of the fur trade (Youngman 1982, 1990). Neither may one discount the status and/or decorative value of predator skins, given the frequent use of carnivore teeth as pendants (Corchón Rodríguez 1986) and other symbolic associations with predators. Examples of the latter category include parietal artwork depicting predators and the carved feline/human head from El Juyo (Freeman *et al.* 1988; Freeman and González Echegaray 1981).

The same species exploited during the Solutrean period continue to be important during the Magdalenian period. *V. vulpes* appears in large amounts at Urtiaga (Altuna 1972) and Rascaño (Altuna 1981). Skeletal part representation at the former site indicates a large proportion of foot elements, but teeth are also highly represented (Altuna 1972:246). Remaining elements are generally present in lower amounts. This pattern of bone representation reflects the relative survivability of the individual elements, with teeth and the small, dense foot elements predominating. No over-representation of foot elements can be assessed here with certainty. Fox skeletal part representation at Rascaño shows a slight emphasis upon foot elements, though the remainder of the skeleton is represented well enough to indicate that whole individuals at least at times were introduced to the site. In (Initial Magdalenian) Level 5 of Rascaño, foot elements make up 14 of 21 (66.7%) identified fox skeletal elements: 1 calcaneus, 1 talus, 8 metapodials and 4 first phalanges. In (Lower Magdalenian) Level 4b, foot elements make up 6 of 29 (20.7%) identified elements (Altuna 1981:255). The natural disposition of these small, dense elements to survive in an identifiable state would explain this part representation pattern without their having been introduced in greater numbers to Rascaño. This site seems to have been utilized as a specialized highland ibex hunting camp, so other specialized hunting practices were perhaps not entirely compatible with the main site function. While fox was exploited in significant numbers at each of these sites, the differential introduction of foot elements transported in furs, with the remaining elements left behind in the field, cannot be detected.

Small mustelids also show an increase in exploitation. In particular, the Magdalenian level of Aitzbitarte yielded 31 remains of *Mustela nivalis* (Altuna 1972:276). This amount is particularly significant given its size, the age of the excavation (begun anew in 1960), and the small size of this species with the concomitant difficulty of its recovery. No foot elements were identified, but these elements of *M. nivalis* are unlikely to be retrieved without the use of very fine mesh screens. One should also note that the small size of this species makes the transport of its entire carcass to a site, only to dispose of most of its mass, less cumbersome. A similar situation is encountered at (Lower Magdalenian) Level V of Erralla, where 135 remains of *M. nivalis* were recovered, including no foot elements (Altuna and Mariezkurrena 1985:112). These more recent excavations at

Erralla (concluded in 1983) perhaps not coincidentally have produced the largest sample of this species from a Cantabrian Paleolithic occupation to date. One must always suspect that the absence or low recovery of the smaller carnivoran taxa from certain occupations results from the antiquity of their excavation.

Table 9.5.1: Teeth of Fur-Bearing Taxa from El Juyo.

Level	*Mustela nivalis*	*Mustela putorius*	*Vulpes vulpes*	*Oryctolagus cuniculus*
12			10	
11	6			
8	3		4	
6	4	2	1	
4+4S	9		5	1

The ongoing excavations at El Juyo, through the use of very fine mesh screens (0.25 mm) have allowed the recovery of multiple foot elements of these taxa (Tables 9.5.1 and 9.5.2). This pattern of recovery is striking in its contrast to the near complete lack of other elements from these taxa, including teeth. Differential survivability therefore is an unlikely explanation for this pattern. This interpretation is reinforced by comparison with recovery of hedgehog (*Erinaceus europaeus*), a species for which many teeth were recovered as well as non-foot postcranial elements (Chapter 2). Hedgehog falls into the same size range as these smaller carnivoran species, yet due to its spiny pelage it cannot be utilized for fur. Since the skeletons of hedgehog and the other taxa discussed here are equivalently susceptible to destruction, postdepositional processes cannot be blamed for loss of hypothetical carnivoran teeth and non-foot postcranial elements. These elements probably were never introduced to El Juyo, the most logical explanation for which is their abandonment at the site of killing/processing.

Identification of these medium carnivore foot elements is extremely difficult, so most can only be assigned to a probable taxon based upon morphology and general size class. The majority of these remains come from the Mustelidae family, falling into the size range of *Mustela putorius* (ferret) with some smaller *M. erminea* (ermine) and larger *Martes* (marten). Specific identification cannot be made in most cases, since *Mustela lutreola* (mink) falls into the same size range as *M. putorius*. Confusion with the foot elements of Canidae or Felidae is very unlikely due to size and morphological differences. The foot elements of *Erinaceus europaeus* are distinctive morphologically from those of medium carnivore species, so confusion with this species' remains may be ruled out (Chapter 3).

The Upper Paleolithic of Cantabrian Spain closes with the Azilian period (Table 9.6, end of chapter). In addition to the usual spectrum of species *Genetta genetta* appears for the first time, in Level I of El Pendo (Fuentes Vidarte 1980). Since the modern northernmost range of this generally African species extends to Iberia and France, its appearance in the Azilian probably reflects climate amelioration and the

Table 9.5.2: Postcranial Remains of Fur-Bearing Taxa from El Juyo.

Level	Element	Size Class/Taxon
12: (n=1)	right front 1st phalanx (III)	*Vulpes vulpes*
10: (n=1)	midshaft phalanx	cf. *Mustela putorius*
9: (n=4)	metapodial fragment	cf. *Mustela erminea*
	3rd phalanx	cf. *Mustela putorius*
	2nd phalanx	"
	right front 2nd phalanx II	*Vulpes vulpes*
8: (n=30)	3rd phalanx	cf. *Mustela erminea*
	prox. 1st/2nd phalanx	"
	1st phalanx	"
	left metapodial (I)	"
	3rd phalanx	cf. *Mustela putorius*
	(8) 2nd phalanx	"
	(4) 1st phalanx	"
	right 1st phalanx (I)	"
	1st phalanx (II/III)	"
	(2) left front metapodial (II)	"
	right front metapodial (II/III/IV)	"
	(2) 1st phalanx	cf. *Martes*
	carpal fragment	small carnivore
	right front metapodial (I)	"
	1st phalanx (II/III)	*Vulpes vulpes*
	right front 1st phalanx (V)	"
	right front 2nd phalanx (III/IV)	"
	unfused 2nd phalanx	"
7: (n=1)	2nd phalanx	cf. *Mustela putorius*
6: (n=18)	prox. 3rd phalanx	cf. *Mustela erminea*
	(2) 2nd phalanx	"
	3rd phalanx	cf. *Mustela putorius*
	(3) 2nd phalanx	"
	left front 1st phalanx	"
	(1)1st phalanx	"
	1st phalanx (II/II/IV)	"
	unfused phalanx frag.	"
	(2) carpals	"
	right prox. radius	"
	(1) 2nd phalanx	cf. *Martes*
	frag. 1st phalanx (III/IV)	*Vulpes vulpes*
	right front metapodial (II)	"
	1st phalanx	cf. *Felis silvestris*
4/4s: (n=17)	(3)3rd phalanx	cf. *Mustela erminea*
	2nd phalanx	"
	1st phalanx	"
	(3) 3rd phalanx	cf. *Mustela putorius*
	(3) 2nd phalanx	"
	frag. 1st/2nd phalanx	"
	frag 1st phalanx	"
	right front metapod. (IV)	"
	right 1st phalanx (I)	*Vulpes vulpes*
	right 1st phalanx (V)	"
	right prox. unfused femur	*Oryctolagus cuniculus*
Total: n=72		

return of availability of this species rather than a shift in faunal exploitation. Rather than an increased intensity of fur acquisition being indicated, the pattern of faunal recovery is more similar to the Solutrean than to the Magdalenian period. While *Meles meles* (badger) is represented by 30 remains at Urtiaga (Altuna 1972) and 18 remains at Ekain Levels III to V (Altuna and Marriezkurrena 1984), faunal representation is otherwise sparse. *V. vulpes* appears in low numbers at many sites: Santimamiñe, Urtiaga, El Pendo, La Riera, Ekain, and Aitzbitarte. Smaller amounts of felids, mustelids and lagomorphs are also present. This lessened retrieval of fur-bearing species from Azilian levels may reflect, as in the Solutrean, the archaeological recovery methods employed at most of the excavations. A real de-emphasis of fur exploitation would also fit with the general climatic amelioration occurring during this terminal Pleistocene tool industry. In a climate more temperate than glacial Spain furs would no longer be as necessary, and clothing could have been made entirely from textiles or the hides of large game animals.

HUNTING/TRAPPING METHODS

Many methods could have been employed to acquire these species, since the ingenuity of Upper Paleolithic peoples seems limited only by technology in other areas of faunal acquisition. These areas include the invention of the bow and arrow. An innovation of this importance could have spread from one part of Europe to another in a short period of time (Bergman 1993). The Cantabrian region itself does have small Solutrean stone points and small antler points suitable for use as arrow armatures, such as those from the Magdalenian occupations of La Paloma (Martinez and Chapa 1980), Rascaño (Barandiarán 1981) and El Juyo (Barandiarán 1987; Pokines 1993). In terms of hunting large game, a variety of projectile tips (Pokines 1993) as well as drive and surround techniques (Klein *et al.* 1981; Freeman 1981; Straus 1977, 1987a, 1987b, 1992) were almost certainly employed. Projectile innovation also extended to rear-facing barbed points during the Solutrean and Lower Magdalenian (Pokines and Krupa, in prep.) and true harpoons during the Upper Magdalenian and Azilian periods (Corchón Rodríguez 1986; Moure Romanillo 1975; Moure Romanillo and Cano Herrera 1976, 1978, 1979) most likely used for fishing. Therefore, the ability of these peoples to contrive efficient methods for fur hunting and trapping is unquestioned.

What these methods were, however, is subject to debate. These taxa present the simultaneous problems in their acquisition of relatively small target size, the ability to defend and conceal themselves, and generally nocturnal activity patterns in which their sensory abilities give them an advantage over humans. They also represent a smaller payoff in terms of usable meat or hide than larger prey such as red deer, horse, or ibex. The large spear points typical to Upper Paleolithic occupations would be unsuitable for use upon all but the largest of these taxa (foxes, badgers and lynx), though arrows if used could have functioned in this role. Smaller taxa remain more problematic. A logical expedient would have been the employment of traps, which operate independently of human activity and thus can acquire nocturnal species. High species diversity also argues for the employment of traps, in that many types of small carnivore would be attracted to the same bait and captured in the same manner. Mortality would extend across a range of species, since specialization upon one would prove difficult. The number of person-hours necessary to acquire these prey items is also greatly reduced, since once placed the traps need be checked only once per day. Adult labor load may have been further reduced, since children could have maintained trap lines with little supervision (personal observation). The pattern of fur exploitation detected at El Juyo and reflected at large in the Cantabrian Upper Paleolithic record shows a trend of increasing human exploitative efficiency over the Pleistocene environment (discussed in Chapter 10).

Traps could have taken many forms, none of which are ordinarily retrievable archaeologically due to their placement away from habitation sites and perishable construction. Snares, pitfalls and deadfalls or any variation upon these main themes could have been employed (Coon 1971). Necessary raw materials are minimally wood, hide thongs or some type of fiber or sinew for binding, and stone tools suitable for sharpening stakes. Implements for digging could have been improvised (simple sticks), or the multiple rib diggers retrieved from El Juyo may have been used in this role (Freeman *et al.* 1988). It remains possible that traps are depicted in some of the geometric motifs in parietal art, but unequivocal representations are lacking.

SUMMARY

The Cantabrian region shows a pattern of increasing utilization of small fur-bearing mammals throughout the Paleolithic. Nearly non-existent in the poorly represented Acheulean levels, these taxa appear sporadically in Mousterian levels. The Early Upper Paleolithic, with the presumed arrival of anatomically modern humans, shows an increase in abundance and diversity of these taxa, including for the first time *Mustela* spp. Skeletal representation of remains from this period does not indicate a pattern of skinning in the field and transporting foot elements (phalanges, metapodials, carpals/tarsals) back to cave sites along with furs. A further shift becomes evident later in the Magdalenian period, where species abundance increases in several sites, and evidence of differential transport of foot elements into sites (El Juyo) has been recovered. The Azilian period shows an apparent decline in hunted mammal species diversity and abundance. This trend perhaps resulted from a de-emphasis of fur acquisition relating to the general climatic amelioration at the close of the Pleistocene.

Table 9.1: Fur-Bearing Taxa from Mousterian Levels.

	Amalda	Axlor	El Castillo	Lezetxiki	
	VII		20	Vb	VI
Mustela nivalis					
Mustela erminea					
Mustela putorius					
Martes martes					
Martes					
Martes foina					
Meles meles					2
Gulo gulo					
Vulpes vulpes	29	[2]	X		9
Alopex lagopus					
Genetta genetta					
Felis					
Felis lynx					2
Felis silvestris		[1]			
Lagomorpha					
Oryctolagus cuniculus					5
Lepus					
Lepus capensis					
Lepus timidus					
Marmota marmota				1	5

Key to Tables:
All numbers denote NISP, except where bracketed [=MNI]
"X" denotes presence

Sources:
Abauntz: (Altuna and Marriezkurrena 1982)
Aitzbitarte: (Altuna 1972)
Altamira: (Altuna 1972; Castro Bernardez 1986)
Amalda (Altuna 1990; Pemán 1990)
Axlor: (Altuna 1989; Straus 1982)
Los Azules: (Fernández-Tresguerres 1980)
Bolinkoba: (Straus 1974a, 1977)
Candamo: (Altuna 1972; Obermaier 1924)
El Castillo: (Cabrera 1984)
La Chora: (González Echegaray et al. 1963)
Coberizas: (Altuna 1972; Obermaier 1924)
Cueto de la Mina: (Altuna 1972; Castro Bernardez 1986)
Ekain: (Altuna and Marriezkurrena 1984)
Ermittia: (Altuna 1972)
Erralla: (Altuna and Marriezkurrena 1985; Pemán 1985)
Lezetxiki: (Altuna 1972)
Cueva Morín: (Altuna 1978)
Otero: (González Echegaray et al. 1966)
La Paloma: (Castaños 1980)
Pasiega: (Straus 1977)
El Pendo: (Fuentes Vidarte 1980)
Piélago: (López-Berges and Valle 1985)
Rascaño: (Altuna 1981)
La Riera: (Altuna 1986)
Santian: (Altuna 1972)
Santimamiñe: (Straus 1974a, 1977; Altuna 1972)
Sofoxo: (Straus 1977)
Tito Bustillo: (Altuna 1976)
Torre: (Altuna 1972)
Urtiaga: (Altuna1972)

Table 9.2: Fur-Bearing Taxa from Early Upper Paleolithic Levels.

	Aitzbitarte	Amalda		Bolinkoba	El Castillo								Cueto de la Mina	Ekain
		VI	V		18	16 Inf.	16 Basal	16	16 Sup.	14	13	12	7	Xa
	Aurig.	Per. V	Ev. Per.	Grav.	Aurig.			Aurig.		Aurig.	Aurig.	Aurig./Per.	Aurig.	Chat.?
Mustela nivalis	7		1				X						12	
Mustela erminea	1					X	X		X		X			
Mustela putorius			2				X							
Martes														2
Martes martes														
Martes foina														
Meles meles											X			
Gulo gulo														
Vulpes vulpes		133	22	[3]	X			X		X		X		9
Alopex lagopus		1												
Genetta genetta														
Felis			1											
Felis lynx														
Felis silvestris												X		
Lagomorpha														
Oryctolagus cuniculus									X					
Lepus									X					
Lepus capensis		3												
Lepus timidus														
Marmota marmota		1												

Table 9.2: (Cont.).

	Lezetxiki							Morín				Otero		El Pendo	Rascaño	Santian	Santimamiñe
	Va	IVc	IVb	IVa	IIIb	IIIa	II	6	5c	5a	4			V	7		
	Ster.	Protoaurig.	Ster.	Aurig.	Ster.	Aurig.	Sol./Grav.	Aurig.	Aurig. II	Grav.	Grav.	Aurig./Moust.	Aurig. V	Grav.	Aurig.?	Inf. Aurig.	Inf. Pergord.?
Mustela nivalis				1													
Mustela erminea															3		
Mustela putorius				1		3											
Martes						1											
Martes martes																	
Martes foina																	
Meles meles						5											
Gulo gulo							1										
Vulpes vulpes	2			1	3	6	7	1	1	1	1	X	X			X	
Alopex lagopus																	
Genetta genetta																	
Felis														1			
Felis lynx											1						
Felis silvestris	2		2			1		1			1						X
Lagomorpha																	
Oryctolagus cuniculus	7				2	2											
Lepus															3		
Lepus capensis									1	1	1						
Lepus timidus																	
Marmota marmota	16	3	5	25	50	59											

Table 9.3: Fur-Bearing Taxa from Solutrean Levels.

	Aitzbitarte	Altamira	Amalda	Bolinkoba	Candamo	El Castillo	Coberizas	C. de la Mina		Ermittia	Pasiega	La Riera									Santimamiñe
			IV			10		h	f			2+3	4	7	8	9	10	14	15	16	
Mustela nivalis	19						2														
Mustela erminea	2									1											
Mustela putorius	1																				
Martes																					
Martes martes																					
Martes foina																					
Meles meles				[7]																	
Gulo gulo																					
Vulpes vulpes	3	X	27	[6]	X	X	1	X	X	15	[2]	2			1	1	1		1	1	[1]
Alopex lagopus																					
Genetta genetta																					
Felis																					
Felis lynx		X								3											
Felis silvestris																					
Lagomorpha																					
Oryctolagus cuniculus													1	11							
Lepus																					
Lepus capensis										2		1						1			
Lepus timidus																					
Marmota marmota																					

Table 9.4: Fur-Bearing Taxa from Magdalenian Levels.

	Abauntz	Aitzbitarte	Altamira	El Castillo	La Chora	Coberizas	Cueto de la Mina		Ekain		Ermittia	Erralla				Lezetxiki	Morín
	e			8			e	c	VII	VI		VI	V	IV	I-III	I	2
	Lower	Mag.	Lower	Lower	Upper	Upper/Azil.	Lower	Upper	Lower	Upp./Fin	Mag.	Ster.	Lower	Ster.	Final	Upper	Final
Mustela nivalis		31							10			3	135	1	3		
Mustela erminea		3							7				13	4	6		
Mustela putorius		1											1				
Martes																	
Martes martes																	
Martes foina																	
Meles meles		1														1	
Gulo gulo																	
Vulpes vulpes	13	6		X	X		X	X	33	13	3		14			1	
Alopex lagopus																	
Genetta genetta																	
Felis												1					
Felis lynx																	
Felis silvestris	1					1											
Lagomorpha																	
Oryctolagus cuniculus															1		
Lepus			X							32			5	2	8		
Lepus capensis	17	9									3						1
Lepus timidus																	
Marmota marmota																	

Table 9.4: (Cont.).

	Otero	La Paloma		Pasiega	El Pendo		Rascaño							La Riera			
		4	8		IIcd	IIab	8	7	5	4b	4a	3	2	21-23	24	25	26
	Upper	Upper	Lower	Lower	Final	Final	Ster.	Ster.	Init.	Lower	Lower	Lower	Upper	Upper	Upper		Upp./Azil.
Mustela nivalis									1			1					
Mustela erminea								3									
Mustela putorius				X													
Martes																	
Martes martes																	
Martes foina																	
Meles meles													2	1			
Gulo gulo																	
Vulpes vulpes	X	3	1		3	2			21	29	8	4	4	1	4	2	1
Alopex lagopus																	
Genetta genetta																	
Felis																	
Felis lynx													2				
Felis silvestris	X																
Lagomorpha																	
Oryctolagus cuniculus	X												1				
Lepus																	
Lepus capensis							2	3	2			2					
Lepus timidus																	
Marmota marmota																	

Table 9.4: (Cont.).

	Sofoxo	Tito Bustillo			Torre	Urtiaga				
		1c	1b	1a						
	Upper	Upper	Upper	Upper	Upper	Sterile	Sterile	Final	Sterile	Final
Mustela nivalis										
Mustela erminea			1					2		4
Mustela putorius								7	1	
Martes										
Martes martes										
Martes foina										
Meles meles	[1]				X				2	8
Gulo gulo										
Vulpes vulpes	[1]	1	4	8	X	7	3	67	52	156
Alopex lagopus										
Genetta genetta										
Felis										
Felis lynx		3						1	1	4
Felis silvestris							2			
Lagomorpha										
Oryctolagus cuniculus				3						
Lepus			1			3	3	2	3	10
Lepus capensis										
Lepus timidus										
Marmota marmota										

Table 9.6: Fur-Bearing Taxa from Azilian Levels.

	Abauntz	Aitzbitarte	Los Azules	Ekain		Ermittia	El Pendo	Piélago						Rascano	La Riera	Santimamiñe	Urtiaga
	d Epipaleolithic		3	III-V	IV		I	1	2	3	4	5	6	1	27		
Mustela nivalis		7		1												X	
Mustela erminea		4														X	
Mustela putorius				2													3
Martes																	
Martes martes	1																
Martes foina																	
Meles meles			1	18		2											30
Gulo gulo																	
Vulpes vulpes		1		2	1	2	1								1	X	18
Alopex lagopus																	
Genetta genetta							2										
Felis																	
Felis lynx										1				1			
Felis silvestris	1		1													X	5
Lagomorpha								9	25	9	1	2	1				
Oryctolagus cuniculus			X														
Lepus				1										1			4
Lepus capensis																	
Lepus timidus																	
Marmota marmota																	

CHAPTER 10
HUMAN ENVIRONMENTAL INTERACTIONS

Upper Paleolithic human-environment interaction in Cantabrian Spain exhibits an accelerating pattern of separation of human actions from the immediate constraints of environment. These changes are visible as early as the Early Upper Paleolithic (Aurignacian, Chatelperronian, Gravettian/Upper Perigordian) and greatly increase throughout the Late Upper Paleolithic (Solutrean, Magdalenian and Azilian). These changes are more remarkable considering the major and minor fluctuations of the climate. These proceeded from vacillating interstadial conditions, through the glacial maximum ca. 22,000-18,000 B.P. and the stepwise retreat/readvance of glaciers before the warm climate of the Holocene was obtained by 10,000 B.P. The multiple environmental zones of coastal Cantabrian Spain are also striking, with littoral, riverine, coastal lowland and montane zones all within easy mutual reach. Variable environment was thus obtainable both over time and through physical travel over the landscape.

Mousterian occupation of Cantabrian Spain was sparse and less intrusive upon the local resources. While surface finds are numerous, Mousterian sites number just over a dozen in a recent summary by Straus (1992). These include levels at El Castillo (Cabrera 1984), Amalda (Altuna *et al.* 1990), Cueva Morín (González Echegaray and Freeman 1978) and El Pendo (González Echegaray 1980). Faunal exploitation for this period is more opportunistic than seems to be the case during at least the Late Upper Paleolithic. While red deer show up in most Mousterian levels and are often the most numerous species, their numbers are sometimes less than or comparable to other food taxa. This situation is illustrated at Cueva Morín Level 17, where *Cervus elaphus* (NISP = 142) is found in similar amounts as large bovids (NISP = 215) and *Equus Caballus* (NISP = 124) (Altuna 1978:203). In Amalda Level VII, *Cervus elaphus* (NISP = 150) are far outnumbered by *Rupicapra rupicapra* (NISP = 536) (Altuna 1990:151). While *C. elaphus* is abundant among the Mousterian levels of Axlor, large bovids, *Capra pyrenaica*, and *Equus caballus* all outnumber it in at least one level (Altuna 1989 cited in Straus 1992).

The pattern of Mousterian prey acquisition indicates more opportunistic hunting was the norm for many later occupations (Freeman 1973, 1981; Straus 1977), with species taken on an individual encounter basis. Such encounter hunting implies a lesser need for group cooperation: fewer hunters are necessary for the kill, nor are there large amounts of meat to process, transport and store at one time. Less reward is balanced by lowered risk. Hunting success is based upon a series of frequent encounters, any one of which may prove successful, rather than large, planned game drives involving many hunters at key seasonal/topographic aggregations of animals.

Mousterian levels also tend to yield substantial numbers of large carnivore remains. This is especially true of Lezetxiki, where *Ursus spelaeus* (cave bear), *U. arctos* (brown bear), *Canis lupus* (wolf) and *Panthera spelaea* (cave lion) are numerous, and their remains sometimes outnumber those of more probable human food taxa (Altuna 1972:141-142). These elevated levels of large carnivores indicate perhaps a larger degree of direct competition with humans, both for cave denning sites and ungulate prey (Stiner 1993, 1994). Hominids seem to have acted in ways analogous to those of other large predators present.

The nature of the Middle/Upper Paleolithic transition in Europe is an enormous topic (Chase and Dibble 1987; Harrold 1980; Gargett 1989; Klein 1989; Mellars 1990; Mellars and Stringer 1989), of which only one aspect is ecological. Early Upper Paleolithic Cantabrians began to shift this ecological pattern and show increasing control over the landscape. Large predators no longer appear in such great numbers among human habitation levels. Some species reach near extinction during this period, including *Ursus spelaeus* and *Panthera spelaeus* (Kurtén 1972, 1976). Others reach near local extinction, such as *Panthera leo* and *P. pardus*. Some very large prey species also declined to or near extinction, a pattern that would repeat across Eurasia. These include now extinct forms of rhinoceros (*Dicerorhinus kirchbergensis* and *Coelodonta antiquitatus*), elephant (*Mammuthus primigenius* and *Elephas antiquus*) and deer (*Megaceros giganteus*). Although debate continues on the proximate cause of megafaunal extinctions around this time, the theory of climate change as cause has less credence in the face of the regular, rapid fluctuations in climate occurring throughout the Pleistocene (CLIMAP 1976; McIntyre *et al.* 1976; Ruddiman and McIntyre 1976; Shackleton and Opdyke 1976; van Donk 1976). Since these species regularly adapted to major and minor glacial fluctuations, their demise on the grounds of these changes alone is implausible. This interpretation is reinforced by their decline and in most cases extinction in Western Europe before the substantially warmer conditions of the Holocene were obtained. Human hunting (and direct competition for resources in the case of large carnivores) therefore is the more logical cause of these species' demise (Martin and Klein 1984). The evidence for Cantabrian Spain during the Early Upper Paleolithic suggests that humans had assumed a greater impact upon the ecosystem than other large predators.

Early Upper Paleolithic site numbers increased over the Mousterian, even though a much shorter period was involved (40,000 to 20,500 B.P.). In a recent summary Straus (1992: 75) indicates as many as 26 sites containing Aurignacian, Chatelperronian, and/or Gravettian/Upper Perigordian levels. These occupations include El Pendo (González Echegaray 1980), El Castillo (Cabrera 1984), Ekain (Altuna and Merino 1984), Amalda (Altuna *et al.* 1990) and Aitzbitarte (Altuna 1972).

Early Upper Paleolithic faunal exploitation centered upon the same ungulate species as during the Mousterian period. The balance among large prey species is illustrated at La Riera Level 1, where 4 taxa have an MNI of 4 or more: *Equus caballus* (5), *Capra pyrenaica* (4), *Bos/Bison* (4) and *Cervus elaphus* (7) (Altuna 1986). This balance among large ungulates is also illustrated among the Aurignacian and Gravettian levels of Cueva Morín, where *Cervus elaphus*, *Capreolus capreolus*, *Bos/Bison*, *Capra pyrenaica* and *Equus caballus* all were taken in substantial amounts (Altuna 1978:203). Other levels from this time period, however, show the beginnings of more intensive use of one or two ungulate species. At El Castillo Aurignacian Level 18, *Cervus elaphus* (MNI = 216) and *Bos* (MNI = 29) dominate the faunal profile (Cabrera 1984:414-415). At Bolinkoba Gravettian Level F is dominated by the remains of *Rupicapra rupicapra* (NISP = 634), while the next highest represented taxon is *Bos/Bison* (NISP = 37) (Straus 1992:262). This trend was to continue and be greatly expanded upon during the ensuing occupation of Cantabrian Spain.

The Late Upper Paleolithic (Solutrean, Magdalenian and Azilian) Cantabrians chose both to harvest previously exploited species with increasing intensity and to broaden their resource base during this time of fluctuating but generally ameliorating climate that ultimately led to the Holocene. Population also seems to have expanded, with about 44 Solutrean, 52 Lower Magdalenian, 45 Upper Magdalenian and 35 Azilian sites (Straus 1992). These sites were also accumulated over shorter periods than during the Mousterian or Early Upper Paleolithic: 4,000 years for the Solutrean, 3,500 for the Lower Magdalenian, 3,000 years for the Upper Magdalenian and 2,000 years for the Azilian. Increasing specialization of site function would also increase the number of sites, but the increase is so great that it must also reflect population increase. These figures can only suggest changing population levels. Since nearly all known Paleolithic sites in Cantabrian Spain are in caves, destroyed or undiscovered open-air sites would if anything raise population estimates. These figures support both an the increased intensity of cave site usage, given the increased number of sites and amount of excavated materials, and an increase in population.

Red deer were the most abundant large mammalian prey and were exploited with increasing intensity (Freeman 1973, 1975, 1981; Straus 1974a, 1974b, 1975, 1976, 1977). This pattern begins in Solutrean times. In surveys of Solutrean faunas by Straus (1977, 1992), red deer are the most abundant large mammal taken in nearly all sites and levels. The exceptions include Bolinkoba and Ermittia, where ibex rank ahead of red deer and Buxú and Amalda where chamois rank ahead.

This trend was intensified during the Magdalenian, when large kills of red deer become more pronounced. As discussed in Chapter 7, red deer dominate the Lower Magdalenian levels of El Juyo. Other Lower Magdalenian occupations with elevated proportions of red deer include Altamira (Straus 1977), La Paloma (Castaños 1980), Cueto de la Mina (Castaños 1982), La Riera (Altuna 1986), El Pendo (Fuentes 1980), Urtiaga (Altuna 1972), Cueva Morín (Altuna 1978), Ekain (Altuna and Mariezkurrena 1984) and Tito Bustillo (Altuna 1976).

Another change in the faunal exploitation pattern is the use of specialized hunting camps, as at the upland site of Rascaño, where Magdalenian and Azilian levels are dominated by ibex (Altuna 1981). Collubil also seems to have served this purpose during this time (González Morales 1974 cited in Straus 1992:132). Ibex are also plentiful at other sites such as Erralla (Altuna and Mariezkurrena 1985), Ermittia and Urtiaga (Altuna 1972), and La Riera (Altuna 1986). Age evidence for ibex from Rascaño indicates most were adults over 2 years old (Altuna 1981:251). Though breakdown by age class was not indicated, a high proportion of adults is more likely to reflect a catastrophic kill. Ibex were most likely harvested using drive and surround techniques, since their ability to escape over rugged terrain makes their individual stalking by hunters without modern firearms extremely difficult (Straus 1987b). This species also comes from the highland altitudinal zone not normally occupied by these hunter-gatherers, especially during episodes of alpine glaciation. This concentrated extractive exploitation of a montane resource indicates further intensity of landscape usage.

The methods of deer harvest also seem to have changed during the Late Upper Paleolithic in Cantabrian Spain. The red deer taken at Level 4/4S and 6 at El Juyo show a catastrophic age profile, a probable indication that drive and surround techniques taking whole herds were practiced (Klein and Cruz Uribe 1987; Klein *et al.* 1981). Random encounter hunting tends to produce a paucity of prime age adults, since these individuals are the best able to avoid predators, while the oldest and youngest are more vulnerable. The taking of prime age adults in proportion with their natural abundance in a population indicates that the technique used to harvest them neutralizes the benefits of their more robust state. Predator avoidance equalizes among individuals in a trapped herd. Logical places for ambushes to occur are in lakes, streams or wetlands where mobility is lessened, in natural traps such as ravines, or in artificial traps such as pens. Trapping and dispatching an entire deer herd also requires a large cooperating group of hunters to block avenues of escape and to process and transport the kill.

This emphasis upon "wild harvesting" (Freeman 1973:39-41, 1975:254-255) at selected seasonal times appears to be reflected as well in the projectile technology of this period. The Solutrean period is defined in one respect by its large lithic projectile points found in multiple forms: laurel leaf, willow leaf and shouldered. Found along with these points are significant numbers of oval-profiled antler projectile tips. By the Lower Magdalenian period, large lithic projectile tips had disappeared and antler tips had assumed this role. The dominant profile shape of these points shifted to quadrangular, a shape that is a hallmark of the Cantabrian Lower Magdalenian. The introduction of the bow and arrow may have also occurred at some point in this period (Bergman 1993). Projectile technologies changed again later with the antler harpoon forms that partially define the Upper Magdalenian and Azilian periods.

Antler projectile tips are more reliable than lithic ones in an engineering sense (Bleed 1986). Experimentation with replicated points (Pokines 1993, in press) indicates that antler projectile tips can withstand repeated impacts with minimal damage. They are also much more likely to withstand lateral stresses after impact while being dragged by a wounded animal. Repair is possible more often and is much quicker, usually involving only the resharpening of a damaged tip. Antler projectiles throughout the Later Upper Paleolithic also show a pattern of more systematic manufacture. Quadrangular forms can be produced more quickly than oval forms, since more of the original shape of the antler blank is retained in the former. Evidence from El Juyo also shows that long blanks (large enough for multiple points) were cut in strips from antler, allowing more efficient projectile production.

This change in projectile tips complements other changes in hunting technique. If hunting success is based upon a few isolated, labor intensive instances where an entire herd is dispatched, then the consequences of failure are much greater. Failure during random encounter hunting leaves several hunters in other locations who may still be successful on that same day. Group drives have a greater penalty for failure, since most of a groups' resources have been invested in one area. Projectile tips that have been more reliably engineered (more likely to withstand impact, less likely to be damaged accidentally, able to withstand repeated use, and easier to repair if damaged) are worth any additional invested manufacture time. Antler tips therefore become a more logical choice in situations where reliability is required and longer periods of down time are available between hunting episodes.

An additional change in projectile technology occurred in harpoon styles. Laterally barbed harpoon heads made of antler appear by definition at the outset of the Upper Magdalenian period. The usual function assigned to these tips is fishing, since they solve the problem of securing aquatic prey that might otherwise float away if speared with an unbarbed projectile tip. They also appear when evidence of fishing tends to increase. Harpoons, however, are preceded by an unbarbed projectile tip that could have functioned in the same role (Pokines and Krupa 1997). This antler point type, which occurs in Solutrean and Lower Magdalenian contexts, has a lateral hafting surface and a curved rear portion ending in a sharp point. When hafted laterally to the end of a shaft, the rear tip projects outward and forms a rear facing barb. These tips function well as fish spears in modern experiments. Their presence in sites predating the Upper Magdalenian period indicates that spear fishing may have gone on more intensively during this earlier period. This interpretation is supported by new evidence from El Juyo, where very abundant fish remains occur in a Lower Magdalenian context. More intensive "wild harvesting" of dense, seasonal resources may have occurred earlier in this period (Davidson 1976a, 1976b, 1983).

Landscape usage changed in other manners. Research by Conkey (1978, 1980) on engraved mobile art motifs from Lower Magdalenian sites suggests that some sites may have served as seasonal aggregation loci for bands that remained dispersed for the rest of the year. Altamira and Cueto de la Mina are two sites identified for this possible role. Their mobile art has a greater variety of engraved motifs than surrounding sites, suggesting that periodic visitors from many parts of the surrounding region left these behind. That smaller nearby sites have lower decorative diversity she interprets as reflecting the fact that they were exclusively used by one or a few groups whose decorative repertoires were limited. This pattern may, however, also be explained by the lumping together of old excavated collections from multiple levels at proposed "aggregation sites" whose deposition spans more time (Straus 1980). Temporal instead of spatial stylistic diversity would thus be reflected. The seasonal aggregation hypothesis is reinforced at least at Altamira by that site's status as a major painted cave whose motifs are centered upon mating bison (Freeman 1987) and through geographic analysis of the region (Freeman 1994). The temporary joining of small dispersed bands for the purposes of hunting massed herds of ungulates meshes with the other evidence of hunting behavior interpreted here. At Altamira there is also evidence that Magdalenian occupations were not especially long-lasting (González Echegaray 1988; Freeman 1988). Further analysis of newly excavated materials with better temporal controls may provide a conclusive resolution.

Marine molluscs were also subject to increasingly selective exploitation throughout this period (Clark 1983; Krupa 1994; Straus *et al.* 1981). While of extremely limited dietary importance during both Mousterian and Early Upper Paleolithic occupations, these assume a significant role during the Solutrean and were accumulated in true shell middens beginning with the Lower Magdalenian. The major occupation levels at El Juyo (4, 6, and 8) have each yielded in excess of 8,000 whole shells and many times that quantity of fragments, almost entirely of limpets (*Patella vulgata*) and periwinkles (*Littorina littorea*). Significant quantities have also been found at sites such as Altamira (Straus 1976-77), El Castillo (Cabrera 1984), and Otero (González Echegaray *et al.* 1966), and Tito Bustillo (Moure Romanillo 1975) during the Upper Magdalenian. This trend was to accelerate, culminating in the true *concheros* of the Asturian period, a regional manifestation of the Mesolithic (Clark 1971, 1983; Straus 1991a).

As discussed in Chapter 7, fish show the same pattern of increasing and concentrated exploitation during this period, primarily Atlantic salmon (*Salmo salar*) (Krupa 1994, 1996). Many other Cantabrian sites have yielded fish remains, though in smaller amounts (probably due to differences in recovery system). The modern systematic excavations of the remnant stratigraphy at La Riera yielded 81 identified *Salmo* remains from its Late Upper Paleolithic sequence (Straus and Clark 1986).

Waterfowl were another resource whose exploitation intensified during the Late Upper Paleolithic in Cantabrian Spain. Magdalenian levels yielding these come from Balmori, Cueto de la Mina, La Riera, Coberizas, Altamira, El Pendo, Otero, La Chora, Lumentxa, Torre, Erralla, Ekain, Urtiaga and Aitzbitarte (summarized by Straus 1992:151). The majority of bird remains from El Juyo await further identification, but represent a significant sample. Migratory waterfowl are a seasonal resource that was perhaps harvested when this humid coastal region was a stopping point on migratory fly-ways or a seasonal refuge area.

Faunal exploitation in this period includes hunting or trapping of small to medium sized fur-bearing mammals specifically for their pelts (Chapter 9). El Juyo so far has yielded 72 identified foot elements, primarily phalanges and metapodials, of carnivores in the size range of ferret (*Mustela putorius*), European mink (*Mustela lutreola*), marten (*Martes* spp.) and possibly wolverine (*Gulo gulo*). The presence of these bones is in striking contrast to the virtual lack of any other skeletal elements, including teeth, from these taxa. Differential survival due to the durability of these foot elements cannot explain this pattern, since the hypothetically "missing" teeth are the most durable skeletal elements of all. Non-carnivores of this size range (such as rabbit or hedgehog) and the smaller, more fragile mammalian taxa do not show this same element representation discrepancy. The author therefore attributes the preponderance of fur-bearing mammal foot elements to their selective introduction to the site. This pattern would be the logical result of foot bones left in a pelt removed from a trapped or hunted animal away from the site, with the remaining carcass discarded in the field. Mustelid meat has a strong, unpleasant flavor and may have been therefore a relatively less desirable food choice, used only in times of hardship. This pattern of fur exploitation has been suggested for other Cantabrian sites (Straus 1977, 1992). Aitzbitarte IV in Basque country has yielded several remains from *Mustela nivalis*, *M. erminea*, and *M. putorius* in its Solutrean and Magdalenian levels (Altuna 1972). Bolinkoba, a Basque site, has also yielded the remains of Mustelids and red fox (*Vulpes vulpes*). The suggested pattern of specialized fur extraction emerges during the Late Upper Paleolithic, in conjunction with other changes in faunal exploitation noted.

The exploitation of plant resources is naturally less visible in the paleolithic archaeological record due to poor chances of preservation. With the advent of new flotation procedures such as those in practice at El Juyo (Freeman *et al.* 1998; Stettler 1995) and at Franchthi Cave in Greece (Hansen 1991), future excavations of earlier occupations may indicate whether plant resource use changed over this period. Plant resources during summer in this interstadial climate would have been much more abundant and would seasonally complement winter meat resources.

In short, sites from the Cantabrian region exhibit interrelated trends in human/environment relationships throughout the Late Upper Paleolithic, a period of generally rigorous and fluctuating climate. Large mammal exploitation became focused upon relatively rapidly renewable, highly productive congregating species which were taken with increased efficiency. Evidence of this trend includes the mass slaughter of red deer and ibex herds, as well as the specific use of some sites (Rascaño) as primarily single-species extractive hunting camps. Salmon runs were also subject to increasing exploitation through Magdalenian times and into the Epipaleolithic (Pokines and Krupa 1997). These runs by their nature provide seasonal and concentrated sources of food, and as such require group cooperation and coordination for their efficient "wild harvesting". Major protein resources, with the exception of marine molluscs, became increasingly concentrated in temporary, perishable surpluses.

Changes in food storage needs represent a logical new problem for the practitioners of mass harvest techniques. Encounter hunting lacks the same urgency for meat storage, since individual animals procured on a semi-daily basis can be consumed before they spoil and killed only when no surplus already exists. The generally cool climate also reduced the urgency for a more planned response. Meat sharing among group members may also have reduced any need for storage: a temporary personal surplus that cannot be consumed before spoiling can be traded among members of the group or between groups for future reciprocation where the situation is reversed. One thus gives up only what cannot be used in return for future resources.

Mass harvesting of perishable resources presents an entirely new problem to solve and with it the potential for greatly changed group behavior. While marine molluscs could be harvested in large amounts only as needed, a mass harvest of red deer, ibex, or salmon produces a highly perishable resource that will rapidly spoil unless measures are taken for its preservation. The option of trading away a perishable resource for future considerations is greatly reduced. Since group cooperation was necessary from the outset, each member has the simultaneous problem of immediate surplus and future need. Since this faunal exploitation pattern holds for the whole Cantabrian region, trading with one's neighbors is less of an option (they are likely to have the same problem and at the same time). Therefore, meat storage becomes a pressing need in a humid climate ill suited for it during most of the year.

The season of occupation for humans at El Juyo is partially clarified by the new data presented here (Chapter 2) concerning the season of occupation of barn owls in the site. While concurrent occupation is possible (and occurs commonly in other settings), the spatial layout of the cave vestibule makes the simultaneous habitation by roosting barn owls and any substantial, semi-permanent human encampment there unlikely. Occupation by barn owls for particular seasons therefore suggests that humans tended not to occupy El Juyo at those same times. The age profiles, where determinable, for barn owl small mammal prey indicates that many of these individuals were very young juveniles, perhaps only days past weaning. These very young individuals would have been present in the local populations only during the late spring/summer/early autumn months, roughly April through September each year (see Chapter 3 for summaries of the breeding season of these small mammal species). Barn owls must have been present in El Juyo at least part of the time during these months, though their presence there during the remainder of the year cannot be discounted by analysis of their diet.

Human occupation at El Juyo therefore may have been concentrated from late autumn through late spring. No clear season of human occupation was determinable through investigation of dental attrition modes and cementum annuli of the red deer teeth (Freeman *et al.* 1988). The presence of cervid foetal bones in the Level 4 Sanctuary feature at El Juyo shows that human occupation lasted at times until late spring/early summer when full-term (or nearly so) pregnant females were present in the red deer population. The presence of mostly unshed antler indicates that significant numbers of adult male red deer were taken during autumn/early winter, since their antlers are shed shortly after the mating season ends (in January). Another indication of human occupation of El Juyo in the autumn is the presence of many seed species, most of which are produced (and are optimally harvested) in the autumn months. Human acquisition, however, is not the only explanation for their entry into the site. Salmon harvest most likely occurred during the late autumn spawning run, which further indicates human occupation of El Juyo at this time of year. Human seasonal cycles in Lower Magdalenian Cantabrian Spain therefore may have included autumn and winter occupation of lowland sites like El Juyo, combined with summer occupation of warm season-accessible upland sites such as Rascaño.

The easiest method of meat preservation is simple freezing during winter. Since interstadial conditions existed throughout most of this period, substantial freezing during the winter months becomes a viable partial solution. An ideal time to harvest red deer is when they congregate for the autumn rutting season, so large meat supply could be acquired just as temperatures dip below freezing. No special effort is necessary other than an even greater need to reduce local scavenger/carnivore populations. Since many of the red deer remains from El Juyo were taken in late spring/early summer when freezing of the meat associated with these skeletal remains would not be feasible, some other method of preservation must have been practiced at that time. An option in this situation is drying/smoking meat. This procedure is necessary to process large amounts of meat acquired during the warmer months of the year and presents no great technological obstacles (Coon 1971). Thin sections of meat may be air dried in sunlight, or larger pieces may be cured by smoking. Meat processed this way can be kept for long periods of time.

Direct archaeological evidence of these procedures is more difficult to obtain. Evidence for the systematic processing of fauna that is a necessary preliminary step in these procedures, however, has been obtained from El Juyo. A large proportion of red deer remains were discarded in the dugout structure in Level 6, with indications of selectivity in body part representation and disposal (Freeman *et al.* 1988). This pattern suggests methodical, wholesale processing of red deer. The orderly pattern of skeletal disposal is difficult to explain if individual deer were acquired in multiple episodes and is reinforced by the catastrophic age profile obtained from crown heights. Red deer were not only harvested methodically, but also processed methodically. The relative lack of bones from meatier body portions in Level 6 is an indication that substantial processing of carcasses took place away from the cave vestibule. Ongoing research into the faunal remains of El Juyo may disclose further reinforcing patterns of butchery or part representation.

Fish remains from El Juyo also suggest patterns of methodical processing of large amounts of fish en masse. This evidence is obtained from skeletal part representation. While all paleolithic levels and features of the site have yielded fragmentary fish remains, almost none of these is a cranial element. No otoliths have been discovered. The elements that have been recovered in greatest amounts from screening site deposits are fragmentary ribs and fin spines, with vertebral elements a far distant second. These elements are the most fragile, so their overrepresentation cannot be explained in terms of differential preservation. This pattern suggests that most cranial elements were discarded away from the site, perhaps at the site of capture. This practice would be efficient and necessary if large amounts of fish were taken at once and the flesh was deboned and preserved. Since preliminary investigation indicates that most of these remains are those of Atlantic salmon, harvesting must have taken place during seasonal spawning runs in nearby rivers. These occur beginning in late autumn, though some fish could have been taken in the principle spawning months of November through January (Netboy 1980).

The necessity for food storage also affects patterns of landscape usage (Bailey and Davidson 1983). Large amounts of stored meat decrease the mobility of a group, especially in this generally rugged terrain. This decreased mobility is reflected in the use of specialized resource camps, such as the Magdalenian ibex hunting camp of Rascaño. Instead of moving the base camp to where the resources are, the resources are brought to a centralized location. El Juyo has yielded ibex remains from Levels 4/4S and 6 (Klein and Cruz Uribe 1987), even though this montane species would have been many kilometers distant from the site. While red deer and bison were available in the environs of the site, marine molluscs were transported up to 10 km to the site in large numbers. The nearest salmon river (Rio Pas) is over 5 km distant. Major food resources that were concentrated at their point of harvest were transported to El Juyo. One must also consider, however, that in Cantabrian Spain distances between all vertically zoned habitats are small, and mobility need not have been great for a single group exploiting all of these. In addition, true storage pits (that might be expected under the circumstances suggested) are virtually absent. Other means of food storage may have been practiced. These include the structural remnants from Level 6 of El Juyo, or archaeologically less visible means such as suspension on platforms or poles away from predators.

During this same period we see a gradual increase in population, as shown by the increasing number of sites and in the intensity of their occupation (Straus 1977, 1992). An increasing population logically contains on average a higher proportion of children. Two other increasingly important areas of faunal exploitation during this period in Cantabrian Spain are the gathering of marine molluscs and fur trapping. Both are highly suitable activities for children, as they respectively involve immobile, easily-harvested molluscs or the maintenance of a trap line where small to medium furbearers are already dead or captured. Their harvest at that point simply involves collection from the traps. Far less strength, speed or critical judgment are necessary at the time of harvest of such species than during the driving, surrounding, and slaying of large fauna. Weight loads for transport to a habitation site are also easily controlled by limiting mollusc harvest and skinning trapped animals in the field, the processing pattern suggested by their proportional skeletal representation. The collection of molluscs by children especially frees adults for more difficult tasks, while supplying a large and reliable supply of protein.

DISCUSSION

The conflation of population increase and resource management at this time of climatic oscillations--general amelioration with intermittent periods of deterioration--indicates a growing separation of human behavior from the direct constraints of an unpredictable natural setting. Microfaunal and other paleoenvironment indicators portray a humid, cold and open landscape moving through multiple cycles of poorer or more favorable conditions, approaching the temperate and more forested Holocene. Red deer, ibex, molluscs, salmon, and perhaps furs were obtained in an increasingly efficient and cooperative manner. These trends are reflected in the faunal remains themselves, as well as the changing patterns of site use and technology. Instead of human resource acquisition following fluctuations caused by changing environment, human control over the harvest of these resources accelerates independent of environmental fluctuations.

While the author in no way suggests that Cantabrian Spain was a hearth area for agriculture (which came to this region rather late), this changing human relationship with environment in many ways presages the changes necessary for agricultural systems to develop (Freeman 1973, 1975). Food production, whether by herding or crop growing, represents separation from the constraints of naturally harvestable resources. Environmental factors in the form of climate, terrain and biogeography of species available for domestication still affect agricultural systems. Agriculturists in many cases rely upon wild species to supplement their own food production, and so remain tied to their natural environment rather closely. They also create artificial ecosystems whose internal workings approach the complexity of natural ecosystems. Cross pollination and hence gene flow occurs between domesticated and wild ancestral forms (as with maize and teosinte in Central America), further linking this relationship. Ultimate resource harvest, however, is largely determined by direct human action and does not rely only upon natural productivity to establish resource yields.

Wild harvesting of resources places human ecological interaction along the spectrum closer to agriculture. The other end of this spectrum is random-encounter collection, where food resources are acquired more in proportion with their natural abundances as found on the landscape. The extreme end of this spectrum is not obtainable, since humans make choices in their food acquisition even to the point of starvation in the presence of culturally unacceptable food resources. Nor is the other end of this spectrum obtainable (total independence of environmental constraints upon food production), since these factors place boundaries that limit the breadth of human decision making. No evidence exists that Upper Paleolithic Cantabrians were acting to replenish natural resources or alter environments that would favor the growth of preferred resource species, other than the reduction of competition from other large predators and the possible deliberate cultivation of marine molluscs (Freeman 1973, 1975). Other possible actions in this regard include territorialism in order to protect important resources from other hunter-gatherer groups. Landscape alteration using fire to favor the habitats of certain ungulate prey species is another possibility, or the selective predation where possible upon weakened or non-reproducing individuals. The allowance of "fallow" periods for areas of mollusc harvest is another conceivable practice. Changes in resource acquisition therefore lie not so much in control of production but in control of harvest.

Increased efficiency in landscape usage, technology and its employment, and staging of the exploitation and processing of resources places a human group in an entirely different relationship with its environment than had previously been the pattern for Cantabrian Spain. As mentioned above, humans had begun in general during the Upper Paleolithic to reduce competition from predators and greatly decrease the numbers and ranges or aid in the extinction of many megafauna species (Martin and Klein 1984). These actions place humans distant from other large predators, whose numbers are largely determined by the availability of prey species.

Another parallel to early changes related elsewhere to the origins of agriculture (Cohen 1977; Reed 1977) is the increased importance of stored food resources and their relationship to sedentism. Agriculture lies at one extreme, where most food resources are literally rooted in place and are harvested during one or a few brief periods. The results of this harvest must be stored throughout the following year to provide most food resources and seed stock for the following year's crops. Hunting and gathering relies far less upon stored resources, since these are in essence stored dispersed throughout a large territory, with many natural backups available in the event of the failure of one resource. Controlled wild harvesting that produces a stored resource pushes a hunting-gathering group towards the constraints of sedentism. More of their food resources are seasonally or periodically concentrated. This efficiency and resource concentration may in turn allow population to grow, forming a positive feedback relationship between increased labor and increased returns. Sedentism may also affect this relationship, since reduced mobility allows shorter birth intervals and an increase in population growth. These factors became very important in later changes associated with advanced agricultural societies: the growth of urbanism, craft specialization and societal stratification.

Efficient wild harvesting also impacts the form of ethnographically known hunter-gatherer societies, such as the chiefdoms of the Northwest Coast Indians of North America. These richly complex, sedentary and highly territorial tribes had as their successful food energy base the rich salmon resources of the North Pacific Ocean (Netboy 1980). This coastal region hardly allows successful agriculture and parallels Late Pleistocene Cantabrian Spain in the suitability of its resources. Intense, controlled seasonal harvesting and storage of a wild resource allowed the growth of settled villages, dense population, elaborate art, intense craft specialization. Territorialism also became viable, since this non-dispersed resource was worth protecting. Elaborate ceremonials at least partly related to temporary surplus resource redistribution (the potlatch) were an additional cultural response to the parameters of this environment.

One factor that may have shaped the growth of the cultural systems of the Pacific Northwest is the geography of major salmon river drainages. The Cantabrian coast has many short rivers flowing directly from the coastal mountains with each draining a relatively tiny area. The amount of salmon re-entering these rivers each year to spawn is proportional to this drainage and its area of spawning streams. The Cantabrian salmon resource is therefore dispersed among many entry points to individual valleys. Larger drainage systems in the Pacific Northwest mean that a greater number of salmon must pass through the bottleneck points of key river mouths. This region also has a richer variety of salmon species (6 species in the genus *Oncorhynchus* inhabit the North Pacific vs. one salmon species in the genus *Salmo* inhabiting the North Atlantic) whose runs, significantly, do not coincide. Atlantic salmon also enter the freshwater system on a more dispersed time frame (Mills 1989). The salmon resources of the Pacific Northwest therefore allowed significantly more concentration of multiple fish species at key geographic points during several seasonal runs and a more prolonged harvest. The Cantabrian region during the Late Upper Paleolithic had a more dispersed and less rich salmon resource. The Cantabrian region also had other rich resources to offer (red deer, ibex, marine molluscs), so that while primary reliance upon salmon was not possible, it was also unnecessary.

In summary, the paleoenvironmental relationships of the Upper Paleolithic Cantabrians show an increasing pattern of separation of human action from the immediate constraints of their environment (Freeman 1973, 1975). Harvesting of wild resources increased in intensity throughout this period despite major climatic fluctuations in this open, humid coastal zone. Instead of abandoning this region during the glacial maximum and other cold periods, the Cantabrians steadily intensified their utilization of the landscape. Subsistence focused upon a few highly productive and concentrated resources, while continuing to expand into ones previously underexploited. These changes in landscape usage (away from more random resource exploitation and towards planned, intensive harvesting at efficient locations and times) in many ways presage the changes necessary for agricultural systems to arise.

CHAPTER 11
SUMMARY AND CONCLUSIONS

The goal of this research is the reconstruction of the paleoenvironmental setting of the Lower Magdalenian site of El Juyo and related sites in Cantabrian Spain. The data central to this project are terrestrial mammalian microfaunal remains. These have been compared with other sources of paleoenvironmental data: chronological synchronization with global ocean level, sedimentology, large terrestrial fauna, aquatic fauna, palynology, and macrobotanical remains. Each of these sources possesses inherent biases that must be taken into account when interpreting past environmental conditions from them. The author has demonstrated that in both the resolution of information contained and the potential to correct the biases incurred in their introduction to a site, terrestrial mammalian microfauna, whenever these are recovered, are a much more satisfactory source of information on past environmental conditions. The taphonomic journey traveled by these remains is far more determinable than that of pollen grains, and their introduction to a site has not passed through a cultural filter as have the remains of large terrestrial fauna, aquatic fauna, or macrobotanical remains. Microfauna are more directly relevant to the reconstruction of local paleoenvironment than are sedimentological or global ocean level extrapolations. An even higher degree of paleoenvironmental resolution of course may be achieved using these multiple sources in conjunction, but microfaunal remains must be considered if information yield and reliability are to be maximized.

This study of paleoenvironment has addressed three related questions. The first is the reconstruction of the Lower Magdalenian habitats surrounding El Juyo and the extrapolation of these environmental reconstructions to the coastal zone, including the numerous other sites in the same time range. These Magdalenian sites include the occupations at Altamira (Altuna and Straus 1976; Straus 1976-77; Freeman 1988; González Echegaray 1988), La Riera (Straus and Clark 1986), El Castillo (Cabrera 1984), Tito Bustillo (Moure Romanillo 1975; Moure Romanillo and Cano Herrera 1976, 1978, 1979), Cueva Morín (Altuna 1978), and Ekain (Zabala 1984). The data for the present reconstructions were the varying proportions of terrestrial mammalian microfauna from El Juyo and related sites. Where possible, analysis was extended beyond the presence/absence of microfaunal species to take into account the role of barn owls and their pattern of prey accumulation.

The second research problem is the comparison of this source of environmental data with other sources already employed at El Juyo and other sites. The microfaunal taxa are representative of certain microhabitats and tend to be far more specialized in their choices than are large fauna (Avery 1982a; Bourlière 1975; Delibes de Castro 1985; Fa *et al.* 1982). A cultural filter was not involved in the introduction of the microfaunal remains, as is the case with large terrestrial fauna, aquatic fauna, cave sediments, pollen and macrobotanical remains. Additionally, the taphonomic trajectory followed by microfaunal remains is better understood and hence more controllable than other sources of paleoenvironmental data. Thus, microfauna are inherently more reliable paleoenvironmental indicators than are other data categories, disregarding the popularity of use of the latter.

The third problem is the nature of the adaptation of the human groups inhabiting Cantabrian Spain at the end of the Pleistocene in relation to this changing environment. Broad changes in food procurement strategies over time are evident in the archaeological levels already excavated. Mousterian, Chatelperronian, Aurignacian and Perigordian/Gravettian levels are characterized by opportunistic hunting and/or scavenging of less dangerous large mammal taxa: horse, bison and red deer (Freeman 1973, 1975, 1981; Straus 1977, 1981, 1992). Beginning with the Solutrean and accelerating through the Magdalenian, Azilian and Asturian (Mesolithic) periods, wild food procurement in Cantabrian Spain shows an intensification in exploitation of certain species. These were primarily red deer, Atlantic salmon and marine molluscs (limpets and periwinkles). Site location and function were diversified and tailored to exploit other specific resources, such as the ibex hunted at Rascaño. Other newly exploited resources include birds and small/ medium mammal pelts. The Lower Magdalenian occupation of El Juyo spanned a period of intensification of some aspects of resource procurement and a broadening of the resource base to include some that were previously underutilized.

These questions were first addressed with a taphonomic analysis of all microfaunal remains from the cave. This analysis was performed to determine the series of factors influencing the passage of these remains from living individuals of the local population to their final recovery through the sorting of ultrafine-screened site deposits. The intermediate stages proceed through the freshly dead individuals culled from the local population, to their defleshed/digested skeletons, their introduction to El Juyo and burial among the archaeological sediments, the preservation of a subset of these elements, the excavation of a portion of the cave vestibule, and the processing of these sediments by flotation and wet screening. Each of these stages represents some loss of elements and hence potential loss of information from the proceeding stage (Davis 1987; Klein and Cruz-Uribe 1984). Significantly, movement between these stages in many cases adds information in terms of marks upon the bones themselves, their pattern and location, and data inherent to the species represented such as age profile, activity patterns, and body size. Factors affecting the final state of the microfaunal remains also leave their marks upon them, frequently allowing the determination of the stages in the winnowing process. The goal of taphonomic research therefore is to backtrack up this trail of intermediate stages in order to reconstruct as nearly as possible the original living populations of animals and their interactions with the human population of a site. Even a partial reconstruction of this passage allows a finer assessment of the biases affecting the proportions of a microfaunal assemblage.

The conclusions of the analysis presented here are that owls were overwhelmingly responsible for the introduction of microfaunal remains into El Juyo and are the most likely vector of introduction for most Cantabrian cave sites. More specifically, barn owls (*Tyto alba*) are the most likely source among the possible species of nocturnal raptor. The evidence for this conclusion came from multiple sources: site setting, the amount and distribution of microfauna in El Juyo, prey age profiles and size distribution, their habitat preferences, activity patterns, and the degree of breakage and digestive corrosion upon the individual remains. All of these considerations correspond with known barn owl behavior and physiology, which are assumed not to have changed significantly since the Lower Magdalenian period. Other taphonomic agencies could all be ruled out on multiple grounds. These include physical agencies such as natural pitfall trapping by a limestone cave, which was not plausible due to the spatial layout of the El Juyo vestibule. Running water was similarly dismissed. Other predators were dismissed by the pattern of bone recovery and breakage. Small carnivorans leave behind prey skeletal remains with relatively high breakage and degrees of digestive corrosion. Diurnal raptors follow a similar pattern. Other species of owl (*Bubo bubo*, *Asio otus*, *A. flammeus* and *Strix aluco*) were eliminated on the grounds of their usual roosting patterns, hunting patterns and prey profiles.

An exception to this pattern was found with the remains of hedgehog (*Erinaceus europaeus*), a larger species of microfauna which probably found its way into El Juyo as a semi-commensal. Also exceptional were the remains of small fur-bearing mammals such as weasel (*Mustela nivalis*) and ferret (*M. putorius*), which the author suggests entered El Juyo as a result of human trapping activity. The majority of remains consist of rodents and insectivores and were introduced as the products of barn owl pellet regurgitation. This identification of the accumulating agency allows a greater resolution to paleoenvironmental interpretation than would the presence or absence of indicator species. Barn owls are a generalist predator whose diet closely reflects the relative proportions of prey species' associated habitats. Relative proportions of small mammal prey species therefore closely reflect relative proportions of habitat types present in the Lower Magdalenian landscape.

The behavioral, morphological, and ecological characteristics of the small mammal species present in Cantabrian Spain during the Late Pleistocene and Holocene were presented in Chapter 3. It is fair to assume that the behavior of these species has not changed significantly in the few thousand intervening years. In almost all cases, this assumption is not radical since these species have undergone no major physical changes in that time. One possible exception is the evolution of commensal behavior on the part of some of these taxa, which necessarily cannot predate the evolution of human-induced impacts upon the landscape: deforestation, agriculture and the extensive construction of habitations. This coevolutionary process could not have begun on a large scale before the Neolithic Period and its resultant changes in landscape usage. One may therefore infer that this process was in an incipient stage during the Upper Paleolithic, particularly with the barn owls responsible for the introduction of small mammal remains to the site. Hedgehog behavior may also fall into this category.

In addition, one species of vole present in Pleistocene Spain (*Pliomys lenki*) has gone extinct, and its presence in El Juyo represents an unexpectedly late local survival. Its behavioral characteristics were therefore surmised from its lineage, morphology, and associated species/habitats.

Reconstructing past ecosystems presented a more difficult problem, since the Pleistocene environments of Cantabrian Spain do not have direct modern analogs. Analogy to modern northern Europe fails to consider differences in solar radiation, which was high and day length more regular even during glacial conditions in Cantabrian Spain. Topography for the northern coast of Spain is also a complicating consideration, due to the proximity of the highland zone so close to the Atlantic coast. Some species interactions no longer occur: the range of *Microtus oeconomus* (arctic vole) no longer overlaps with endemic Iberian or Pyrenean vole species such as *Pitymys lusitanicus* and *P. pyrenaicus*. The presence of other species of vole may have affected ecological behavior and habitat exploitation, as is the case today with certain species interactions (Ligtvoet and Van Wijngaarden 1994). Some ecological conditions of the Late Pleistocene are no longer directly observable even in modern northern climates, which reduces the possibility and usefulness of direct analogy with modern conditions.

The modern microfauna of the Cantabrian region and the surrounding Iberian peninsula were presented to serve as a baseline for comparison to the paleomicrofauna. Data concerning the contemporary fauna came from both trapping studies and the analysis of raptor pellet samples, with new data from each of these sources derived from the immediate environs of El Juyo. Relative population densities, compositions, and variation across climatic, altitudinal and latitudinal clines of these microfaunal species were presented for Spain. The type of traps used or the species of predator whose meal remains are analyzed greatly affects the relative proportions of prey species. The source of mortality for an archaeological or paleontological sample of small mammals is of extreme importance in its interpretation. The same is logically true for a modern sample, since the major difference between them is the recentness of mortality. Additionally, examination of the modern small mammal fauna of a region and its relationship to the contemporary climate and environment can illuminate the relationship between Pleistocene faunas and their associated paleoenvironments.

This analysis indicated that barn owls today are extremely effective samplers of the terrestrial small mammals within their hunting territory. The biomass of small mammals taken from each major habitat type within this range (forest, wetlands, human habitations, shrubland, dry grassland, humid grassland, etc.) was shown to reflect the general areal proportions of these habitats. While some small mammal species will be underrepresented due to their activity patterns (especially arboreal and fossorial species), other species from their general habitat (forest or humid meadow) will represent the presence of that biome. The patterns detected were used to interpret the microfauna from the Pleistocene levels of El Juyo.

The smaller microfaunal remains (Rodentia, Insectivora and *Mustela nivalis* of Carnivora) from El Juyo and the reconstruction of paleoenvironment from their relative proportions was presented in Chapter 5. The Lower Magdalenian levels are not characterized by a complete microfaunal turnover when compared with modern Cantabrian Spain: the changes reflect a shift in emphasis. While microtines (*Microtus*, *Arvicola* and *Pitymys*) are a significant component of the modern microfauna, they dominate the Lower Magdalenian levels. This dominance comes at the expense of murids (*Apodemus* and *Micromys*, plus the introduced modern commensals *Mus* and *Rattus*).

Shifts also occurred among the microtine species. While *Microtus nivalis* (snow vole) is present in the Cantabrian Cordillera today, its range normally does not extend into the coastal lowlands. The presence of this alpine species in the lowland zone indicates a significant downward altitudinal shift concurrent with climatic deterioration at the end of Lower Magdalenian occupation at El Juyo (Level 4/4S). *M. oeconomus* is not found today in Spain, yet during the Pleistocene it was a regular part of the microfauna. Its presence there indicates cliserial shift of habitat zones southward through Europe. This shift also indicates significant cooling of the local climate, as well as humid and open conditions. *Arvicola terrestris* dominates the Lower Magdalenian microfauna, but is present in lower proportions in barn owl pellet samples from contemporary Cantabrian Spain. *Clethrionomys glareolus* was not present in the Lower Magdalenian microfauna. This microtine species is common in Cantabrian Spain today and is associated with forest habitats. The Cantabrian lowlands are more densely populated by *Pitymys lusitanicus*, while the El Juyo microfauna is dominated by *P. pyrenaicus*. This indicates another altitudinal shift in species representation, since *P. pyrenaicus* density correlates with increasing altitude (Delibes de Castro 1985).

The insectivores present in the Lower Magdalenian levels of El Juyo (*Sorex minutus*, *S. coronatus*, *Neomys fodiens*, *N. anomalus*, *Talpa europaeus*, *T. occidentalis*, and *Erinaceus europaeus*) are common to modern Cantabrian Spain. *Crocidura russula* and *C. suaveolens* were not found among the El Juyo Lower Magdalenian microfauna, and both are found rarely in Cantabrian Upper Paleolithic levels. Each species is associated with dry meadow conditions, and their complete absence from El Juyo further reinforces direct indications of humid conditions. *Galemys pyrenaicus* (Pyrenean desman) is also absent from the El Juyo microfauna. This alpine species is not found today in the coastal lowlands, but is found in some Pleistocene levels of other lowland sites (Chapter 8).

Microfauna among the Lower Magdalenian levels of El Juyo changed very little over their accumulation. Open, cold, and humid habitat species dominate the total microfauna biomass through all levels. Small amounts of forest and dry grassland species biomass are present. Forest covered much more of the heavily deforested landscape of modern Cantabrian Spain. The most significant shift is in the appearance of the alpine species *Microtus nivalis* in Level 4/4S, indicating deterioration of climate during its accumulation.

Some elements of the El Juyo microfaunal sample were examined metrically (Chapter 6). Measurement of selected elements aids taxonomic identification (Ruprecht 1971), separation among the species of a genus, and quantifies microevolutionary changes in populations (Berry 1978). Biometrical analysis indicated that the *Arvicola* remains present in El Juyo primarily belong to *A. terrestris*, though the presence of some individuals of *A. sapidus* cannot be discounted. The former species has smaller tooth dimensions than the latter among Pleistocene and Holocene samples. The presence of *A. terrestris* is also more likely due to its more northerly modern range. Analysis also indicated that the remains of *Apodemus* more likely belong to the smaller *A. sylvaticus* than to *A. flavicollis*. Both species are associated with forest habitats. The presence of both *T. europaeus* and *T. occidentalis* among the remains assigned to *Talpa* was indicated by measurement of humerus length. This element is larger in the former species, with no overlap with the latter species in most modern samples. Measurement of ${}_1M_1$ identified to *Microtus agrestis/arvalis* reconfirmed that the latter species was more prevalent at El Juyo. Estimation from identification of the ${}^2M^2$ of *M. agrestis* was that this species contributed about 25% of the tooth remains attributable to these two species combined.

Tooth length also reconfirmed identification in some cases, such as those of *Mustela nivalis*, whose teeth are similar to morphologically but smaller than those of *M. erminea*. The identification of the extinct species *Pliomys lenki* was also reconfirmed from tooth length.

The multiple sources of paleoenvironmental data from El Juyo were considered together in Chapter 7 to give a more complete picture of the waning Pleistocene landscape. Sea core foraminifera (CLIMAP 1976; McIntyre *et al.* 1976; Ruddiman and McIntyre 1976; Shackleton 1987; Shackleton and Opdyke 1976; van Donk 1976) indicate that the North Atlantic surface ocean water temperatures at 42^0 North latitude (the Cantabrian coast) averaged 10^0 C colder at 18,000 B.P. than at present. This temperature indicates polar water conditions with attendant pack ice. By 14,000 B.P. (the occupation of El Juyo), ocean levels were again lowering as continental ice caps expanded again after a period of retreat. Ocean level fluctuated somewhere intermediate to the -100 m glacial maximum lowpoint and modern sea level that was reached shortly after the Holocene began. Surface ocean water temperatures and accompanying climate were therefore somewhere between those of the glacial maximum and the warmer Atlantic climate of the Holocene. Local Cantabrian climate need not have followed precisely changes in global sea level. Changes in ocean circulation and weather patterns also allow great variation in local climate: hence the need to examine other more direct evidence of Cantabrian Pleistocene conditions.

Sedimentological analysis of the El Juyo deposits suggests that colder conditions returned during the time of deposition of the upper habitation Levels 6-4 (Freeman *et al.* 1988:11-12), following an earlier period of relatively drier, temperate conditions in Levels 7-13. Humidity or the seasonal concentration of precipitation had increased by the time of deposition of Level 6 such that the water table of the cave reached this level, depositing clays. Both Level 5 (a sterile flowstone layer) and Level 4 show an increase in *éboulis*, which may indicate an increase in cold regime due to spalling of the cave ceiling. Level 3 was formed by the collapse of the main entrance followed by further sealing of the vestibule deposits with a layer of travertine, suggesting a climate regime of high humidity followed by a period of cold, dry conditions.

The large mammal, marine mollusc and fish remains introduced to El Juyo do not give a definite picture of paleoenvironment. This situation is due to the wide environmental tolerances of these species, almost all of which are found today (or were extirpated locally in historical times) in Cantabrian Spain. El Juyo's large mammal fauna (in descending order of importance) consist of red deer, roe deer, bison, horse, ibex, boar, lion and wolf. These species are now or were formerly found throughout Europe. Red deer thrive on mixed forest/open habitats with their edge effects and resultant browsing foliage. Both roe deer and boar, found in much smaller amounts than red deer, prefer forest habitats. Horses and bison are open grassland/parkland species, and each is found in small amounts throughout the site levels. Ibex are an upland species that were within transportable range of El Juyo even during warmer climatic episodes. The large fauna therefore indicate a range of habitats present in unknown proportions. Molluscan fauna were similarly inconclusive, since the two primary harvest species *Patella vulgata* and *Littorina littorea* are found throughout a wide range of Atlantic coastlines, both north and south of Cantabrian Spain. Warm and cold indicator species, retrieved in very small amounts, provide conflicting or inconclusive evidence. The many salmonid remains recovered are also equivocal indicators of paleoenvironment, since modern Spain and Portugal are at the southern edge of their spawning river range (which extends to modern Scandinavia).

Palynological analyses offer a finer resolution of the paleoenvironment of El Juyo, but suffer from the limitations inherent to this type of research. Previous interpretation (Boyer-Klein and Ar. Leroi-Gourhan 1987; Boyer-Klein 1988; Ar. Leroi-Gourhan 1994) indicated that the lower levels (12-7) of El Juyo could be correlated with the relatively temperate Pre-Bölling climatic phase. This interpretation is based upon the presence of moderate levels of tree and steppe pollen and the dominance of Ericaceae (heather) pollen. The upper levels of El Juyo's Lower Magdalenian occupation were assigned to the Dryas I climatic phase, since a major deterioration of conditions was indicated by the dominance of steppic pollens at the expense of tree and shrubs. Sanchez Goñi (1993) reinterpreted this pollen data as still reflecting the cold, dry conditions of Dryas I in the upper levels (6-3) of the site, with the lower Levels 12-7 not assignable to the Pre-Bölling. The lower levels therefore also may have been deposited during Dryas conditions, which is in accordance with radiocarbon dating from the site.

The macrobotanical remains from El Juyo give similarly mixed paleoenvironmental signals, with a varied group of species present. Level 8 has yielded a proportionally high abundance of marsh-favoring plants, indicating wetter conditions during its formation or the more intensive exploitation of the nearby marshes and *rias* (estuaries). Subalpine taxa appear during the deposition of Level 4, reinforcing the interpretation that colder conditions returned in the later occupation of El Juyo (though the more intensive exploitation of alpine resources is an alternative possibility for the introduction of some taxa). Throughout all levels, taxa favoring disturbed habitats are abundant. The presence of these taxa indicates the selected nature of the macrobotanical remains introduced to the site by deliberate and accidental human activities, since human occupation itself produces disturbed local habitats. Thus, a cultural filter has in large part affected the relative abundances of these remains and lessened the resolution of their paleoenvironmental information.

Using these many data sources in conjunction with the microfauna sample, the paleoenvironmental picture of El Juyo is far clearer. Temperatures were generally cool, intermediate between the glacial maximum of 22,000-18,000 B.P. and the temperate Holocene period. Large mammal indicator species agree with the interpretation that open conditions were mixed with at least some areas of forest cover and grassland among heath. The relative amounts of these vegetation types are best reflected by the microfauna, whose accumulation did not pass through a cultural filter. High humidity is reflected by the microfauna through all levels. Moist habitat microfauna are accompanied by macrobotanical remains of marsh plants in occupations below Level 4. Sedimentology, palynology and macrobotanical analysis all agree with the microfaunal data that some deterioration of climate occurred during the deposition of Level 4, indicating that some cooling and drying of the climate occurred. During this time the alpine zone intruded to lower altitudes, with some plant and microfauna taxa present in this lowland site.

The smaller microfaunal remains (Rodentia, Insectivora, and Carnivora) from Cantabrian sites and the interpretation of paleoenvironment from their relative proportions were presented in Chapter 8. Some levels from the Mesolithic period and other later environments were presented for purposes of paleoenvironmental comparison across the Pleistocene/Holocene transition. Some sites from just outside the Cantabrian region (Pyrenean France, Navarra province in Spain) were similarly included. The remains from these sites reinforce interpretations of the pattern derived from El Juyo.

The Middle and Upper Paleolithic environments of Cantabrian Spain were dominated by open, humid-associated species. Microtines are the most numerous rodents in nearly every level from all sites, primarily *Arvicola* cf. *terrestris* and *Microtus oeconomus*. At many older excavations, only the relatively large taxa *Arvicola* and *Talpa* were recovered. Some dry grassland microtines were present (*M. arvalis* and *M. agrestis*), though generally in lower proportions. Insectivores were also typically dominated by humid meadow and water edge species (*Sorex coronatus*, *S. minutus*, *Neomys fodiens* and *N. anomalus*) at the expense of dry grassland species (*Crocidura russula* and *C. suaveolens*). Colder episodes are indicated by the appearance of alpine/tundra taxa (*Microtus nivalis*, *Galemys pyrenaicus* and *Microtus gregalis*) in the coastal lowland zone. Warmer episodes and the return of forested conditions with the onset of the Holocene are indicated by the increased proportion of *Apodemus* at many sites, reinforced by the presence of *Glis glis* and *Clethrionomys glareolus*. Modern Cantabrian region faunas are characterized by an increased importance of murid species, including the introduced commensal/agriculturally associated taxa (*Rattus*, *Mus musculus* and *Micromys minutus*).

An important subcategory of small mammal was recovered in significant amounts from the fine screen deposits of El Juyo: the remains of medium sized fur-bearing species, most of which belong to Order Carnivora. While some of these species fall below the upper boundary of 5 kg used in delimiting microfauna, many do not (at least as adults). These larger species, while fitting under the category of micromammals, are generally larger than the typical rodent and insectivore fauna falling prey to barn owls. As such, these taxa experience different taphonomic processes, including source of mortality, manner of transport to the site, pattern and degree of skeletal element destruction, and deposition within the site. While it is possible that some of these individuals (especially the smallest species such as *Mustela nivalis* and *M. erminea* or juveniles of the larger species such as *Oryctolagus cuniculus*) were introduced as the prey of owls, the pattern of skeletal element recovery suggests that these were brought into the cave as the direct action of humans. The depositional pattern of these fur-bearing mammals is also in direct contrast to that of hedgehog (*Erinaceus europaeus*) remains recovered from El Juyo. While this species is of a similar size to some fur-bearers (about 1 kg adult mass), it is not a usable source of fur due to its protective spines.

El Juyo has yielded to date 72 foot elements, primarily phalanges and metapodials of carnivores in the size range of ermine (*Mustela erminea*), ferret (*M. putorius*), and marten (*Martes* spp.). The presence of these elements is in striking contrast to the virtual lack of *any* other elements, including teeth, from these taxa (Chapter 5). Non-carnivores of this size range (such as rabbit or hedgehog) do not show this same element representation discrepancy, nor do smaller, more fragile rodent and small insectivore taxa. Differential destruction of non-foot elements of the fur-bearing species therefore may be ruled out. The author attributes the preponderance of fur-bearing mammal foot elements to their selective introduction to the site by a human agency. Such a pattern is the logical result of foot bones left in a pelt that was skinned from an animal trapped or hunted away from the site, with the remaining carcass including all other skeletal elements discarded in the field. Mustelid meat has a strong, unpleasant flavor due to their strong scent glands (personal observation) and may have been therefore a less desirable food choice. These species were thus probably acquired wholly for their pelts, with no other archaeologically recoverable portion of them reaching the site.

This pattern of increasing exploitation of fur-bearing species is reflected throughout the Cantabrian Upper Paleolithic. These taxa are nearly non-existent in the few Acheulean levels and appear only sporadically in Mousterian levels. The Early Upper Paleolithic (Aurignacian, Chatelperronian and Upper Perigordian/Gravettian) with its presumed arrival of anatomically modern humans has an increased abundance and diversity of these species, including for the first time *Mustela* spp. Skeletal part representation for this period does not indicate a pattern of skinning in the field and transporting foot elements (phalanges, metapodials, carpals/tarsals) back to cave sites along with furs, though sample sizes are generally quite small. The presence of these species in human habitation levels during this period therefore may be due as much to intermittent occupation of sites by these small predators or their introduction as the prey of larger predators as to sporadic use by humans.

The Late Upper Paleolithic (Solutrean, Magdalenian and Azilian) in Cantabrian Spain is characterized by a further increase in fur exploitation. This pattern becomes most evident in the Magdalenian period, where species abundance and diversity increases in several sites, and body part representation (at El Juyo) virtually rules out other explanations. The Azilian period shows an apparent decline in hunted/trapped species diversity and abundance. This trend perhaps resulted from a de-emphasis of fur acquisition relating to the general climatic amelioration at the close of the Pleistocene. It may also reflect decreased densities of small and medium predators in an increasingly human-altered environment.

Upper Paleolithic human-environment interaction in Cantabrian Spain exhibits an accelerating pattern of separation of human actions from the direct and immediate constraints of environment. This interaction includes the controlled harvest of red deer, ibex, salmon and marine molluscs by cooperative human action. Other changes include the expansion into previously less important resources (avifauna and fur trapping) and more organized and specialized landscape use. These changes are visible as early as the Early Upper Paleolithic and greatly increase throughout the Late Upper Paleolithic. This shift is more remarkable considering the major and minor fluctuations of the Pleistocene climate, proceeded from vacillating interstadial conditions, through the intense cold of the glacial maximum ca. 22,000-18,000 B.P. and the stepwise retreat/readvance of glaciers before the temperate conditions of the Holocene, which began about 10,000 B.P. The proximity of Cantabrian Spain's multiple and diverse environmental zones is also striking: littoral, riverine, coastal lowland and montane zones are within easy mutual reach. Access to very different environmental resources was available over space by physical travel over the landscape and through changes in environment over time.

The Upper Paleolithic Cantabrians show an ability or propensity to free themselves progressively from the vagaries of changing environmental conditions. Wild resource harvesting increased in intensity throughout this period despite major climatic fluctuations in this open, humid coastal zone. Instead of abandoning the region during the glacial maximum and other cold peaks, the Cantabrians steadily improved and refined their landscape utilization. Subsistence focused upon continually or seasonally concentrated resources and expanded into ones previously underexploited, and subsistence and settlement systems shifted with local or seasonal shifts in resource abundance. These changes in resource use (away from more random exploitation and towards planned, intensive harvesting at efficient locations and times with appropriately engineered technology) in many ways presages the changes necessary for agricultural systems to arise. These changes in Cantabrian Spain, though not directly leading to agriculture, do indicate locally the possible evolution of increased social complexity. Further evidence of this pattern is found in the art (Conkey 1978, 1980; Freeman *et al.* 1987) and landscape use patterns (Freeman 1994) of the period.

Paleolithic environmental study must be planned as part of a research program before excavation begins. Recovery of necessary data, especially in the cases of microfauna, pollen and macrobotanical remains cannot take place retroactively (Freeman *et al.* 1998). Sediment samples for pollen analysis must be derived from freshly exposed deposits to minimize the risk of contamination and preferably taken at multiple points over the exposed surface of a level to adjust for spatial variations in pollen flux. Macrobotanicals must be obtained through flotation of sediments and therefore are obtained during the initial processing of excavated sediment lots. Microfauna are similarly obtained through fine screening. These studies therefore cannot be carried out retroactively upon already excavated collections, as can meaningful analyses of large fauna or molluscs. Modern excavations in Cantabrian Spain have begun to incorporate these procedures from the outset of excavation planning, as at El Juyo (Freeman *et al.* 1988; Freeman *et al.* 1998) and to some extent at La Riera (Straus and Clark 1986), Ekain (Altuna and Merino 1984), Amalda (Altuna *et al.* 1990) and Eralla (Altuna *et al.* 1995).

Future research should follow several paths (Brothwell and Jones 1978). One area is the establishment of further criteria for identifying closely related species by isolated teeth alone. Such separation is necessary for archaeological samples, where complete crania are rare and most measurements typically used for species identification are not possible. Taxa that should be researched in this manner are *Arvicola*, *Talpa* and *Apodemus*. Small mammals are also potentially identifiable by their postcranial remains (Graham and Saunders 1978). These data may help to identify accumulating predators by their patterns of prey postcranial destruction (Pokines and Kerbis Peterhans 1998), which in turn aids in paleoenvironmental interpretation. Ageing criteria for small mammal species by tooth wear or postcranial fusion will also aid in predator identification, yet research into this topic has been highly sporadic. The continually growing molars of microtines represent a particular problem in this regard. Prey age profiles are a further criterion in the identification of accumulating predator. Additional owl pellet and trapping studies of the altitudinal relationships of Cantabrian microfauna would also better illuminate the altitudinal shifts in species ranges that accompany episodes of alpine glaciation. Further investigation of the relationships between local species biometrics over different habitats and climates in their modern ranges (Klein 1991) may also allow interpretation of past changes in environmental parameters (temperature or rainfall).

While barn owl behavior is well researched, that of other potential accumulating agencies has been less so. More research into the biases they may cause in paleoecological interpretation by affecting relative proportions of prey remains is needed. Taphonomic reanalysis of already excavated microfaunal collections might also prove illuminating, allowing a finer degree of paleoenvironmental interpretation. Samples that may benefit from such analysis exist from the sites of Erralla (Pemán 1985), Aitzbitarte (Altuna 1972), Amalda (Pemán 1990) and Ekain (Zabala 1984).

At El Juyo, microfaunal age profile analysis indicates that barn owls were regularly present during the warmer months of the year, though their presence during the colder months cannot be discounted. Barn owls and humans are unlikely to have occupied the cave at the same time. This pattern, along with indications of human food acquisition during the colder months of the year reinforces the interpretation of seasonal human occupation of this lowland site. Similar analyses of microfauna from other sites may further illuminate patterns of human site utilization and seasonal movement, as counterpoised against patterns of non-human animal activities.

Future research at El Juyo will concentrate upon the acquisition of larger microfaunal samples from the lower site levels (11-8), as well as the identification of bird, reptile and amphibian remains already excavated. This information will be combined with new large fauna, fish, sedimentological, and macrobotanical data as they accumulate continually during the ongoing excavations.

Parallel paths of research include analysis of the relationships between animal depictions and faunal remains (González Sainz 1988) or animal depictions and site function/hierarchy (Conkey 1978, 1980; Freeman 1994; Freeman *et al.* 1987). Research of this type examines the cognitive role in human paleoecological relationships and how human perceptions of the environment may have affected patterns of resource and landscape use. Continuing research into changing patterns of Paleolithic resource exploitation or overexploitation (Klein and Cruz Uribe 1987; Krupa 1994) will allow further interpretation of human interaction with the changing Cantabrian environment.

Paleoenvironmental research is relevant to more than the natural history and evolution of the small mammals themselves, or to human ecological interaction specific to extinct hunter-gatherer groups in a glacial environment that no longer exists. The evolutionary path that we are on did not begin with the Industrial Age or even with the "Neolithic Revolution", but tens of thousands of years earlier as our environmental relationships began to deviate from those of other large social predators. This deviation can even be seen to have begun much earlier with the evolution of *Homo* and preceding/concurrent developments such as bipedalism, increasing relative brain size, and tool use. The human evolutionary niche therefore may be seen as the development of the ability to adapt the natural environment to human needs (the construction of a cultural environment) rather than adaption to the natural environment itself. This evolutionary trend does have limitations. Humans must operate within the parameters set by their contemporary environment, alter it as they will, and are discovering this fact with increasing immediacy as biodiversity shrinks (Connell 1978; Huston 1979; Stork 1988; Wilson 1989) and environmental crises grow (Soulé 1992). The origins and trajectory of this path of increasing ability to alter the biosphere becomes directly relevant to understanding this increased modern propensity to damage the biosphere and learning how to control it.

APPENDIX A
METRICAL DATA

Table A.1: Maximum Length (mm) of $_1M_1$ of *Arvicola*.

				El Juyo Level										El Juyo	
11	9	8	7	6			4R	4S					4	Quarry Site 2	
4.02	3.44	3.74	3.60	3.77	3.81	4.11	3.38	3.67	3.55	3.64	3.78	3.91	3.90	3.46	3.55
	3.49	3.94	3.61	3.36	3.62	3.47	3.23	3.76	3.22	3.71	3.68	3.69	3.62	3.40	3.67
	3.45	3.97	3.67	3.36	3.76	3.58	3.89	3.68	3.60	3.44	3.90	3.73	3.79	3.55	3.45
	4.01	3.56	4.09	3.89	3.91	3.46	3.34	3.90	3.55	3.85	4.24	3.73	3.63	3.24	3.66
		4.65		3.61	3.75		3.23	3.78	3.51	3.61	3.43	3.69	3.78	3.46	3.56
		4.21		4.03	3.85		3.70	3.28	3.83	3.97	3.95	3.40	3.69	3.61	3.47
		3.96		4.05	3.82		3.75	3.91	3.56	3.27	3.67	4.06	3.87	3.60	3.44
		3.51		3.91	3.56		4.07	3.85	3.69	4.01	3.73	3.70	4.16	3.87	4.03
		3.47		3.86	3.56		3.96	3.83	3.57	4.16	4.01	3.96	3.70	3.73	3.35
		3.57		3.79	3.62		4.05	3.62	3.65	3.84	3.59	3.47	3.52	3.83	3.50
				3.67	4.01		4.24	3.74	3.87	3.58	3.63	3.77	3.51	3.59	3.47
				3.61	4.42		3.60	3.75	3.85	3.68	3.85	3.85	3.52	3.55	3.65
				3.57	3.98		3.66	3.52	3.50	3.63	4.01	3.91	3.55	3.67	3.41
				3.60	3.96			3.92	3.71	3.60	3.47			3.26	3.31
				3.66	3.77			3.84	3.83	3.65	3.47			3.51	3.39
				3.66	3.81			4.41	3.76	3.45	3.35			3.31	3.47
				3.60	3.60			3.11	3.58	3.86	3.61			3.59	3.31
				4.18	3.81			3.85	3.51	3.97	4.43			3.31	
				3.54	3.61			3.42	3.32	3.82	3.84			3.52	
				3.98	3.68			3.75	4.11	3.61	4.00			3.71	

Table A.2: Maximum Length (mm) of $^3M^3$ of *Arvicola*.

				El Juyo Level												El Juyo	
11	9	8	7	6		4R	4S						4		2/4	Quarry Site 2	
2.80	2.38	2.22	2.52	2.33	2.59	2.54	2.33	2.65	2.20	2.39	2.87	2.18	2.56	2.39	2.14	2.31	2.37
	2.22	2.31	2.43	2.29	2.32	2.40	2.47	2.45	2.25	2.07	2.31	2.69	2.25	2.30		2.56	2.66
	2.57	2.35		2.48	2.61	2.21	2.26	2.55	2.42	2.14	2.50	2.34	2.20	2.51		1.92	2.09
	2.19	1.97		2.16	2.42	2.46	2.58	2.26	2.21	2.53	2.46	2.15	2.28	2.28		2.26	2.28
	2.26	2.22		2.55	2.29	2.06	2.21	2.45	2.60	2.13	2.43	2.61	2.28			2.24	2.33
	2.27	2.10		2.71	2.70	2.26	2.06	2.96	2.41	2.47	2.09	2.61	2.05			2.26	2.20
	2.72	2.30		2.04	1.92	2.41	2.44	2.57	2.31	2.65	2.36	2.26	2.08			2.50	
		2.37		2.48	2.29	2.43	2.39	2.53	2.45	2.45	2.46	2.44	1.98			2.03	
		2.41		1.94	2.41	2.16	2.37	2.26	2.60	2.37	2.52	2.19	2.01			2.02	
		2.57		2.32	2.51	2.62	2.21	2.11	2.36	2.47	2.16	2.39	2.31			2.11	
		2.26		2.66	2.23	2.36	2.51	2.19	2.26	2.39	2.40	2.26	2.11			2.05	
				2.12	2.58	2.45	2.20	2.55	2.60	2.52	2.51	2.40	2.11			2.13	
				2.27	2.30		2.31	2.57	2.51	2.44	2.26	2.30	2.57			2.09	
				2.02	2.13		2.16	2.63	2.99	2.57	2.23	2.59	2.37			2.22	
				2.22	2.23		2.32	2.24	2.17	2.29	2.06	2.25	2.86			1.96	
				2.50	2.25		2.16	2.26	2.23	2.33	2.11	2.55	2.45			2.10	
				2.60			2.61	2.27	2.25	2.52	1.98	2.50	2.21			2.09	
				2.43			2.57	2.41	2.36	2.52	2.28	2.34	2.82			2.15	
				2.72			2.32	2.18	2.31	2.50	2.41	2.66	2.04			1.85	
				2.39			2.40	2.34	2.65	2.62	2.21	2.65	2.42			2.26	
				2.76			2.30	2.26	2.07	2.70	2.21	2.65	2.00			2.35	
				2.42			2.50	2.35	2.33	2.31	2.37	1.89	2.70			2.23	
				2.54			2.40	2.23	2.45	2.18	2.35	2.06	2.39			2.13	
				2.54			2.56	2.45	2.21	2.77	2.27	2.40	2.40			2.43	
				2.12			2.23	2.15	2.42	2.62	2.31	2.15	2.80			2.32	
				2.79			2.46	2.16	2.63	2.76	2.19		2.34			2.25	

Table A.3: Maximum Length (mm) of $_1M_1$ of *Microtus oeconomus*.

El Juyo Level														
11	9	8	7	6/8	6		4R	4S				4		2/4
2.40	2.27	2.52	2.57	2.57	2.51	2.84	2.59	2.42	2.86	2.65	2.92	2.26	2.62	2.47
		2.46			2.56	2.53	2.22	2.50	2.70	2.53	2.59	2.45	2.51	2.55
		2.82			2.55	2.47	2.59	2.72	2.70	2.45	2.58	2.23		
		2.46			2.54	2.59	2.48	2.74	2.53	2.67	2.45	2.37		
		2.36			2.44	2.77	2.58	2.74	2.65	2.81	2.67	2.49		
		2.56			2.55	2.38	2.63	2.70	2.41	2.46	2.69	2.59		
		2.46			2.37	2.43	2.35	2.65	2.67	2.68	2.08	2.59		
					2.56	2.38	2.35	2.68	2.50	2.48	2.61	2.72		
					2.20	2.53	2.81	2.55	2.50	2.49	2.79	2.42		
					2.54	2.61	2.72	2.72	2.54	2.62	2.57	2.29		
					2.55	2.37	2.61	2.60	2.70	2.55	2.56	2.47		
					2.57	2.56	2.37	2.58	2.70	2.45	2.61	2.33		
					2.66	2.57	2.71	2.52	2.26	2.41	2.50	2.61		
					2.47	2.48		2.57	2.46	2.51	2.58	2.71		
					2.66	2.48		2.65	2.88	2.33	2.37	2.46		
					2.50	2.51		2.35	2.65	2.72	2.54	2.43		
					2.60	2.46		2.58	2.66	2.58	2.82	2.43		
					2.50	2.40		2.30	2.46	2.57	2.51	2.61		
					2.39	2.63		2.37	2.76	2.45	2.36	2.66		
					2.42			2.40	2.70	2.48	2.52	2.56		
					2.46			2.59	2.52	2.51	2.58	2.53		
					2.55			2.51	2.84	2.25	2.44	2.74		
					2.56			2.65	2.81	2.47	2.44	2.62		
					2.36			2.55	2.47	2.69	2.70	2.58		
					2.37			2.52	2.62	2.57	2.67	2.45		

Table A.4: Maximum Length (mm) of $_1M_1$ of *Pitymys* spp.

El Juyo Level										
11	9	8	7	6	4R		4S			4
2.63	2.36	2.57	2.41	2.51	2.73	2.51	2.31	2.38	2.39	2.68
		2.22		2.70	2.30	2.30	2.63	2.38	2.52	2.53
		2.48		2.25		2.56	2.47	2.52	2.35	2.35
		2.66		2.63		2.72	2.56	2.56	2.61	2.72
		2.42		2.69		2.65	2.42	2.60	2.54	2.43
		2.40		2.50		2.56	2.60	2.76	2.71	2.64
		2.42		2.55		2.38	2.64	2.48	2.46	2.44
		2.50		2.60		2.58	2.65	2.50	2.40	2.72
		2.52				2.55	2.72	2.27	2.64	2.60
						2.62	2.60	2.45	2.54	2.58
						2.57	2.65	2.37	2.52	2.47
						2.71	2.50	2.46		

Table A.5: Maximum Length (mm) of $_2M_2$ of *Pitymys* spp.

El Juyo Level									
11	8	7	6	4S				4	
1.42	1.37	1.38	1.40	1.46	1.39	1.48	1.38	1.44	1.37
	1.44		1.49	1.40	1.33	1.25	1.25	1.31	1.34
	1.47		1.39	1.40	1.30	1.47	1.35	1.43	1.26
	1.24		1.51	1.42	1.45	1.31	1.29	1.39	1.30
	1.40			1.45	1.42	1.31	1.38	1.38	1.45
	1.42			1.35	1.48	1.29	1.39	1.37	1.25
				1.40	1.44	1.29	1.40	1.22	1.37
				1.41	1.43	1.40	1.20	1.43	1.37
				1.30	1.44	1.30		1.46	1.39
				1.50	1.26	1.36			
				1.50	1.29	1.38			

Table A.6: Maximum Length (mm) of $^3M^3$ of *Pitymys lusitanicus*.

El Juyo Level			El Juyo Quarry Site 2				
8	4R	4S					
1.84	1.83	1.97	1.66	1.51	1.53	1.52	1.44
		1.87	1.59	1.51	1.41	1.62	1.35
		1.75	1.61	1.42	1.31	1.37	1.68
		1.44	1.43	1.51	1.57	1.42	1.46
		1.43	1.42	1.75	1.51	1.71	1.41
		1.53	1.46	1.59	1.51	1.41	1.56
		1.47	1.50	1.45	1.64	1.47	1.42
		1.47	1.60	1.44	1.60	1.64	1.40
		1.40	1.69	1.53	1.38	1.68	1.46
		1.44	1.50	1.48	1.55	1.64	

Table A.7: Maximum Length (mm) of $^3M^3$ of *Pitymys pyrenaicus*.

El Juyo Level						
9	8	6		4S		4
1.76	1.66	1.81	1.62	1.59	1.75	1.60
	1.51	1.65	1.67	1.66	1.76	1.75
	1.82	1.62	1.93	1.65	1.78	1.67
	1.88	1.98	1.85	1.59	1.67	1.77
	1.87	1.84	1.75	1.64	1.76	1.64
	1.75	1.72	1.45	1.69	1.90	1.81
	1.61	1.52	1.59	1.80	1.71	1.61
	1.70	1.45	1.55	1.48	1.76	1.42
	1.69	1.47	1.85	1.76	1.67	1.62
	1.71	1.45	1.92	1.38	1.58	1.42
	1.77	1.91	1.70	1.85	1.84	1.54
	1.74	1.76	1.65	1.79	1.66	1.50
		1.77	1.70	1.86	1.65	1.74
			1.61	1.64	1.64	1.70
			1.90	1.50	1.55	1.63
			1.75	1.49		1.75

Table A.8: Maximum Length (mm) of $_1M_1$ of *Microtus agrestis/arvalis*.

El Juyo Level												El Juyo Quarry Site 2
10	9	8	7	6	4R	4S			4		2/4	
2.47	2.42	2.84	2.58	2.59	2.35	2.71	2.66	2.74	2.61	2.82	2.61	2.71
	2.64	2.61	2.40	2.46	2.79	2.32	2.34	2.58	2.46	2.53	2.81	2.57
	2.79	2.76	2.74	2.47		2.62	2.36	3.02	2.50	2.42		2.65
	2.79	2.41		2.41		2.67	2.26	2.76	2.55	2.48		2.69
		2.69		2.69		2.64	2.73	2.81	2.60	2.35		2.57
		2.78		2.77		2.50	2.37	2.86	2.57	2.51		2.72
		2.86		2.73		2.73	2.67	2.90	2.81	2.37		2.75
		2.54		2.70		2.59	2.55	3.01	2.86	2.38		2.82
		2.74		2.78		2.36	2.67	2.90	2.64	2.73		2.74
				2.71		2.65	2.56	2.75	2.65	2.90		2.59
				2.77		2.41	2.54	2.78	2.74	2.37		3.03
				2.59		2.47	2.74	2.82	2.76	2.72		2.68
				2.62		2.63	2.54	2.82	2.74	2.41		2.74
				2.66		2.55	2.65	2.63	2.59	2.81		2.75
				2.68		2.64	2.44	2.70	2.73	2.77		2.87
				2.93		2.72	2.60	2.91	2.51	2.77		2.76
				2.92		2.53	2.75	2.57	2.68	2.69		2.94
				2.76		2.74	2.73	2.65	2.63	2.57		2.78
						2.71	2.70	2.70	2.84	2.68		2.80
						2.70	2.56	2.81	2.86	2.73		
						2.47	2.68	2.76	3.16			
						2.62	2.69	2.84	3.22			
						2.82	2.74	2.81	2.73			
						2.68	2.70	2.57	2.78			
						2.44	2.76	2.90	2.78			
						2.70	2.79	2.86	2.82			
						2.78	2.67	2.88	2.84			
						2.89	2.77	2.91	3.23			
						2.81	3.00	3.07	2.27			
						2.74	2.77	2.71	2.45			
						2.67	2.76	2.87	2.44			
						2.60	2.74		2.37			

Table A.9: Maximum Length (mm) of $^2M^2$ of *Microtus agrestis.*

El Juyo Level										El Juyo Quarry Site 2	
9	8	7	6	4R	4S			4	2		
1.59	1.61	1.55	1.59	1.59	1.61	1.60	1.62	1.71	1.56	1.66	1.64
	1.52		1.70		1.59	1.50	1.54	1.76		1.63	1.69
	1.52				1.45	1.65	1.60	1.57		1.65	1.70
	1.69				1.72	1.66	1.68	1.66		1.56	1.59
	1.66				1.61	1.46	1.45			1.61	1.61
					1.62	1.59	1.56			1.61	
					1.65	1.62	1.58			1.53	
					1.72	1.66	1.60			1.60	
					1.62	1.64	1.60			1.64	
					1.56	1.61				1.63	

Table A.10: Maximum Length (mm) of $_1M_1$ of *Microtus nivalis.*

El Juyo (All Levels)
2.61
2.96
3.27

Table A.11: Maximum Length (mm) of $_1M_1$, $_2M_2$, $^1M^1$and $^2M^2$ of *Apodemus.*

El Juyo Level														
8	6	4S		4	8	4S	4	3	4S		4	4S	4
M1 low	M1 low	M1 low		M1 low	M2 low	M2 low	M2 low	M2 low	M1 up		M1 up	M2 up	M2 up
1.71	1.84	2.00	1.70	1.90	1.26	1.33	1.17	1.38	1.87	2.10	1.96	1.41	1.27
1.90		1.86	1.90	1.95		1.28			1.95	1.99		1.25	1.28
		1.74	1.99	1.90		1.28			1.83	1.59		1.11	
		1.85	1.75	1.86		1.25			1.85			1.41	
		1.71	1.80	1.82		1.20			1.78			1.23	
		1.73	1.78			1.26			1.95			1.25	
		1.76				1.24			1.96			1.27	
		1.75				1.19			1.95			1.17	
		1.72							1.75			1.23	
		1.66							1.80				
		1.71							1.96				
		1.72							1.81				

El Juyo Quarry Site 2							
A. flavicollis				*A. sylvaticus*			
M1 low	M2 low	M1 up	M2 low	M1 low	M2 low	M1 up	M2 low
1.92	1.25	1.91	1.22	1.78	1.22	1.89	1.22
1.77	1.26	1.97	1.31	1.70	1.18	1.76	1.11
1.74	1.17	1.76	1.14	1.81	1.26	1.78	1.16
1.82	1.26	1.95	1.20	1.68	1.18	1.77	1.26
1.89	1.37	1.94	1.32	1.76	1.23	1.90	1.19
1.85	1.34	2.02	1.31	1.80	1.22	1.98	1.32
1.81	1.26	1.89	1.28	1.82	1.27	1.91	1.27
				1.73	1.17	1.72	1.20
				1.74	1.17	1.79	1.19
				1.84	1.28	1.86	1.26
				1.72	1.11	1.76	1.13
				1.75	1.23	1.80	1.22

Table A.12: Maximum Ramus Height and Length of $_1M_1$, $_2M_2$ and $_3M_3$(mm) of *Sorex coronatus*.

El Juyo Level

8	6	4S	4	8	6	4S	4	8	6	4S	4	8	6	4S	4
ramus	ramus	ramus	ramus	M1	M1	M1	M1	M2	M2	M3	M2	M3	M3	M3	M3
3.88	4.44	4.11	4.14	1.67	1.50	1.45	1.34	1.32	1.16	1.23	1.27	1.06	0.98	1.05	1.04
4.36	4.18	4.34	4.25		1.40	1.42	1.45	1.27	1.27	1.22	1.24	1.08	1.01	0.90	
	4.38				1.43	1.48	1.37		1.30	1.24	1.24		1.01		
	4.00				1.55	1.48	1.42		1.42	1.26			0.99		
	4.26				1.47	1.50			1.28				1.07		
					1.49				1.31						
					1.44				1.22						
					1.39				1.39						
									1.28						

El Juyo Quarry Site 2

Ramus		M1		M3		M3	
4.27	4.50	1.38	1.32	1.22	1.20	0.97	1.06
4.50	4.26	1.40	1.30	1.27	1.24	1.07	1.08
4.44	4.65	1.36	1.44	1.30	1.31	1.00	1.05
4.41	4.52	1.35	1.40	1.25	1.34	0.91	1.05
4.57	4.50	1.50	1.50	1.33	1.28	1.03	1.01
4.28	4.40	1.42	1.43	1.25	1.21	1.04	1.05
4.43	4.47	1.45	1.46	1.31	1.30	1.05	1.10
4.50	4.32	1.40	1.45	1.32	1.28	1.06	1.00
4.65	4.54	1.38	1.45	1.27	1.32	1.00	1.06
4.46	4.55	1.41	1.36	1.28	1.26	1.03	1.10
4.56	4.53	1.42	1.42	1.31	1.25	1.02	1.02
4.42	4.52	1.49	1.45	1.31	1.26	1.00	1.00
4.69	4.56	1.42	1.38	1.31	1.25	1.03	1.04
4.50	4.78	1.46	1.46	1.31	1.35	1.09	0.97
4.45	4.56	1.42	1.43	1.30	1.29	1.01	0.98
4.36	4.45	1.50	1.37	1.34	1.31	1.00	1.00
4.50	4.39	1.41	1.46	1.29	1.27	1.08	
4.54	4.56	1.45	1.45	1.29	1.30	1.00	
4.57	4.40	1.48	1.40	1.31	1.32	1.05	
4.47	4.37	1.49	1.36	1.39	1.23	1.03	
4.57	4.40	1.40	1.32	1.25	1.21	1.00	
4.50	4.42	1.40	1.40	1.32	1.25	1.00	
4.56	4.13	1.44	1.42	1.30	1.34	0.95	
4.43	4.59	1.40	1.51	1.28	1.29	1.04	
4.54	4.26	1.40	1.42	1.28	1.30	1.00	
4.74	4.21	1.43	1.43	1.25	1.27	1.01	
4.55	4.54	1.42	1.37	1.23	1.31	1.00	
4.42	4.50	1.50	1.35	1.28		1.04	
4.73	4.20	1.41	1.32	1.28		0.95	
4.45	4.26	1.43	1.46	1.32		0.97	
4.44	4.32	1.41		1.25		1.10	
4.35	4.43	1.45		1.30		1.00	

Table A.13: Maximum Ramus Height and Length of $_1M_1$, $_2M_2$ and $_3M_3$(mm) of *Sorex minutus*.

El Juyo (All Levels)

Ramus	M1	M2	M3
2.97	1.28	1.10	0.92
3.12	1.27	1.14	0.95
3.00	1.41	1.15	0.93
3.26	1.27	1.13	0.89
3.20	1.24	1.14	0.92
3.15		1.09	0.98
2.71		1.04	0.92
3.14			

Table A.14: Maximum Length (mm) of $_1M_1$ and $^1M^1$ of *Erinaceus europaeus*.

El Juyo (All Levels)	
1 lower	1 upper
5.84	5.60
6.00	5.74
6.53	6.00
6.84	6.34
7.25	

Table A.15: Maximum Length (mm) of $_1M_1$, $_4P_4$ and $^4P^4$ of *Mustela nivalis*.

El Juyo (All Levels)		
M1 low	P4 up	P4 low
4.60	3.39	1.89
3.90		2.02
4.56		

REFERENCES

Adamczewska-Andrzejewska, K. A., 1967. Age reference model for *Apodemus flavicollis* (Melchior, 1834). *Ekologia Polska Seria A* 15, 787-790.

Alibhai, S.K., 1980. An X-ray technique for ageing bank voles (*Clethrionomys glareolus*) using the first mandibular molar. *Journal of Zoology* 191, 418-423.

Almaça, C., 1973. Sur la structure des populations des *Pitymys* Ibériques. *Revista da Facultad de Ciencias de Lisboa* 17, 383-426.

Altuna, J., 1972. Fauna de mamíferos de las yacimientos prehistóricos de Guipúzcoa. *Munibe* 24, 1-464.

Altuna, J., 1976. Los mamíferos del yacimiento prehistórico de Tito Bustillo (Asturias). In *Excavaciones en la Cueva de "Tito Bustillo" (Asturias)*, eds. J. A. Moure Romanillo and M. Cano-Herrera. Oviedo: Instituto de Estudios Asturianos, pp. 151-193.

Altuna, J., 1978. Los mamíferos de Cueva Morín. In *Vida y Muerte en Cueva Morín*, eds. J. González Echegaray & L. G. Freeman. Santander: Institución Cultural de Cantabria, pp. 201-209.

Altuna, J., 1981. Restos oseos del yacimiento prehistórico del Rascaño. In *El Paleolitico Superior de la Cueva del Rascaño (Santander)*, eds. J. González Echegaray & I. Barandiarán Maestu. Santander: Centro de Investigacíon y Museo de Altamira, Monografia No. 3, pp. 221-269.

Altuna, J., 1985a. Introducción. *Munibe*, 37:7-9.

Altuna, J., 1985b. Dataciones del C14. *Munibe* 37, 25-28.

Altuna, J., 1986. The mammalian faunas from the prehistoric site of La Riera. In *La Riera Cave*, eds. L. G. Straus & G. A. Clark. Arizona State University, Anthropological Research Papers No. 36, pp. 237-274

Altuna, J., 1989. Subsistence d'origine animale pendant le Moustérien dans la région cantabrique. In *L'Homme de Néanderthal, Volume* 6, ed. M. Otte. Etudes et Recherches Archéologiques de l'Université de Liège, pp. 31-44.

Altuna, J., 1990. Caza y alimentación procedente de macromamíferos durante el Paleolítico de Amalda. In *La Cueva de Amalda (Zestoa, Pais Vasco)*, eds. J. Altuna, A. Baldeón & K. Mariezkurrena. San Sebastián: Sociedad de Estudios Vascos, pp. 149-192.

Altuna, J., A. Baldeón & K. Mariezkurrena, eds., 1990. *La Cueva de Amalda (Zestoa, Pais Vasco)*. San Sebastián: Sociedad de Estudios Vascos.

Altuna, J, & K. Mariezkurrena, 1982. Restos oseos del yacimiento prehistórico de Abauntz (Arraiz, Navarra). *Trabajos de Arqueologia Navarra* 3, 347-353.

Altuna, J, & K. Mariezkurrena, 1984. Bases de subsistencia de origen animal en el yacimiento de Ekain. In *El Yacimiento de la Cueva de Ekain (Deba, Guipúzcoa)*, eds. J. Altuna & J.M. Merino. San Sebastián: Sociedad de Estudios Vascos, Serie B1, pp. 211-280.

Altuna, J, & K. Mariezkurrena, 1985. Bases de subsistencia de los pobladores de Erralla: Macromamíferos. *Munibe* 37, 87-117.

Altuna, J. & J. M. Merino, eds., 1980. *El Yacimiento de la Cueva de Ekain (Deba, Guipúzcoa)*. San Sebastián: Sociedad de Estudios Vascos, Serie B1.

Altuna, J. & L. G. Straus, 1976. The Solutrean of Altamira: The artifactual and faunal evidence. *Zephyrus* 26-27, 175-182.

Amores, F., 1980. Feeding habits of the stone martens, *Martes foina* (Erxleben, 1777), in southwestern Spain. *Säugetier- kundlichen Mitteilungen* 54, 316-322.

Anderson, S. & C. A. Long, 1961. Small mammals in pellets of barn owls from Miñaca, Chihuahua. *American Museum Novitates* 2052, 1-3.

Andrews, P., 1990, *Owls, Caves and Fossils*. Chicago: University of Chicago Press.

Andrews, P., J. M. Lord, & E. M. Nesbit Evans, 1979. Patterns of ecological diversity in fossil and modern mammalian Faunas. *Biological Journal of the Linnean Society* 11, 177-205.

Andrews, P. & E. M. Nesbit Evans, 1983. Small mammal bone accumulations produced by mammalian carnivores. *Paleobiology* 9, 289-307.

Angersbjörn, A. & J. E.C. Flux, 1995. *Lepus timidus. Mammalian Species* 495, 1-11.

Avery, D. M., 1982a. Micromammals as paleoenvironmental indicators and an interpretation of the Late Quaternary in the Southern Cape Province, South Africa. *Annals of the South African Museum* 85, 183-374.

Avery, D. M., 1982b. The micromammalian fauna from Border Cave, Kwazulu, South Africa. *Journal of Archaeological Science* 9, 187-204.

Avery, D. M., 1987. Late Pleistocene coastal environment of the Southern Cape Province of South Africa: Micromammals from Klasies River Mouth. *Journal of Archaeological Science* 14, 405-421.

Avery, D. M., 1992. The environment of early modern humans at Border Cave, South Africa: Micromammalian evidence. *Palaeogeography, Palaeoclimatology, Palaeoecology* 91, 71-87.

Avery, D. M., 1993. Ecological data on micromammals collected by barn owls *Tyto alba* in the West Coast National Park, South Africa. *Israel Journal of Zoology* 38, 385-397.

Avery, D. M., 1995a. Micromammalian studies: Information from the past of relevance to the future. *Transactions of the Royal Society of South Africa* 50, 41-47.

Avery, D. M., 1995b. Physical environment and site choice in South Africa. *Journal of Archaeological Science* 22, 343-353.

Bailey, G., 1983. Economic change in Late Pleistocene Cantabria. In *Hunter-Gatherer Economy in Prehistory* ed. G. Bailey. Cambridge: Cambridge University Press, pp. 149-165.

Bailey, G. & I. Davidson, 1983. Site exploitation territories and topography: Two case studies from Palaeolithic Spain. *Journal of Archaeological Science* 10, 87-115.

Barandiarán Maestu, I., 1981. Industria osea. In *El Paleolitico Superior de la Cueva del Rascaño (Santander)*, eds. J. González Echegaray & I. Barandiarán Maestu. Santander: Centro de Investigación y Museo de Altamira, Monografías No.3, pp. 97-164.

Barandiarán Maestu, I., 1987. Industria osea paleolítica de la Cueva del Juyo, excavaciones de 1978 y 1979. In *Excavaciones en la Cueva Del Juyo*, eds. I. Barandiarán Maestu, L. G. Freeman, J. González Echegaray & R. G. Klein. Santander: Centro de Investigación y Museo de Altamira, Monografías No. 14, pp. 161-194

Barandiarán, I., L. G. Freeman, J. González Echegaray & R. G. Klein, eds., 1987. *Excavaciones en la Cueva Del Juyo*. Santander: Centro de Investigación y Museo de Altamira, Monografías No. 14.

Barash, D. P., 1989. *Marmots: Social Behavior and Ecology*. Stanford: Stanford University Press.

Bartolomei, G., J. Chaline, O. Fejfar, D. Jánossy, M. Jeannet, W. von Koenigswald & K. Kowalski, 1975. *Pliomys lenki* (Heller 1930) (Rodentia, Mammalia) en Europe. *Acta Zooligica Cracoviensia* 20, 393-467.

Behre, K.-E., 1986. *Anthropogenic Indicators in Pollen Diagrams*. Rotterdam: A.A. Balkema.

Behrensmeyer, A. K., 1978. Taphonomic and ecologic information from bone weathering. *Paleobiology* 4, 150-162.

Behrensmeyer, A. K., 1991. Terrestrial vertebrate accumulations. In *Taphonomy: Releasing the Data Locked in the Fossil Record*, eds. P. A. Allison & D. E. G. Briggs. New York: Plenum Press, pp. 291-335.

Beltrán, J. F. & M. Delibes, 1993. Physical characteristics of Iberian lynxes (*Lynx pardinus*) from Doñana, Southwestern Spain. *Journal of Mammalogy* 74, 852-862.

Bergman, C. A., 1993. The development of the bow in Western Europe: A technological and functional perspective. In *Hunting and Animal Exploitation in the Later Palaeolithic and Mesolithic of Eurasia*, eds. G. Larsen Peterkin, H. M. Bricker & P. Mellars. Archeological Papers of the American Anthropological Association, Number 4, pp.95-105.

Berry, R. J., 1978. Micro-evolutionary studies in animals: Their relevance to archaeology. In *Research Problems in Zooarchaeology*, eds. D. R. Brothwell, K. D. Thomas & J. Clutton-Brock. London: Institute of Archaeology, Occasional Publication No. 3, pp. 1-8.

Bischoff, J. L., J. Francisco García & L. G. Straus, 1991. Uranium-series isochron dating at El Castillo cave (Cantabria, Spain): The "Acheulean"/"Mousterian" question. *Journal of Archaeological Science* 19, 49-62.

Bishop, I. R. & M. J. Delany, 1963. Life histories of small mammals in the Channel Islands in 1960-61. *Proceedings of the Zoological Society of London* 141, 515-526.

Bleed, P., 1986. The optimal design of hunting weapons: Maintainability or reliability. *American Antiquity* 51, 737-747.

Blumenschine, R. J., 1986. Carcass consumption sequences and the archaeological distinction of scavenging and hunting. *Journal of Human Evolution* 15, 639-659.

Boonstra, R. & C. J. Krebs, 1978. Pitfall trapping of *Microtus townsendii*. *Journal of Mammalogy* 59, 136-148.

Bordes, F., 1968. *The Old Stone Age*. New York: McGraw-Hill.

Borghi, C.E., S. M. Giannoni & J.P. Martínez-Rica, 1994. Habitat segregation of three sympatric fossorial rodents in the Spanish Pyrenees. *Zeitschrift für Säugetierkunde* 59, 52-57.

Bourlière, F., 1975. Mammals, small and large: The ecological implications of size. In *Small Mammals: Their Productivity and Population Dynamics*, eds. F. B. Golley, K. Petrusewicz & L. Ryszkowski. Cambridge: Cambridge University Press, pp. 1-8.

Boyer-Klein, A. 1976. Análisis polínico de la Cueva de Tito Bustillo (Asturias). In *Excavaciones en la Cueva de Tito Bustillo (Asturias). Excavaciones de 1975*,. eds. A. Moure Romanillo & M. Cano Herrera. Oviedo: Instituto de Estudios Asturianos, pp. 203-206

Boyer-Klein, A., 1981. Análisis polínico de la cueva del Rascaño. In *El Paleolítico Superior de la Cueva del Rascaño (Santander)*, eds. I. Barandiarán Maestu & J. González Echegaray. Centro de Investigación y Museo de Altamira, Monografías No. 3, pp. 217-220.

Boyer-Klein, A., 1984. Analyses polliniques cantabriques au Tardiglaciaire. *Revue de Paléobiologie*, Vol. spécial, pp. 33-39.

Boyer-Klein, A., 1985. Analyse pollinique de la grotte d'Erralla. *Munibe* 37, 45-48.

Boyer-Klein, A. & Ar. Leroi-Gourhan, 1987. Análisis palinologico de la Cueva del Juyo. In *Excavaciones en la Cueva Del Juyo*, eds. I. Barandiarán Maestu, L. G. Freeman, J. González Echegaray & R. G. Klein. Santander: Centro de Investigación y Museo de Altamira, Monografías No. 14, pp. 55-61

Brain, C. K., 1974. The use of microfaunal remains as habitat indicators in the Namib. *South African Archaeological Society Goodwin Series* 2, 55-60.

Brain, C. K., (1980) Some criteria for the recognition of bone-collecting agencies in African Caves. In *Fossils in the Making: Vertebrate Taphonomy and Paleoecology*, eds. A. K. Behrensmeyer & A. P. Hill. Chicago: The University of Chicago Press, pp. 107-130.

Brain, C. K., 1981. *The Hunters or the Hunted? An Introduction to African Cave Taphonomy.* Chicago: The University of Chicago Press.

Brain, C.K. & V. Brain, 1977. Microfaunal remains from Mirabib: Some evidence of palaeo-ecological changes in the Namib. *Madoqua* 10, 285-293.

Brothwell, D. R. & R. Jones, 1978. The relevance of small mammal studies to archaeology. In *Research Problems in Zooarchaeology*, eds. D. R. Brothwell, K. D. Thomas & J. Clutton-Brock. London: Institute of Archaeology, Occasional Publication No. 3, pp. 47-57

Brown, J. C. & G. J. Twigg, 1969. Studies on the pelvis in British Muridae and Cricetidae (Rodentia). *Journal of Zoology, London* 158, 81-132.

Brown, J. C. & G. J. Twigg, 1971. Mammalian prey of the barn owl (*Tyto alba*) on Skomer Island, Pembrokeshire. *Journal of Zoology, London* 165, 527-530.

Brunet-Lecomte, P., 1990. Evolution morphologique de la première molaire inférieure des campagnols souterrains d'Europe (Arvicolidae, Rodentia). *Zeitschrift für Säugetier-kunde* 55, 371-382.

Brunet-Lecomte, P. , G. Brochet, J. Chaline & M. Delibes, 1987. Morphologie dentaire comparée de *Pitymys lusitanicus* et *Pitymys duodecimcostatus* (Arvicolidae, Rodentia) dans le nord-ouest de l'Espagne. *Mammalia* 51, 145-158.

Brunet-Lecomte, P. & M. Delibes, 1984. Alimentación de la lechuza comun *Tyto alba* en la Cuenca del Duero, España. *Doñana Acta Vertebrata* 11, 213-229.

Brunet-Lecomte, P., T. Lode & P. Pailley, 1995. Morphométrie comparée de la première molaire inférieure du campagnol de Gerbe, *Microtus (Terricola) pyrenaicus gerbei* (Rodentia, Arvicolidae). *Mammalia* 59, 249-254.

Brunet-Lecomte, P., P. Thouy & J. Chaline, 1994. Etude comparée de populations actuelles et fossiles de *Microtus (Terricola) pyrenaicus* (Rodentia, Arvicolidae). *Bulletin de la Société Préhistorique Française* 119, 37-49.

Buchalczyk, T. & J. Raczynski, 1961. Taksonomiczna wartose niektorych pomiarow czaszki krajowych przedstawicicli rodzaju *Sorex* Linnaeus, 1758 i *Neomys* Kaup, 1829. *Acta Theriologica* 5, 115-124.

Buckland, W., 1823. *Reliquiae Diluvianae.* London: John Murray.

Buckley, J., 1977. Barn owl predation on the harvest mouse. *Mammal Review* 7, 117-121.

Buckley, J. & J. G. Goldsmith, 1975. The prey of the barn owl (*Tyto alba alba*) in East Norfolk. *Mammal Review* 5, 13-16.

Buhler, P., 1964. Zur Gattungs- und Artbestimmung von *Neomys* Schodeln. *Zeitschrift für Säugetierkunde* 29, 65-93.

Bunn, D. S., 1972. Regular daylight hunting by barn owls. *British Birds* 65, 26-30.

Bunn, D. S., A. B. Warburton & R. D. S. Wilson, 1982. *The Barn Owl.* Vermillion, SD: Buteo Books.

Burton, J. A., ed., 1973. *Owls of the World.* London: Peter Lowe.

Butzer, K. W., 1980. Investigación preliminar de la geología de la cueva de El Pendo. In *El Yacimiento de la Cueva de "El Pendo" (Excavaciones 1953-1957)*, ed. J. González Echegaray. Madrid: Bibliotheca Praehistorica Hispana, Vol. 17, pp. 199-213.

Butzer, K. W., 1983. Global sea level stratigraphy: An appraisal. *Quaternary Science Reviews* 2, 1-15

Butzer, K. W., 1986. Paleolithic adaptations and settlement in Cantabrian Spain. In *Advances in World Archaeology*, Vol. 5, eds. F. Wendorf & A. E. Close. Orlando: Academic Press, pp. 201-252.

CLIMAP, 1976. The Surface of the Ice Age Earth. *Science* 191, 1131-1137.

Cabrera, A., 1907. Three new Spanish insectivores. *Annals and Magazine of Natural History* Ser. 7, 20, 212.

Cabrera, A., 1908. On *Muscardinidae* from the Iberian Peninsula. *Annals and Magazine of Natural History* Ser. 8, 1, 188.

Cabrera, A., 1914. *Fauna Ibérica: Mamíferos*. Madrid: Museo Nacional de Ciencias Naturales.

Cabrera Valdés, V., 1984. *El Yacimiento de la Cueva de "El Castillo" (Puente Viesgo, Santander)*. Madrid: Bibliotheca Praehistorica Hispanica, No. 22.

Cabrera Valdes, V. & M. Gimenez La Rosa, 1981. Sobre la industria osea de Altamira. In *Espacio, Tiempo y Forma*, Series 1, Prehistoria, Vol. 4. Madrid: Universidad Nacional de Educación a Distancia, pp. 93-110

Camacho Muñoz, I., 1975. La alimentición de *Tyto alba* (Scop.) en la Vega de Granada. *Cuadernos de Ciencias Biologicas, Universidad de Granada* 4, 111-121.

Campos, F., 1978. Regimen alimenticio de *Tyto alba* en las Provencias de Salamanca y Zamora. *Ardeola* 24, 105-119.

Candau, P. & F. J. González Minero, 1992. Relationship between vegetation and pollen spectrum in Southwest Spain. *Israel Journal of Botany* 41, 285-295.

Cartailhac, E., 1902. Les cavernes ornées de dessins. La grotte d'Altamira, Espagne: 'Mea Culpa d'un sceptique'. *L'Anthropologie* 13, 348-354.

Castaños, P., 1980. La macrofauna de la Cueva de La Paloma (Pleistoceno Terminal de Asturias). In *La Cueva de la Paloma*, eds. M. Hoyos, M. I. Martínez Navarrete, T. Chapa Brunet, P. Castaños & F. B. Sanchíz. Excavaciones Arqueológicas en España No. 116, pp. 65-100.

Castién, E. & J. Gosálbez, 1992a. Distribución de micromamíferos (Insectivora y Rodentia) en Navarra. *Miscellanea Zoologica* 16, 183-195.

Castién, E. & J. Gosálbez,(1992b) Distribución geografica y habitats ocupados por *Galemys pyrenaicus* (Geoffrey, 1811) (Insectivora: Talpidae) en los Pirineos Occidentales. *Doñana, Acta Vertebrata* 19, 37-44.

Castién, E. & J. Gosálbez, 1995a. Diet of *Sorex coronatus* in the western Pyrenees. *Acta Theriologica* 40, 113-121.

Castién, E. & J. Gosálbez, 1995b. Reproductive cycle of *Microtus pyrenaicus* De Sélys-Longchamps, 1847 (Mammalia, Rodentia) in the Western Pyrenees. *Zeitschrift für Säugetierkunde* 60, 170-175.

Castro Bernardez, D., 1986. *Estudio de la microfauna Pleistocena Superior yacimientos de Altamira, El Juyo y Cueto de la Mina (Region Cantabrica)*. Unpublished M.S. Thesis, Universidad de Santiago.

Cendrero, O., 1915. *Resumen de los bastones perforados (bastones de mando) hallados en la provincia de Santander, y noticia sobre uno nuevo de la Caverna "El Pendo"*. Madrid: Comisión de Investigaciónes Paleontológicas y Prehistóricas, No. 1.

Chaline, J. 1972. *Les Rongeurs du Pléistocène Moyen et Supérieur de France*. Paris: Centre National de la Recherche Scientifique.

Chaline, J., 1974. Esquisse de l'evolution morphologique, biometrique et chromosomique du genre *Microtus* (Arvicolidae, Rodentia) dans le Pleistocene de l'hemisphere Nord. *Bulletin de la Société Géologique de France* ser. 7, 16, 440-450.

Chaline, J., 1985. *Histoire de l'homme et des climats au Quaternaire*. Paris: Doin Editeurs.

Chaline, J. & J.-D. Graf, 1988. Phylogeny of the Arvicolidae (Rodentia): Biochemical and paleontological evidence. *Journal of Mammalogy* 69, 22-33.

Chase, P. G. & H. L. Dibble, 1987. Middle Paleolithic symbolism: A review of current evidence and interpretations. *Journal of Anthropological Archaeology* 6, 262-296.

Chitty, D., 1938. A laboratory study of pellet formation in the short-eared owl (*Asio flammeus*). *Proceedings of the Zoological Society of London* Ser. A, 108, 267-287.

Clark, G. A., 1971. *The Asturian of Cantabria: A Re-evaluation*. Unpublished Ph.D. Dissertation, Department of Anthropology, University of Chicago.

Clark, G. A., 1974-75. Excavations in the Late Pleistocene cave site of Balmori, Asturias (Spain). *Quaternaria* 18, 383-426.

Clark, G. A., 1983. *The Asturian of Cantabria: Early Holocene Hunter-Gatherers in Northern Spain*. Tuscon: University of Arizona Press. Anthropological Papers of the University of Arizona, No. 41.

Clark, G. A. & L. R. Richards, 1978. Late and post-Pleistocene industries and fauna from the cave site of La Riera (Province of Asturias, Spain). In *Views of the Past*,. ed. L. G. Freeman. The Hague: Mouton Publishers, pp. 117-152

Clark, G. A. & L. G. Straus, 1983. Late Pleistocene hunter-gatherer adaptations in Cantabrian Spain. In *Hunter-Gatherer Economy in Prehistory*, ed. G. Bailey. Cambridge: Cambridge University Press, pp. 131-148.

Clark, R. 1981. Cold climate features of the Cordillera Cantabrica, northern Spain. *Biuletyn Peryglacjalny* 28, 5-13.

Clark, R. J., 1972. Pellets of the Short-eared owl and Marsh hawk compared. *Journal of Wildlife Management* 36, 962-964.

Clutton-Brock, T. H., F. E. Guiness & S. D. Alban, 1982. *Red Deer: Behavior and Ecology of Two Sexes*. Chicago: The University of Chicago Press.

Cohen, M. N., 1977. *The Food Crisis in Prehistory*. New Haven: Yale University Press.

Conkey, M. W., 1978. *An Analysis of Design Structure: Variability among Magdalenian Engraved Bones from North Coastal Spain*. Unpublished Ph.D. Dissertation, Department of Anthropology, University of Chicago.

Conkey, M. W., 1980. The identification of prehistoric hunter-gatherer aggregation sites: The case of Altamira. *Current Anthropology* 21, 609-630.

Connell, J. H., 1978. Diversity in tropical rain forests and coral reefs. *Science* 199, 1302-1310.

Coon, C. S., 1971. *The Hunting Peoples*. New York: Nick Lyons.

Corchón, M. Soledad, 1981. *Cueva de las Caldas*. Madrid: Excavaciones Arqueológicas en España No. 115.

Corchón, M. Soledad, 1986. *El Arte Mueble Paleolítico Cantábrico*. Santander: Centro de Investigación y Museo de Altamira, Monografías No. 16.

Corbet, G. B., 1966. *The Terrestrial Mammals of Western Europe*. London: G.T. Foulis & Co.

Corbet, G. B., 1978. *The Mammals of the Palearctic Region: A Taxonomic Review*. London: Cornell University Press.

Corbet, G. B., 1984. *The Mammals of the Palearctic Region: A Taxonomic Review (Supplement)*. Dorchester: Henry Ling.

Corbet, G. B., J. Cummins, S. R. Hedges & W. Krzanowski, 1970. The taxonomic status of British Water voles, genus *Arvicola*. *Journal of Zoology, London* 161, 301-316.

Corbet, G. B. & S. Harris, eds., 1991. *The Handbook of British Mammals*. Oxford: Blackwell Scientific Publications.

Corbet, G. B. & D. Ovenden, 1982. *Manual de los Mamíferos de España y de Europa*. Barcelona: Ediciones Omega.

Crowcroft, P. 1966. *Mice All Over*. London: G.T. Foulis.

Crowe, W. T., 1987. Tecnicas de recuperación integral de los datos obtenidos en los sedimentos de yacimientos prehistóricos. In *Excavaciones el la Cueva del Juyo*,. eds. I. Barandiarán, L. G. Freeman, J. González Echegaray & R. G. Klein. Centro de Investigación y Museo de Altamira, Monografías No. 14, pp. 63-74.

Cummings, J. H., G. E. Duke & A. A. Jegers, 1976. Corrosion of bone by solutions simulating raptor gastric juice. *Raptor Research* 10, 55-57.

Da Luz Madureira, M. 1983. On the identification of West-European *Microtus* of the sub-genera *Microtus* and *Chionomys* (Mammalia, Rodentia) a principal component analysis. *Revista de Biologia* 12, 615-632.

Davidson, I., 1976a. Les Malaetes and Monduver: The economy of a human group in prehistoric Spain. In *Problems in Economic and Social Archaeology*, eds. G. de G. Sieveking, I. H. Longworth & K. E. Wilson. London: Duckworth, pp. 483-499.

Davidson, I., 1976b. Seasonality in Spain. *Zephyrus* 26-27, 175-182.

Davidson, I., 1983. Site variability and prehistoric economy in Levante. In *Hunter-Gatherer Economy in Prehistory*, ed. G. Bailey. Cambridge: Cambridge University Press, pp. 79-95.

Davis, D. H. S., 1958. Notes on some small mammals in the Kalahari Gemsbok National Park, with special reference to those preyed upon by barn owls. *Koedoe* 1, 184-188.

Davis, D. H. S., 1959. The barn owl's contribution to ecology and palaeoecology. In *Proceedings of the First Pan-African Ornithological Conference*, eds. M. K. Rowan. *Ostrich*, Supplement No. 3, pp. 143-153.

Davis, M. B., 1968. Pollen grains in sake sediments: Redeposition caused by seasonal water circulation. *Science* 162, 796-799.

Davis, S. J. M., 1987. *The Archaeology of Animals*. New Haven: Yale University Press.

Day, M. G., 1966. Identification of hair and feather remains in the gut and faeces of stoats and weasels. *Journal of Zoology* 148, 201-217.

Dayan, T. & E. Tchernov, 1988. On the first occurrence of the common weasel (*Mustela nivalis*) in the fossil record of Israel. *Mammalia* 52, 165-168.

de Paz, O. & J. Benzal, 1989. Clave para la identificación de los murciélagos de la Península Ibérica (Mammalia, Chiroptera). *Miscellania Zoologica* 13, 153-176.

Dean, W. R. J,. 1973. Age distribution of *Praomys natalensis* prey in *Tyto alba* pellets. *Zoologica Africana* 8, 140.

Delany, M. J. & P. E. Davis, 1961. Observations on the ecology and life history of the Fair Isle field-mouse *Apodemus sylvaticus fridariensis* (Kinnear). *Proceedings of the Zoological Society of London* 136, 439-452.

Delattre, P., 1983. Density of weasel (*Mustela nivalis* L.) and stoat (*Mustela nivalis* L.) in relation to water vole abundance. *Acta Zoologica Fennica* 174, 221-222.

Delibes, M., 1983. Interspecific competition and the habitat of the stone marten *Martes foina* (Erxleben 1777) in Europe. *Acta Zoologica Fennica* 174, 229-231.

Delibes, M., P. Brunet-Lecomte & M. Manez, 1984. Datos sobre la alimentación de la lechuza comun (*Tyto alba*), el buho chico (*Asio otus*) y el mochuelo (*Athene noctua*) en una misma localidad de Castilla la Vieja. *Ardeola* 30, 57-63.

Delibes de Castro, J., 1985. Distribution and abundance of small mammals in a gradient of altitude. *Acta Zoologica Fennica* 173, 53-56.

de Mortillet, G. 1883. *Le Préhistorique*. Paris: Reinwald. Bibliothèque des Sciences Contemporaine, No. 8.

Denys, C. & M. Mahboubi. 1992. Alterations structurales et chimiques des elements squelettiques de pelotes de regurgitations d'un rapace diurne. *Bulletin du Museum National d'Histoire Naturelle* 4th series, section A, no. 1, 14, 229-249.

Derting, T. L. & J. A. Cranford, 1989. Physical and behavioral correlates of prey vulnerability to barn owl (*Tyto alba*) predation. *American Midland Naturalist* 121, 11-20.

Dice, L. R., 1945. Minimum intensities of illumination under which owls can find dead prey by sight. *American Naturalist* 79, 385-416.

Dickman, C. R., M. Predavec & A. J. Lyman, 1991. Differential predation of size and sex classes of mice by the barn Owl, *Tyto alba*. *Oikos* 62, 67-76.

Dienske, H., 1969. Notes on differences between some external and skull characteristics of *Microtus arvalis* (Pallas, 1778) and of *Microtus agrestis* (Linnaeus, from the Netherlands. *Zoologische Mededelingen* 44, 83-108.

Diputación Regional de Cantabria, 1994. Disposiciones generales. No. 260, December 30. Consejería de Ganadería, Agricultura y Pesca, Santander, Spain.

Dodson, P., 1973. The significance of small bones in paleoecological interpretation. *Contributions to Geology, University of Wyoming* 12, 15-19.

Dodson, P. & D. Wexlar, 1979. Taphonomic investigations of owl pellets. *Paleobiology* 5, 275-284.

Donazár, J. A. & O. Ceballos, 1989. Selective predation by eagle owls *Bubo bubo* on rabbits *Oryctolagus cuniculus*: Age and sex preferences. *Ornis Scandinavica* 20, 117-122.

Donazár, J. A. & F. Hiraldo, M. Delibes & R. R. Estrella, 1989. Comparative food habits of the eagle owl *Bubo bubo* and the great horned owl *Bubo virginianus* in six Palearctic and Nearctic biomes. *Ornis Scandinavica* 20, 298-306.

Duke, G. E., O. A. Evanson, & A. A. Jegers, 1976. Meal to pellet intervals in 14 species of captive raptors. *Comparative Biochemistry and Physiology* 53A, 1-6.

Duke, G. E., A. A. Jegers, G. Loff & O. A. Evanson, 1975. Gastric digestion in some raptors. *Comparative Biochemistry and Physiology* 50A, 649-656.

Duke, G. E. & D. D. Rhoades, 1977. Factors affecting meal to pellet intervals in Great horned owls. *Comparative Biochemistry and Physiology* 56A, 283-286.

Dunmire, W. W., 1955. Sex dimorphism in the pelvis of rodents. *Journal of Mammalogy* 36, 356-361.

Dunwiddie, P. W., 1987. Macrofossil and pollen representation of coniferous trees in modern sediments from Washington. *Ecology* 68, 1-11.

Dupré, M., 1984. Palinología de los niveles VII a II de la cueva de Ekain. In *El Yacimiento Prehistórico de la Cueva de Ekain (Deva, Guipúzcoa)*, eds. J. Altuna & J. M. Merino. San Sebastián: Sociedad de Estudios Vascos Serie B1, pp. 61-63.

Dupré, M., 1990. Análisis polínico de la cueva de Amalda. In *La Cueva de Amalda (Zestoa, Pais Vasco)*,. eds. J. Altuna, A. Baldeón & K. Mariezkurrena. San Sebastián: Sociedad de Estudios Vascos, pp. 49-51

Eastham, A., 1995. La microfaune. In *Les Derniers Chasseurs de Rennes du Monde Pyrénéen. L'Abri Dufaure: Un Gisement Tardiglaciare en Gascogne*, ed. L. G. Straus. Paris: Société Préhistorique Française, Mémoire 22, pp. 235-245.

Efremov, J. A., 1940. Taphonomy: New branch of paleontology. *Pan-American Geologist* 74, 81-93.

Einarsen, A. S., 1956. Determination of some predator species by field signs. *Oregon State Monographs Studies in Zoology* 10, 1-34.

Elias, S. A., 1987. Paleoenvironmental significance of Late Quaternary insect fossils from packrat middens in South-Central New Mexico. *The Southwestern Naturalist* 32, 383-390.

Epperson, C., 1976. Food remains from a barn owl nest in Nebraska. *Nebraska Bird Review* 34, 54-57.

Erlinge, S., 1987. Why do European stoats *Mustela erminea* not follow Bergmann's rule? *Holarctic Ecology* 10, 33-39.

Evans, F. C. & J. T. Emlen, Jr., 1947. Ecological notes on the prey selected by a barn owl. *Condor* 49, 3-9.

Fa, J. E., E. Shaw & A. Santana, 1992. Habitat associations of small mammals in a southern Spanish fir forest. *Mammalia* 56, 478-481.

Faegri, K. & J. Iversen, 1989. *Textbook of Modern Pollen Analysis* (4th ed.). Chichester: John Wiley & Sons.

Fall, P. L., 1987. Pollen taphonomy in a canyon stream. *Quaternary Research* 28, 393-406.

Farner, D. S., 1960. Digestion and the digestive system. In *Biology and Comparative Physiology of Birds*, Vol. 1, eds. A. J. Marshall. New York: Academic Press, pp. 411-467.

Farrand, W. R., 1975. Analysis of the Abri Pataud sediments. In *Excavation of the Abri Pataud, Les Eyzies (Dordogne)*, ed. H. L. Movius, Jr. Cambridge: Harvard University Press, pp. 27-68.

Fast, S. J. & W. A. Harrison, 1975. Prey preference and hunting habitat selection in the barn owl. *American Midland Naturalist* 96, 503-507.

Fedyk, A. & A. L. Ruprecht, 1971. Taxonomic value of M_1 measurements in *Microtus agrestis* (Linnaeus, 1761) and *Microtus arvalis* (Pallas, 1779). *Acta Theriologica* 16, 408-412.

Felten, H., A. Helfricht & G. Storch, 1973. Die Bestimmung der europaischen Fledermause nach der distalen Epiphyse des Humerus. *Senckenbergiana Biologica* 54, 291-297.

Fernández-Tresguerres, J., 1981. *El Aziliense en las Provincias de Asturias y Santander.* Santander: Centro de Investigación y Museo de Altamira, Monografías No. 2.

Fisher, D. C., 1981a. Crocodilian scatology, microvertebrate concentrations, and enamel-less teeth. *Paleobiology* 7, 262-275.

Fisher, D. C., 1981b. Mode of preservation of the Shotgun Local Fauna (Paleocene, Wyoming) and its implication for the taphonomy of a microvertebrate concentration. *Contributions from the Museum of Paleontology, the University of Michigan* 25, 247-257.

Fisher, D. C., 1981c. Taphonomic interpretation of enamel-less teeth in the Shotgun Local Fauna (Paleocene, Wyoming). *Contributions from the Museum of Paleontology, The University of Michigan* 25, 259-275.

Freeman, L. G., 1973. The significance of mammalian faunas from paleolithic occupations in Cantabrian Spain. *American Antiquity* 38, 3-44.

Freeman, L. G., 1975. By their works you shall know them: Cultural developments in the Paleolithic. In *Hominisation und Verhalten*, eds. G. Kurth & I. Eibl-Eibesfeldt. Stuttgart: G. Fischer, pp. 234-261.

Freeman, L. G., 1981. The Fat of the land: Notes on Paleolithic diet in Iberia. In *Omnivorous Primates*, eds. R. S. O. Harding & G. Teleki. New York: Columbia University Press, pp. 104-165.

Freeman, L. G., ed., 1987. *Altamira Revisited.* Chicago: Institute for Prehistoric Studies.

Freeman, L. G., 1988. The stratigraphic sequence at Altamira, 1980-1981. *Espacio, Tiempo y Forma*, Series 1, Prehistoria, Vol. 1,. Madrid: Universidad Nacional de Educación a Distancia, pp. 149-163

Freeman, L. G., 1994. Paleolithic polygons: Voronoi tesserae and settlement hierarchies in Cantabrian Spain. *Espacio, Tiempo y Forma*, Series 1, Prehistoria, Vol. 7, Madrid: Universidad Nacional de Educación a Distancia, pp. 43-61.

Freeman, L. G. & J. González Echegaray, 1981. El Juyo: A 14,000-year-old sanctuary from northern Spain. *History of Religions* 21, 1-19.

Freeman, L. G. & J. González Echegaray, 1982. Magdalenian mobile art from El Juyo (Cantabria). *Ars Praehistorica* 1, 161-167.

Freeman, L. G. & J. González Echegaray, 1995. The Magdalenian site of El Juyo (Cantabria, Spain): Artistic documents in context. *Bollettino del Centro Camuno de Studi Preistorici* 27, 25-42.

Freeman, L. G., J. González Echegaray, R. G. Klein & W. T. Crowe, 1988. Dimensions of research at El Juyo, an earlier Magdalenian site in Cantabrian Spain. In *Upper Pleistocene Prehistory of Western Eurasia*, eds. H. L. Dibble & A. Montet-White. Philadelphia: University of Pennsylvania Museum, pp. 1-39.

Freeman, L. G., J. González Echegaray, J. Pokines, H. Stettler & M. Krupa, 1998. Tamisage ultra fin et récupération de l'outillage: Observations réalisées a El Juyo (Espagne Cantabrique). *L'Anthropologie* 102, 35-44.

Frey, H. & W. Walter, 1986. Zur Ernährung des Uhus, *Bubo bubo* (Linnaeus 1758), Aves, an einem alpinen Brutplatz in den Hohen Tauern (Salzberg, Österreich). *Annalen die Naturhistorischen Museums en Wien* 88/89, 91-99.

Fuentes Vidarte, C., 1980. Estudio de la Fauna de El Pendo. In *El Yacimiento de la Cueva de "El Pendo" (Excavaciones 1953-1957)*, ed. J. González Echegaray. Madrid: Bibliotheca Praehistorica Hispanica, No. 17, pp. 215-237.

Fulk, G. K., 1976. Owl predation and rodent mortality: A case study. *Mammalia* 40, 423-427.

Gargett, R. H., 1989. Grave shortcomings: The evidence for Neanderthal burial. *Current Anthropology* 30, 157-190.

Garzon-Heydt, J. & J. Castroviejo, 1970. Notas preliminares sobre la distribución de algunos micro-mamíferos en el norte de España. *Säugetierkundliche Mittelungen* 19, 217-222.

Gilliard, E. T., 1958. *Tyto alba*, Part I. *Natural History* 67, 238-245.

Glading, B., D. F. Tillotson & D. M. Selleck, 1943. Raptor pellets as indicators of food habits. *California Fish and Game* 29, 92-121.

Glue, D. E., 1967. Prey taken by the barn owl in England and Wales. *Bird Study* 14, 169-183.

Glue, D. E., 1970. Avian predator pellet analysis and the mammalogist. *Mammal Review* 1, 53-62.

Glue, D. E., 1974. Food of the barn owl in Britain and Ireland. *Bird Study* 21, 200-210.

Glue, D. E., 1975. Harvest mice as barn owl prey in the British Isles. *Mammal Review* 5, 9-12.

González Echegaray, J., 1953. Les oeuvres d'art de la Grotte de las Chimeneas. *Bulletin de la Société Préhistorique de l'Ariège* 8, 75-77.

González Echegaray, J., 1960. El Magdaleniense III de la costa Cantábrica. *Boletín de Seminario de Estudios de Arte y Arqueología*, Universidad de Valledolid 26, 69-100.

González Echegaray, J., 1974. *Cueva de las Chimeneas*. Madrid: Excavaciones Arqueológicas en España No 21.

González Echegaray, J., ed., 1980. *El Yacimiento de la Cueva de "El Pendo" (Excavaciones 1953-57)*. Madrid: Bibliotheca Praehistorica Hispana, No. 17.

González Echegaray, J., 1987. Introducción geografica e histórica. In *Excavaciones en la Cueva Del Juyo*, eds. I. Barandiarán Maestu, L. G. Freeman, J. González Echegaray & R. G. Klein. Santander: Centro de Investigación y Museo de Altamira, Monografías No. 14, pp. 7-14.

González Echegaray, J., 1988. El Magdaleniense de Altamira. *Espacio, Tiempo y Forma*, Series 1, Prehistoria, Vol. 1, Madrid: Universidad Nacional de Educación a Distancia, pp. 165-175.

González Echegaray, J. & I. Barandiarán Maestu, eds,. 1981. *El Paleolitico Superior de la Cueva del Rascaño (Santander)*. Santander: Centro de Investigación y Museo de Altamira, Monografías No. 3.

González Echegaray, J. & L. G. Freeman, 1971. *Cueva Morín: Excavaciones 1966-1968*. Santander: Patronato de las Cuevas Prehistóricas.

González Echegaray, J. & L. G. Freeman, 1973. *Cueva Morín: Excavaciones 1969*. Santander: Patronato de las Cuevas Prehistóricas.

González Echegaray, J. & L. G. Freeman, 1978. *Vida y Muerte en Cueva Morín*. Santander: Institución Cultural de Cantabria.

González Echegaray, J. & L. G. Freeman, 1998. *Le Paléolithique Inférieur et Moyen d'Espagne*. Préhistoire d'Europe 6. Grenoble: Editions Jerome Millon.

González Echegaray, J., M. A. García Guinea, A. Begines Ramirez & B. Madariaga de la Campa, eds., 1963. *Cueva de la Chora*. Excavaciones Arqueológicas en España No. 26.

González Echegaray, J., M. A. García Guinea & A. Begines Ramirez, eds., 1966. *Cueva del Otero*. Excavaciones Arqueológicas en España No. 53.

González Morales, M. R., 1974. *La Cueva de Collubil*. Licenciatura thesis, Departamiento de Prehistoria y Arqueología, Universidad de Oviedo.

González Morales, M. R., 1986. La Riera bone and antler artifact assemblages. In *La Riera Cave*, eds. L. G. Straus & Geoffrey A. Clark. Arizona State University Anthropological Research Papers No. 36, pp. 209-218.

González Sainz, C., 1988. Le fait artistique à la fin du Paléolithique: Quelques réflexions. *Bulletin de la Société Préhistorique Ariège-Pyrénées* 43, 35-62.

Goodman, S. M., 1990. The food habits of the eagle owl (*Bubo bubo ascalaphus*) in Kharga Oasis, Egyptian Western Desert. *Journal of Arid Environments* 18, 217-220.

Goodwin, R. G., 1988. Pollen taphonomy in Holocene glaciolacustrine sediments, Glacier Bay, Alaska: A cautionary note. *Palaios* 3, 606-611.

Goszczynski, J., 1974. Studies on the food of foxes. *Acta Theriologica* 19, 1-18.

Goszczynski, J., 1976. Composition of the food of martens. *Acta Theriologica* 21, 527-534.

Goszczynski, J., 1977. Connections between predatory birds and mammals and their prey. *Acta Theriologica* 22, 399-430.

Graham, I. & A. Saunders, 1978. A multivariate statistical analysis of small mammal bones. In *Research Problems in Zooarchaeology*, eds. D. R. Brothwell, K. D. Thomas & J. Clutton-Brock. London: Institute of Archaeology, Occasional Publication No. 3, pp. 59-67.

Graham, R. W., 1979. Paleoclimates and Late Pleistocene faunal provinces in North America. In *Pre-Llano Cultures of the Americas: Problems and Paradoxes*, eds. R. L. Humphrey & D. Stanford. Washington, D.C.: Washington Anthropological Society, pp. 49-69.

Gray, J. A., 1989. Effects of environmental stress on species rich assemblages. *Biological Journal of the Linnean Society* 37, 19-32.

Grayson, D. K., 1981. A critical view of the use of archaeological vertebrates in paleoenvironmental reconstruction. *Journal of Ethnobiology* 1, 28-38.

Grayson, D. K., 1983. *The Establishment of Human Antiquity.* New York: Academic Press.

Greenwood, R. J., 1979. Relating residue in raccoon feces to food consumed. *American Midland Naturalist* 102, 191-193.

Grimm, R. J. & W. M. Whitehouse, 1963. Pellet formation in a Great horned owl: A roentgenographic study. *Auk* 80, 301-306.

Gromov, I. M. & I. Ya. Polyakov, 1992 *Voles (Microtinae).* Fauna of the USSR, Mammals, Vol. 3, No. 8. Washington: Smithsonian Institution.

Hallam, A., 1989. *Great Geological Controversies* (2nd ed.). Oxford: Oxford University Press.

Hanney, P., 1963. Observations upon the food of the barn owl (*Tyto alba*) in southern Nyasaland, with a method of ascertaining population dynamics of rodent prey. *Annals and Magazine of Natural History* Series 13, 6, 305-313.

Hansen, J. M., 1991. *The Palaeoethnobotany of Franchthi Cave.* Bloomington: Indiana University Press.

Harrold, F. B., 1980. A comparative analysis of Eurasian Paleolithic burials. *World Archaeology* 12, 195-211.

Havinga, A. J., 1967. Palynology and pollen preservation. *Review of Palaeobotany and Palynology* 2, 81-98.

Hausser, J. & D. Jammot, 1974. Étude biométrique des machoires chez les *Sorex* due groupe *araneus* en Europe continentale (Mammalia, Insectivora). *Mammalia* 33, 630-658.

Haynes, G., 1980. Prey bones and predators: Potential ecologic information from analysis of bone sites. *Ossa* 7, 75-97.

Haynes, G., 1982. Utilization and skeletal disturbances of North American prey carcasses. *Arctic* 35, 266-281.

Heim de Balsac, H. & F. de Beaufort, 1969. Contribution á l'Étude des Micromamiferes du Nord-Ouest de l'Espange. *Mammalia* 33, 630-658.

Heinrich, W.-D., 1982. Zur Evolution und Biostratigraphie von *Arvicola* (Rodentia, Mammalia) im Pleistozan Europas. *Zeitschrift für Geologische Wissenschraften* 10, 683-735.

Heinrich, W.-D., 1987. Neue Ergebnisse zur Evolution und Biostratigraphie von *Arvicola* (Rodentia, Mammalia) im Quartar Europas. *Zeitschrift fur Geologische Wissenschraften* 15, 389-406.

Herrera, C. M., 1973. Régimen alimenticio de *Tyto alba* en España sudoccidental. *Ardeola* 19, 359-394.

Herrera, C. M. & F. M. Jaksic, 1980) Feeding ecology of the barn owl in central Chile and southern Spain: A comparative study. *Auk* 97, 760-767.

Herrero, J., J. Canut, D. García-Ferre, R. García-González & R. Hidalgo, 1992. The alpine marmot (*Marmota marmota* L.) in the Spanish Pyrenees. *Zeitschrift für Säugetierkunde* 57, 211-215.

Hevly, R. H., 1981. Pollen production, transport and preservation: Potentials and limitations in archaeological palynology. *Journal of Ethnobiology* 1, 39-54.

Hillson, S. 1986. *Teeth.* Cambridge: Cambridge University Press.

Hinton, M. A. C., 1911. The British fossil shrews. *Geological Magazine*, New Series 8, 529-529.

Hinton, M. A. C., 1926. *Monograph of Voles and Lemmings*, Vol. I. London: Trustees of the British Museum.

Hoffman, R., 1983. The contribution of raptorial birds to patterning in small mammal assemblages. *Paleobiology* 14, 81-90.

Holmes, P. L., 1990. Differential transport of spores and pollen: A laboratory study. *Review of Palaeobotany and Palynology* 64, 289-296.

Howard, W. A., 1958. Food intake and pellet formation of a Horned owl. *Wilson Bulletin* 70, 145-150.

Howes, C. A., 1978. Notes on the food of foxes at Gibraltar Point, Lincolnshire. *Naturalist, Leeds* 103, 25-26.

Huston, M., 1979. A general hypothesis of species diversity. *The American Naturalist* 113, 81-101.

Hutton, J., 1795. *The Theory of the Earth.* London: Messrs. Cadell, Jr. Reprinted in 1959, ed. J. Cramer. Weinheim: H. R. Engelmann.

Ibañez-Angulo, M., 1991. *Paleoethnobotanical Analysis: Problems and Methodology. A Case Study: The Cave of El Juyo (Levels 4 and 8).* Unpublished M.A. Thesis, Department of Anthropology, University of Chicago.

Jaksic, F. M., R. L. Seib & C. M. Herrera, 1982. Predation by the barn owl (*Tyto alba*) in Mediterranean habitats of Chile, Spain and California: A comparative approach. *American Midland Naturalist* 107, 151-162.

Janossy, D., 1975. Mid-Pleistocene microfaunas of continental Europe and adjoining areas. In *After the Australopithecines,*. eds. K. W. Butzer & G. Ll. Isaac. The Hague: Mouton Publishers, pp. 375-397.

Janossy, D. & E. Schmidt, 1975. Extreme Varianten des M_1 der Feldmaus (*Microtus arvalis* Pallas) in Ungarn. *Zeitschrift für Säugetierkunde* 40, 34-36.

Janssens, P. & J. González Echegaray, 1958. *Memoria de las Excavaciones de la Cueva del Juyo.* Santander: Patronato de las Cuevas Prehistóricas de la Provincia de Santander (España).

Jedrzejewski, W., B. Jedrzejewski & L. Szymura, 1995. Weasel population response, home range, and predation on rodents in a deciduous forest in Poland. *Ecology* 76, 179-195.

Jiménez, R., M. Burgos & R. Diaz de la Guardia, 1984. Karyotype and chromosome banding in the mole (*Talpa occidentalis*) from the southeast of the Iberian Peninsula: Implications on its taxonomic position. *Caryologia* 37, 253-258.

Jiménez, R., M. Burgos, A. Sánchez & R. Diaz de la Guardia, 1990. The reproductive cycle of *Talpa occidentalis* in the southeastern Iberian Peninsula. *Acta Theriologica* 35, 165-169.

Jones, J. K., Jr., 1952. Notes on the food habits of the great horned owl in Cherry County, Nebraska. *Nebraska Bird Review* 20, 10-11.

Jordá Cerdá, F., 1958. *Avance al Estudio de la Cueva de la Lloseta.* Oviedo: Diputación Provincial de Asturias.

Keller, B. L., 1985. Reproductive patterns. In *Biology of New World* Microtus, ed. R. H. Tamarin. The American Society of Mammalogists, Special Publication No. 8, pp. 725-778.

Kerbis Peterhans, J. C., 1990. *The Roles of Porcupines, Leopards, and Hyenas in Ungulate Carcass Dispersal: Implications for Paleoanthropology.* Unpublished Ph.D. Dissertation, Department of Anthropology, University of Chicago.

Kerbis Peterhans, J. C. & L. Kolska Horwitz, 1992. A bone assemblage from a striped hyaena (*Hyaena hyaena*) den in the Negev Desert, Israel. *Israel Journal of Zoology* 37, 225-245.

King, C. M., 1991. Body size--prey size relationships in European stoats in *Mustela erminea*: A test case. *Holarctic Ecology* 14, 173-185.

Klein, R. G., 1989. *The Human Career.* Chicago: The University of Chicago Press.

Klein, R. G., 1991. Size variation in the Cape Dune molerat (*Bathyergus suillus*) and Late Quaternary climatic change in the Southwestern Cape Province, South Africa. *Quaternary Research* 36, 243-256.

Klein, R. G. & K. Cruz-Uribe, 1984 *The Analysis of Animal Bones from Archeological Sites.* Chicago: University of Chicago Press.

Klein, R. G. & K. Cruz-Uribe, 1987. La fauna mamífera del yacimiento de la Cueva de "El Juyo". Campañas de 1978 y 1979. In *Excavaciones en la Cueva Del Juyo*, eds. I. Barandiarán Maestu, L. G. Freeman, J. González Echegaray & R. G. Klein. Santander: Centro de Investigación y Museo de Altamira, Monografías No. 14, pp. 97-120.

Klein, R. G., C. Wolf, L. G. Freeman & K. Allwarden, 1981. The use of dental crown heights for constructing age profiles of red deer and similar species in archaeological samples. *Journal of Archaeological Science* 8, 1-31.

Korth, W. W., 1979. Taphonomy of microvertebrate fossil assemblages. *Annals of the Carnegie Museum* 48, 235-285.

Kotler, B. P., J. S. Brown & O. Hasson, 1991. Factors affecting gerbil foraging behavior and rates of owl predation. *Ecology* 72, 2249-2260.

Kotler, B. P., J. S. Brown, R. J. Smith & W. O. Wirtz II, 1988. The effects of morphology and body size on rates of owl predation on desert rodents. *Oikos* 53, 145-152.

Kowalski, K., 1966. The stratigraphic importance of rodents in the studies on the European Quaternary. *Folia Quaternaria* 22, 1-16.

Kratochvil, J., 1970. *Pitymys*-Arten aus der Hohen Tatra (Mammalia, Rodentia). *Acta Sc. Nat. Brno* 4, 1-63.

Krupa, M., 1994. *Limpets and Other Marine Molluscs: An Evaluation of Their Archeological Significance at El Juyo and Other Paleolithic Sites in Northern Spain.* Unpublished M. A. Thesis, Department of Anthropology, University of Chicago.

Krupa, M., 1996. Effects of processes of deposition, taphonomy and excavation on the interpretation of fish remains from Upper Paleolithic sites in western Europe. Paper presented at the 61st Annual Meeting of the Society for American Archaeology, April 11, New Orleans, LA.

Krystufek, B., 1986. Variability of enamel tooth patterns in *Microtus arvalis* from Dalmatia (Rodentia, Mammalia). *Bioloski Vestnik* 34, 37-42.

Krystufek, B., 1990. Geographic variation in *Microtus nivalis* (Martins, 1842) from Austria and Yugoslavia. *Bonner Zoologische Beiträge* 41, 121-139.

Krystufek, B. & B. Kovacic, 1989. Vertical distribution of the Snow vole *Microtus nivalis* (Martins, 1842) in Northwestern Yugoslavia. *Zeitschrift für Säugetierkunde* 54, 153-156.

Kuniholm, P. I., 1990. Archaeological evidence and non-evidence for climatic change. *Philosophical Transactions of the Royal Society of London* A 330, 645-655.

Kurtén, B., 1972. The Cave bear. *Scientific American* 226, 60-72.

Kurtén, B., 1976. *The Cave Bear Story*. New York: Columbia University Press.

Kusmer, K. D., 1990 Taphonomy of owl pellet deposition. *Journal of Paleontology* 64, 629-637.

Kwiatowski, A. & K. Lubliner-Mianowska, 1957. Badanie skladu chemicznego pulku, 2. Badania blon pylkowich. *Acta Soc. Botan. Polon.* 26, 501-514.

Le Jacques, D. & T. Lode, 1994. L'alimentation de la Genette d'Europe, *Genetta genetta* L., 1758, dans un bocage de l'ouest de la France. *Mammalia* 58. 383-389.

Le Louarn, H. & G. Janeau, 1975. Répartition et biologie du campagnol des neiges *Microtus nivalis* Martins dans la région de Briançon. *Mammalia* 39, 589-604.

Leroi-Gourhan, Ar., 1959. Résultats de l'analyse pollinique de la grotte d'Isturitz. *Bulletin de la Société Préhistorique Française* 56, 619-624.

Leroi-Gourhan, Ar., 1966. Análisis polínico de la Cueva del Otero. In *Cueva del Otero*, eds. J. González Echegaray, M. A. García Guinea & A. Begines Ramírez. Excavaciones Arqueológicos en España, No. 53, pp. 83-85.

Leroi-Gourhan, Ar., 1978. Palinología. In *Vida y Muerte en Cueva Morín*, eds. J. González Echegaray & L. G. Freeman. Santander: Institución Cultural de Cantabria, pp. 199-200.

Leroi-Gourhan, Ar., 1980. Análisis polinico de El Pendo. In *El Yacimiento de la Cueva de "El Pendo" (Excavaciones 1953-1957)*, ed. J. González Echegaray. Madrid: Bibliotheca Praehistorica Hispana, Vol. 17, pp. 263-266

Leroi-Gourhan, Ar., 1986. The palynology of La Riera cave. In *La Riera Cave*, eds. L. G. Straus & G. A. Clark. Arizona State University, Anthropological Papers No. 36, pp. 59-64.

Leroi-Gourhan, Ar., 1989. Stratigraphie du Würm depuis 35,000 ans par la palynologie. In *Quaternary Type Sections*, eds. J. Rose & C. Schlüchter. Rotterdam: A. A. Balkema, pp. 91-92.

Leroi-Gourhan, Ar., 1994. Le temps d'El Juyo. In *Homenaje al Dr. Joaquín González Echegaray*, ed. J. A. Lasheras. Museo y Centro de Investigación de Altamira, Monografías No. 17, pp. 123-129.

Levinson, M., 1982. Taphonomy of microvertebrates from owl pellets to cave breccia. *Annals of the Transvaal Museum* 33, 115-120.

Ligtvoet, W. & A. Van Wijngaarden, 1994. The colonization of the Island of Noord-Beveland (The Netherlands) by the Common vole *Microtus arvalis*, and its consequences for the Root vole *M. oeconomus*. *Lutra* 37, 1-28.

Lloyd, J. A., 1975. Social structure and reproduction in two freely-growing populations of house mouse (*Mus musculus* L.). *Animal Behavior* 23, 413-424.

Lockie, J. D., 1959. The estimation of the food of foxes. *Journal of Wildlife Management* 23, 224-227.

Longland, W. S. & S. H. Jenkins, 1987. Sex and age affect vulnerability of desert rodents to owl predation. *Journal of Mammalogy* 68, 746-754.

López, P., 1981. Análisis polínico del yacimiento de Los Azules (Cangas de Onís, Oviedo). *Botánica Macaronésica* 8-9, 243-248.

López-Berges, M. A. & M. Valle, 1985. Estudio osteológico de la Cueva de Piélago II. *Sautuola* 4, 113-121.

López-Fuster, M .J., E. Castién & J. Gosàlbez, 1988. Reproductive cycle and population structure of *Sorex coronatus* Millet, 1828 (Insectivora, Soricidae) in the northern Iberian Peninsula. *Bonner Zoologische Beigtrage* 39, 163-170.

López-Fuster, M. J., 1989. Reproductive strategy of the Millet's shrew (*Sorex coronatus* Millet, 1928) versus the common shrew (*Sorex araneus* L., 1758) in the northeast of the Iberian Peninsula. *Zoologische Abhandlungen* 44, 143-148.

López-Fuster, M. J. & J. Ventura, 1995. Somatometry and reproductive data on the northern water shrew, *Neomys fodiens* (Pennant, 1771) from the Aran Valley (Spanish Pyrenees). *Mammalia* 59, 156-159.

López-Gordo, J. L., E. Lazaro & A. Fernández-Jorge, 1977. Comparación de las dietas de *Strix aluco*, *Asio otus* y *Tyto alba* en un mismo biotopo de la Provincia de Madrid. *Ardeola* 23, 189-221.

Lovari, S., A. Renzoni & R. Fondi, 1976. The predatory habits of the barn owl (*Tyto alba* Scopoli) in relation to the vegetation cover. *Bollettino de Zoologia* 43, 173-191.

Lubinski, P. M., 1996. Fish heads, fish heads: An experiment on differential bone preservation in a salmonid fish. *Journal of Archaeological Science* 23, 175-181.

Lyell, C., 1830-33. *Principles of Geology* (3 Vols.). London: J. Murray. Reprinted facsimile edition (1991), ed. M. J. S. Rudwick. Chicago: University of Chicago Press.

Maddock, A. H., 1992. Comparison of two methods for trapping rodents and shrews. *Israel Journal of Zoology* 38, 333-340.

Madsen, A. B. & A. M. Rasmussen, 1985. Reproduction of the stone marten *Martes foina* in Denmark. *Natura Jutlandica* 21, 145-148.

Marti, C. D., 1969. Some comparisons of the feeding ecology of four owls in North-Central Colorado. *The Southwestern Naturalist* 14, 163-170.

Marti, C. D., 1973a. Ten years of barn owl prey from a Colorado nest site. *Wilson Bulletin* 85, 85-86.

Marti, C. D., 1973b. Food consumption and pellet formation rates in four owl species. *Wilson Bulletin* 85, 178-181.

Martin, P. S. & R. G. Klein, eds., 1984. *Quaternary Extinctions*. Tuscon: University of Arizona Press.

Martínez Navarette, M. I. & Y. T. Chapa Brunet, 1980. La industria prehistórica de la cueva de la Paloma (Soto de las Regueras, Asturias). In *La Cueva de la Paloma*, eds. M. Hoyos, M. I. Martínez Navarette, T. Chapa Brunet, P. Castaños & F. B. Sanchez. Excavaciones Arqueológicas en España, No. 116, pp. 115-204.

Mayhew, D. F., 1977. Avian predators as accumulators of fossil mammal material. *Boreas* 6, 25-31.

McIntyre, A., N. G. Kipp, A. W. H. Be, T. Crowley, T. Kellogg, J. V. Gardner, W. Prell, & W. F. Ruddiman, 1976. Glacial North Atlantic 18,000 years ago: A CLIMAP reconstruction. In *Investigation of Late Quaternary Paleoceanography and Paleoclimatology*, eds. R. M. Cline & J. D. Hays. The Geological Society of America, Memoir 145, pp. 43-76.

Mellars, P., 1990. *The Emergence of Modern Humans*. Ithaca: Cornell University Press.

Mellars, P. & C. Stringer, eds., 1989. *The Human Revolution*. Edinburgh: University of Edinburgh Press.

Mellet, J. S., 1974. Scatological origin of microvertebrate fossil accumulations. *Science* 185, 349-350.

Michalak, I., 1987. Growth and postnatal development of the European water shrew. *Acta Theriologica* 32, 261-288.

Mikkola, H., 1983. *Owls of Europe*. Calton: Poyser Press.

Mills, D., 1989. *Ecology and Management of Atlantic Salmon*. London, Chapman and Hall.

Moon, E. L., 1940. Notes on hawk and owl pellet formation and identification. *Transactions of the Kansas Academy of Science* 43, 457-466.

Moreno, E. & A. Barbosa, 1992. Distribution patterns of small mammal fauna along gradients of latitude and altitude in northern Spain. *Zeitschrift für Säugertierkunde* 57, 169-175.

Moreno, S., 1988. Reproduction of garden dormouse *Eliomys quercinus lusitanicus*, in southwest Spain. *Mammalia* 52, 401-407.

Moreno, S. & E. Collado, 1989 Growth of the garden dormouse (*Eliomys quercinus* Linnaeus, 1766) in southwestern Spain. *Zeitschrift für Säugertierkunde* 54, 100-106.

Moure Romanillo, A., 1975. *Excavaciones en la Cueva de "Tito Bustillo" (Asturias)*. Oviedo: Instituto de Estudios Asturianos.

Moure Romanillo, A. & M. Cano Herrera, 1976. *Excavaciones en la Cueva de "Tito Bustillo" (Asturias)*. Oviedo: Instituto de Estudios Asturianos.

Moure Romanillo, A. & M. Cano Herrera, 1978. Magdalenian habitation structure at Tito Bustillo Cave (Asturias, Spain). *Current Anthropology* 19, 392-394.

Moure Romanillo, A. & M. Cano Herrera, 1979. Tito Bustillo Cave (Asturias, Spain) and the Magdalenian of Cantabria. *World Archaeology* 10, 280-289.

Moure Romanillo, A., C. González Sainz, F. Bernaldo de Quirós & V. Cabrera Valdés, 1996. Dataciones absolutas de pigmentos en cuevas Cantábricas: Altamira, El Castillo, Chimeneas y Las Monedas. In *"El Hombre Fósil" 80 Años Después*, ed. A. Moure Romanillo. Santander: Universidad de Cantabria, Fundación Botín, and Instituto para Investigaciones Prehistóricas, pp. 295-324.

Nadachowski, A., 1982. Late Quaternary rodents of Poland with special reference to morphotype dentition analysis of voles. *Polska Akademia Nauk* 1982, 1-108.

Nadeau, J. H., 1985. Ontogeny. In *Biology of New World Microtus*, ed. R. H. Tamarin. The American Society of Mammalogists, Special Publication No. 8, pp. 254-285.

Nader, I. A., 1969. Animal remains in pellets of the barn owl *Tyto alba* from the vicinity of An-Najaf, Iraq. *Bulletin of the Iraq Natural History Museum* 4, 1-7.

Nagorsen, D. W. & R. L. Peterson, 1980. *Mammal Collectors' Manual*. Life Sciences Miscellaneous Publications, Royal Ontario Museum. Toronto: Alger Press.

Neet, C. R. & J. Hausser, 1990. Habitat selection in zones of parapatric contact between the common shrew *Sorex araneus* and Millet's shrew *S. coronatus*. *Journal of Animal Ecology* 59, 235-250.

Netboy, A., 1980. *Salmon*. Tulsa: Winchester Press.

Niethammer, J. & F. Krapp, eds., 1978. *Handbuch der Säugetiere Europas. Band 1: Nagetiere I.* Wiesbaden: Akademische Verlagsgesellschaft.

Niethammer, J. & F. Krapp, eds., 1982. *Handbuch der Säugetiere Europas. Band 2/1: Nagetiere II.* Wiesbaden: Akademische Verlagsgesellschaft.

Niethammer, J. & F. Krapp, eds., 1990. *Handbuch der Säugetiere Europas. Band 3/I: Insektenfresser, Herrentiere.* Verlag Wiesbaden: AULA.

Niethammer, J. & F. Krapp, eds., 1993. *Handbuch der Säugetiere Europas. Band 5/II: Raubsäuger.* Verlag Wiesbaden: AULA.

Noyes, R. F., G. W. Barrett & D. H. Taylor, 1982. Social structure of feral house mouse (*Mus musculus* L.) populations: Effects of resource partitioning. *Behavioral Ecology and Sociobiology* 10, 157-163.

Obermaier, H., 1924. *Fossil Man in Spain.* New Haven: Yale University Press.

Ognev, S. I., 1962. *Mammals of Eastern Europe and Northern Asia. Vol. II: Carnivora (Fissipedia).* Jerusalem: Israel Program for Scientific Translations.

Ognev, S. I., 1964. *Mammals of Eastern Europe and Northern Asia. Vol. VII: Rodents.* Jerusalem: Israel Program for Scientific Translations.

Oksanen, T., 1983. Prey-caching in the hunting strategy of small mammals. *Acta Zoologica Fennica* 174, 197-199.

Orsini, P., 1985. Le régime alimentaire du hibou grand-duc *Bubo bubo* en Provence. *Alauda* 53, 11-28.

Palmeirim, J. M. & R. S. Hoffman, 1983. *Galemys pyrenaicus. Mammalian Species* 207, 1-5.

Palomo, L. J., J. M. Vargas & M. P. Jiménez-Gómez, 1994. Distribution patterns in Iberian Peninsula rodents. *Polish Ecological Studies* 20, 497-502.

Pasitschniak-Arts, M. & S. Larivière, 1995. *Gulo gulo. Mammalian Species* 499, 1-10.

Payne, R. S., 1971. Acoustic location of prey by barn owls (*Tyto alba*). *Journal of Experimental Biology* 54, 535-573.

Payne, R. S. & W. H. Drury, Jr., 1958. Marksman of the darkness: *Tyto alba* Part II. *Natural History* 67, 316-323.

Pemán, E., 1985. Aspectos climáticos y ecológicos de los micromamíferos del yacimiento de Erralla. *Munibe* 37, 49-57.

Pemán, E., 1990. Los micromamíferos de la Cueva de Amalda y su significado. In *La Cueva de Amalda (Zestoa, Pais Vasco)*, eds. J. Altuna, A. Baldeón & K. Mariezkurrena. San Sebastián: Sociedad de Estudios Vascos, pp. 225-238.

Peñalba, M. C., 1994. The history of the Holocene vegetation in northern Spain from pollen analysis. *Journal of Ecology* 82, 815-832.

Perrin, M. R., 1982. Prey specificity of the barn owl, *Tyto alba*, in the Great Fish River Valley of the Eastern Cape Province. *South African Journal of Wildlife Research* 12, 14-26.

Pitelka, F. A., P. Q. Tomich & G. W. Treichel, 1955. Ecological relations of jaegers and owls as lemming predators near Barrow, Alaska. *Ecological Monographs* 25, 85-117.

Pokines, J. T., 1993. *Antler Points from El Juyo (Santander, Spain): Form, Manufacture and Parameters of Use.* Unpublished M.A. Thesis, Department of Anthropology, University of Chicago.

Pokines, J. T., 1997. *Microfaunal Indicators of Paleoenvironment from El Juyo (Santander, Spain).* Unpublished Ph.D. Dissertation, Department of Anthropology, University of Chicago.

Pokines, J. T., in press. Experimental replication and use of Cantabrian Lower Magdalenian antler projectile points. *Journal of Archaeological Science.*

Pokines, J. T. & J. Kerbis Peterhans, 1998. Barn owl (*Tyto alba*) taphonomy in the Negev Desert, Israel. *Israel Journal of Zoology* 44.

Pokines, J. T. & M. Krupa, 1997. Self-barbed antler spearpoints and evidence of fishing in the Late Upper Paleolithic of Cantabrian Spain. In *Projectile Technology*, ed. H. Knecht. New York, Plenum Publishing, pp. 241-262.

Polunin, O. & M. Walters, 1985. *A Guide to the Vegetation of Britain and Europe.* Oxford: Oxford University Press.

Prentice, I. C., 1986. Forest-composition calibration of pollen data. In *Handbook of Holocene Palaeoecology and Palaeohydrology*, ed. B. E. Berglund. New York: Wiley-Interscience, pp. 799-816.

Pucek, Z. & V. P. W. Lowe, 1975. Age criteria in small mammals. In *Small Mammals: Their Productivity and Population Dynamics*, eds. F. B. Golley, K. Petrusewicz & L. Ryszkowski. Cambridge: Cambridge University Press, pp. 55-72.

Reed, C. A., ed., 1977. *Origins of Agriculture.* The Hague: Mouton Press.

Raczynski, J. & A. L. Ruprecht, 1974. The effect of digestion on the osteological composition of owl pellets. *Acta Ornithologica* 14, 26-38.

Reed, C. I. & B. P. Reed, 1928. The mechanism of pellet formation in the great horned owl (*Bubo virginianus*). *Science* 68, 359-360.

Reichstein, H., 1963. Beitrag zur systematischen Gliederung des Genus *Arvicola* Lacépède 1799. *Zeitschrift für zoologische Systematik und Evolutions-forschung* 1, 155-204.

Robertson, P. A. & J. Whelan, 1987. The food of the Red fox (*Vulpes vulpes*) in Co. Kildare, Ireland. *Journal of Zoology, London* 213, 740-743.

Rodríguez, J. & G. Flor, 1979. Estudio del yacimiento prehistórico de Bañugues y su médio de deposito. *Zephyrus*, 29:205-222.

Ruddiman, W. F. & A. McIntyre, 1976. Northeast Atlantic paleoclimatic changes over the past 600,000 Years. In *Investigation of Late Quaternary Paleoceanography and Paleoclimatology*, eds. R. M. Cline & J. D. Hays. The Geological Society of America, Memoir 145. pp. 111-146.

Ruprecht, A. L., 1971. Taxonomic value of mandible measurements in Soricidae (Insectivora). *Acta Theriologica* 16, 341-357.

Saint Girons, M.-C., 1973. L'age des micromamifères dans le regime de deux rapaces nocturnes, *Tyto alba* et *Asio otus*. *Mammalia* 37, 439-456.

Sanchez Goñi, M. F., 1993. *De la Taphonomie Pollinique à la Reconsitution de l'Environment: L'Exemple de la Région Cantabrique*. British Archaeological Reports, International Series 586.

Santos Reis, M., 1983. Status and distribution of the Portuguese mustelids. *Acta Zoologica Fennica* 174, 213-216.

Saunders, Jr., J. K., 1963. Food habits of the lynx in Newfoundland. *Journal of Wildlife Management* 27, 384-390.

Scott, T. G., 1941. Methods and computation in fecal analysis with reference to the red fox. *Iowa State College Journal of Science* 15, 279-285.

Selleck, D. M. & B. Glading, 1943. Food habits of nesting barn owls and Marsh hawks at Dune Lakes, California, as determined by the "Cage Nest" method. *California Fish and Game* 29, 122-131.

Shackleton, N. J., 1987. Oxygen isotopes, ice volume and sea level. *Quaternary Science Reviews*, 6:183-190.

Shackleton, N. J. & N. D. Opdyke, 1976. Oxygen-isotope and paleomagnetic stratigraphy of Pacific core V28-239 Late Pliocene to Latest Pleistocene. In *Investigation of Late Quaternary Paleoceanography and Paleoclimatology*, eds. R. M. Cline & J. D. Hays. Geological Society of America, Memoir 145, pp. 449-464.

Shaffer, B. S. & J. L. J. Sanchez, 1994. Comparison of 1/8"- and 1/4"-mesh recovery of controlled samples of small-to-medium-sized mammals. *American Antiquity* 59, 525-530.

Sheffield, S. R. & C. M. King, 1994. *Mustela nivalis*. *Mammalian Species* 454, 1-3.

Shipman, P., 1981. *Life History of a Fossil: An Introduction to Taphonomy and Paleoecology*. Cambridge: Harvard University Press.

Shipman, P. & A. Walker, 1980. Bone-collecting by harvester ants. *Paleobiology* 6, 496-502.

Short, H. L. & L. C. Drew, 1962. Observations concerning behavior, feeding, and pellets of short-eared owls. *American Midland Naturalist* 67, 424-433.

Skinner, J. D. & R. H. N. Smithers, 1990. *The Mammals of the Southern African Subregion*. Pretoria: University of Pretoria.

Sleeman, D. P., 1993. Habitats of the Irish stoat. *Irish Naturalist's Journal* 24, 318-321.

Smith, C. R. & M. E. Richmond, 1972. Factors influencing pellet egestion and gastric pH in the barn owl. *Wilson Bulletin* 84, 179-186.

Soulé, M. E., 1992. Conservation: Tactics for a constant crisis. *Science* 253, 744-750.

Southern, H. N., 1954. Tawny owls and their prey. *Ibis* 96, 384-410.

Spitz, F., 1978. Étude crâniométrique du genre *Pitymys*. *Mammalia* 42, 267-304.

Stettler, H. K., 1995. *Macrobotanical Remains from the Site of El Juyo (Santander), Spain Spatial Analysis and Paleoenvironmental Implications*. Unpublished M.A. Thesis, Department of Anthropology, University of Chicago.

Stiner, M. C., 1993. Small animal exploitation and its relation to hunting, scavenging, and gathering in the Italian Mousterian. In *Hunting and Animal Exploitation in the Later Palaeolithic and Mesolithic of Eurasia*, eds. G. Larsen Peterkin, H. M. Bricker & P. Mellars. Archeological Papers of the American Anthropological Association, No. 4, pp. 107-125.

Stiner, M. C., 1994. *Honor Among Thieves*. Princeton: Princeton University Press.

Stork, N. E., 1988. Insect diversity: Facts, fiction and speculation. *Journal of the Linnean Society of London* 20, 321-338.

Straus, L. G., 1974a. Le solutréen du Pays Basque Espagnol: Une esquisse des données. *Munibe* 26, 173-181.

Straus, L. G., 1974b. Notas preliminares sobre el Solutrense de Asturias. *Boletin del Instituto de Estudios Asturianos* 82, 483-504.

Straus, L. G., 1975 El Solutrense de las cuevas del Castillo y Hornos de la Peña (Santander) en el Museo Arqueológico Nacional de Madrid. *Trabajos de Prehistoria* 32, 9-19.

Straus, L. G., 1976. Análisis arqueológico de la fauna paleolítica del norte de la Penísula Ibérica. *Munibe* 28, 227-285.

Straus, L. G., 1976-77. The Upper Paleolithic cave site of Altamira (Santander, Spain). *Quaternaria* 19, 135-148.

Straus, L. G., 1977. Of deerslayers and mountain men: Paleolithic faunal exploitation in Cantabrian Spain. In *For Theory Building in Archaeology*, ed. L. R. Binford. New York: Academic Press, pp. 41-76.

Straus, L. G., 1980. Reply to Conkey. *Current Anthropology* 21, 624-625.

Straus, L. G., 1982. Carnivores and cave sites in Cantabrian Spain. *Journal of Anthropological Research* 38, 75-98.

Straus, L. G., 1985. Stone Age prehistory of northern Spain. *Science* 230, 501-507.

Straus, L. G., 1987a. Hunting in Late Upper Paleolithic Western Europe. In *The Evolution of Human Hunting*, eds. M. H. Nitecki & D. V. Nitecki. New York: Plenum Press, pp. 147-176.

Straus, L. G.,1987b. Upper Paleolithic ibex hunting in southwest Europe. *Journal of Archaeological Science* 14, 163-178.

Straus, L. G., 1990. The original arms race: Iberian perspectives on the Solutrean phenomenon. In *Feuilles de Pierre*, ed. J. Kozlowski. Etudes et Recherches Archéologiques de l'Université de Liège, pp. 425-447.

Straus, L. G., 1991a. Epipaleolithic and Mesolithic adaptations in Cantabrian Spain and Pyrenean France. *Journal of World Prehistory* 5, 83-104.

Straus, L. G., 1991b. Southwestern Europe at the last glacial maximum. *Current Anthropology* 32, 189-199.

Straus, L. G., 1992. *Iberia before the Iberians*. Albuquerque: University of New Mexico Press.

Straus, L. G., 1993. Upper Paleolithic hunting tactics and weapons in Western Europe. In *Hunting and Animal Exploitation in the Later Palaeolithic and Mesolithic of Eurasia*, eds. G. Larsen Peterkin, H. M. Bricker & P. Mellars. Archeological Papers of the American Anthropological Association, No. 4, pp. 83-93.

Straus, Lawrence G., ed., 1995. *Les Derniers Chasseurs de Rennes du Monde Pyrénéen. L'Abri Dufaure: Un Gisement Tardiglaciare en Gascogne*. Paris: Société Préhistorique Française, Mémoire 22.

Straus, L. G., J. Altuna, G. A. Clark, M. González, H. Laville, Ar. Leroi-Gourhan, M. Menendez & J. A. Ortea, 1981. Paleoecology at La Riera (Asturias, Spain). *Current Anthropology* 22, 655-674.

Straus, L. G. & G. A. Clark, 1978. Prehistoric investigations in Cantabrian Spain. *Journal of Field Archaeology* 5, 289-317.

Straus, L. G. & G. A. Clark, eds., 1986. *La Riera Cave*. Arizona State University Anthropological Research Papers No. 36.

Straus, L. G., G. A. Clark, J. Altuna & J. A. Ortea, 1980. Ice-Age subsistence in northern Spain. *Scientific American* 242, 142-152.

Stuart, A .J., 1980. The vertebrate fauna from the interglacial deposits at Sugworth, near Oxford. *Philosophical Transactions of the Royal Society of London,* B 289, 87-97.

Stuart, A .J., 1981. A comparison of the Middle Pleistocene mammal faunas of Voigstedt (Thuringia, GDR) and West Runton (Norfolk, England). *Quartarpalaontologie* 4, 155-163.

Stuart, A .J., 1982. *Pleistocene Vertebrates in the British Isles*. London: Longman.

Stuart, C. T., 1977. Analysis of *Felis libyca* and *Genetta genetta* scats from the Central Namib Desert, South West Africa. *Zoologica Africana* 12, 239-241.

Sutcliffe, A. J., 1970. Spotted hyaena: Crusher, gnawer, digester and collector of bones. *Nature* 227, 1110-1113.

Taylor, I., 1994. *Barn Owls*. Cambridge: Cambridge University Press.

Tchernov, E., 1968. *Succession of Rodent Faunas during the Upper Pleistocene of Israel*. Hamburg: Verlag Paul Parey.

Tchernov, E., 1975. Rodent faunas and environmental changes in the Pleistocene of Israel. In *Rodents in Desert Environments*, eds. I. Prakash & P. K. Ghosh. Monographiae Biologicae, Vol. 28. The Hague: Dr. W. Junk, pp. 331-362.

Tchernov, E., 1979. Quaternary fauna. In *The Quaternary of Israel*, ed. A. Horowitz. New York: Academic Press, pp. 247-290.

Tchernov, E., 1982. Faunal responses to environmental changes in the eastern Mediterranean during the last 20,000 years. In *Palaeoclimates, Palaeoenvironments and Human Communities in the Eastern Mediterranean Region in Later Prehistory*, eds. J. L. Bintliff & W. Van Zeist. Oxford: British Archaeological Reports International Series, 133I, pp. 105-127.

Tessier, L., J.-L. de Beaulieu, M. Conteaux, J.-L. Edouard, P. Ponel, C. Rolando, M. Thinon, A. Thomas & K. Tobolski, 1993). Holocene palaeoenvironments at the timberline in the French Alps--a multidisciplinary approach. *Boreas* 22, 244-254.

Thackeray, J. F., 1987. Late Quaternary environmental changes inferred from small mammalian fauna, Southern Africa. *Climatic Change* 10, 285-305.

Thackeray, J. F. & D. M. Avery, 1990. A comparison between temperature indices for Late Pleistocene sequences at Klasies River and Border Cave, South Africa. *Palaeoecology of Africa* 21, 311-316.

Torre, I. & J. L. Tella, 1994. Distribution of the Cabrera water shrew (*Neomys anomalus*) in northeastern Spain. *Zeitschrift für Säugetierkunde* 59, 282-288.

Traverse, A., 1990. Studies of pollen and spores in rivers and other bodies of water in terms of source-vegetation and sedimentation, with special reference to Trinity River and Bay, Texas. *Review of Palaeobotany and Palynology* 64, 297-303.

Trout, R. C., 1978. A review of studies on populations of wild harvest mice (*Micromys minutus* Pallas). *Mammal Review* 8, 143-158.

Tumlison, R., 1987. *Felis lynx*. *Mammalian Species* 269, 1-8.

Turni, H. & E. F. Müller, 1996. Unterscheidung der Spitzmausarten *Sorex araneus* L., 1758 und *Sorex coronatus* Millet, 1828 mit Hilfe einer neuen Diskriminanzfunktion. *Zeitschrift für Säugetierkunde* 61, 73-92.

Utrilla, P., 1981. *El Magdaleniense Inferior y Médio en la Costa Cantábrica*. Santander: Centro de Investigacíon y Museo de Altamira, Monografia No. 4.

Valladas, H., H. Cachier, P. Maurice, F. Bernaldo de Quirós, J. Clottes, V. Cabrera Valdés, P. Uzquiano & M. Arnold, 1992. Direct radiocarbon dates for prehistoric paintings at the Altamira, El Castillo and Niaux caves. *Nature* 357, 68-70.

van Donk, J., 1976. O^{18} Record of the Atlantic Ocean for the entire Pleistocene epoch. In *Investigation of Late Quaternary Paleoceanography and Paleoclimatology*, eds. R. M. Cline & J. D. Hays. Geological Society of America, Memoir 145, pp. 147-163.

Vega del Sella, Conde de la, 1916. *Paleolitico de Cueto de la Mina*. Madrid: Comisión de Investigaciones Paleontológicas y Prehistóricas, Memoir No. 13.

Veiga, J. P., 1980. Alimentacion y relaciones troficas entre la luchuza comun (*Tyto alba*) y el buho chico (*Asio otus*) en la Sierra de Guadarrama (España). *Ardeola* 25, 113-142.

Veiga, J. P., 1986. Interannual fluctuations of three microtine populations in mediterreanean environments: The effect of rainfall. *Mammalia* 50, 114-116.

Ventura, J., 1988. Revisíon de la corología de *Arvicola terrestris* (Linnaeus, 1758) (Rodentia, Arvicolidae) en la Península Ibérica. *Miscellania Zoologica* 12, 319-328.

Ventura, J., 1991. Morphological characteristics of the molars of *Arvicola terrestris* (Rodentia, Arvicolidae) in its southwestern distribution area. *Zoologische Anzeiger* 226, 64-70.

Ventura, J., 1992. Morphometric data on the scapula and limb long bones of *Arvicola terrestris* (Linnaeus, 1758) (Rodentia, Arvicolidae). *Revue Suisse Zoologie* 99, 629-636.

Ventura, J. & J. Gosálbez, 1986. Características morfológicas y criterios de discriminación sexual a nivel del coxal en *Arvicola sapidus* (Rodentia, Arvicolidae). *Publicaciones del Departmento de Zoología de Barcelona* 12, 117-128.

Ventura, J. & J. Gosálbez, 1987. Reproductive biology of *Arvicola sapidus* (Rodentia, Arvicolidae) in the Ebro delta, Spain. *Zeitschrift für Säugetierkunde* 52, 364-371.

Ventura, J. & J. Gosálbez, 1989. Taxonomic review of *Arvicola terrestris* (Linnaeus, 1758) (Rodentia, Arvicolidae) in the Iberian Peninsula. *Bonner Zoologische Beiträge* 40, 227-242.

Ventura, J. & J. Gosálbez, 1990. Reproductive cycle of *Arvicola terrestris* (Rodentia, Arvicolidae) in the Aran Valley, Spain. *Zeitschrift für Säugetierkunde* 55, 383-391.

Virgós, E., J. G. Casanovas & T. Blázquez, 1994. Genet (*Genetta genetta* L., 1758) diet shift in mountains of central Spain. *Zeitschrift für Säugetierkunde* 61, 221-227.

Von Koenigswald, W., 1973. Veranderungen in der Kleinsaugerfauna von Mitteleuropa zwischen Cromer und Eem (Pleistosan). *Eiszeitalter und Gegenwart* 23-24, 159-167.

Von Koenigswald, W. & F. Nicolaevich Golenishev, 1979. A method for determining growth rates in continuously growing molars. *Journal of Mammalogy* 60, 397-400.

Voorhies, M. R., 1969. Taphonomy and population dynamics of an Early Pliocene vertebrate fauna, Knox County, Nebraska. *University of Wyoming, Contributions to Geology, Special Papers*, No. 1.

Weaver, J. L. & S. W. Hoffman, 1979. Differential detectability of rodents in coyote scats. *Journal of Wildlife Management* 43, 783-786.

Webster, J. A., 1973. Seasonal variation in mammal contents of barn owl castings. *Bird Study* 20, 185-196.

Western, D., 1980. Linking the ecology of past and present mammal communities. In *Fossils in the Making: Vertebrate Taphonomy and Paleoecology*, eds. A. K. Behrensmeyer & A. P. Hill. Chicago: University of Chicago Press, pp. 41-54.

Willis, J .C., 1973. *A Dictionary of the Flowering Plants and Ferns* (8th ed.). Cambridge: Cambridge University Press.

Wilson, E. O., 1989. Threats to biodiversity. *Scientific American* 30, 108-116.

Wilson, M. V. H., 1987. Predation as a source of fish fossils in Eocene lake sediments. *Palaios* 2, 497-504.

Wolff, R. G., 1973. Hydrodynamic sorting and ecology of a Pleistocene mammalian assemblage from California (U.S.A.). *Palaeogeography, Palaeoclimatology, Palaeoecology* 13, 91-101.

Youngman, P. M., 1982. Distribution and systematics of the European mink *Mustela lutreola* Linnaeus 1761. *Acta Zoologica Fennica* 166, 1-48.

Youngman, P. M., 1990. *Mustela lutreola. Mammalian Species* 362, 1-3.

Zabala, J., 1983. Primera cita de *Arvicola sapidus* (Miller 1908, Mammalia, Rodentia) para la provincia de Guipúzcoa. *Munibe* 35, 109-114.

Zabala, J., 1984. Los micromamíferos del yacimiento prehistórico de Ekain (Guipúzcoa). In *El Yacimiento Prehistórico de la Cueva de Ekain (Deba, Guipúzcoa)*, eds. J. Altuna & J. M. Merino. San Sebastián: Sociedad de Estudios Vascos, Serie B1, pp. 317-330.

www.ingramcontent.com/pod-product-compliance
Lightning Source LLC
LaVergne TN
LVHW071209210926
848482LV00015B/762